Lecture Notes in Information Systems and Organisation

Volume 29

Lecture Notes in Information Systems and Organization—LNISO—is a series of scientific books that explore the current scenario of information systems, in particular IS and organization. The focus on the relationship between IT, IS and organization is the common thread of this collection, which aspires to provide scholars across the world with a point of reference and comparison in the study and research of information systems and organization. LNISO is the publication forum for the community of scholars investigating behavioral and design aspects of IS and organization. The series offers an integrated publication platform for high-quality conferences, symposia and workshops in this field. Materials are published upon a strictly controlled double blind peer review evaluation made by selected reviewers. LNISO is abstracted/indexed in Scopus

More information about this series at http://www.springer.com/series/11237

Fred D. Davis · René Riedl
Jan vom Brocke · Pierre-Majorique Léger
Adriane B. Randolph
Editors

Information Systems and Neuroscience

NeuroIS Retreat 2018

Editors
Fred D. Davis
Information Systems
and Quantitative Sciences (ISQS)
Texas Tech University
Lubbock, TX, USA

René Riedl
University of Applied Sciences
Upper Austria
Steyr, Oberösterreich, Austria

and

Johannes Kepler University Linz
Linz, Austria

Jan vom Brocke
Department of Information Systems
University of Liechtenstein
Vaduz, Liechtenstein

Pierre-Majorique Léger
Department of Information Technology
HEC Montréal
Montréal, QC, Canada

Adriane B. Randolph
Department of Information Systems
Kennesaw State University
Kennesaw, GA, USA

ISSN 2195-4968 ISSN 2195-4976 (electronic)
Lecture Notes in Information Systems and Organisation
ISBN 978-3-030-01086-7 ISBN 978-3-030-01087-4 (eBook)
https://doi.org/10.1007/978-3-030-01087-4

Library of Congress Control Number: 2018955917

This Springer imprint is published by the registered company Springer Nature Switzerland AG
The registered company address is: Gewerbestrasse 11, 6330 Cham, Switzerland

Preface

NeuroIS is a field in information systems (ISs) that makes use of neuroscience and neurophysiological tools and knowledge to better understand the development, adoption, and impact of information and communication technologies. The NeuroIS Retreat is a leading academic conference for presenting research and development projects at the nexus of IS and neurobiology (see http://www.neurois.org/). This annual conference has the objective to promote the successful development of the NeuroIS field. The conference activities are primarily delivered by and for academics, though works often have a professional orientation.

Since 2018, the conference is taking place in Vienna, Austria, one of the world's most beautiful cities. In 2009, the inaugural conference was organized in Gmunden, Austria. Established on an annual basis, further conferences took place in Gmunden from 2010 to 2017. The genesis of NeuroIS took place in 2007. Since then, the NeuroIS community has grown steadily. Scholars are looking for academic platforms to exchange their ideas and discuss their studies. The NeuroIS Retreat seeks to stimulate these discussions. The conference is best characterized by its workshop atmosphere. Specifically, the organizing committee welcomes not only completed research, but also work in progress. A major goal is to provide feedback for scholars to advance research papers, which then, ultimately, have the potential to result in high-quality journal publications.

This year is the fourth time that we publish the proceedings in the form of an edited volume. A total of 32 research papers are published in this volume, and we observe diversity in topics, theories, methods, and tools of the contributions in this book. The 2018 keynote presentation entitled "Translational Behavioral Neuroscience: The Use of Neuroscientific Insights to Improve Public Welfare" was given by Bernd Weber, Director of the Center for Economics and Neuroscience, University of Bonn, Germany. Moreover, Christian Montag, Professor for Molecular Psychology at Ulm University in Germany, gave a hot topic talk entitled "The neuroscience of smartphone and social media usage and the growing need to include methods from Psychoinformatics."

Altogether, we are happy to see the ongoing progress in the NeuroIS field. More and more IS researchers and practitioners have been recognizing the enormous potential of neuroscience tools and knowledge.

This year, in 2018, we celebrated the 10 years anniversary of the NeuroIS Retreat. We had a great conference and foresee a prosperous NeuroIS future!

Lubbock, USA — Fred D. Davis
Steyr, Austria — René Riedl
Vaduz, Liechtenstein — Jan vom Brocke
Montréal, Canada — Pierre-Majorique Léger
Kennesaw, USA — Adriane B. Randolph
June 2018

Contents

NeuroIS: A Survey on the Status of the Field

Thomas Fischer, Fred D. Davis and René Riedl

Abstract NeuroIS has emerged as a research field in the Information Systems (IS) discipline over the past decade. Since the inaugural NeuroIS Retreat in 2009, 166 individuals participated at this annual academic conference to discuss research and development projects at the nexus of IS and neuroscience research. Motivated by the fact that the NeuroIS Retreat celebrates its 10-year anniversary in 2018, we invited all 166 former participants of the NeuroIS Retreat to state their opinions in an online survey on the development of the field and its future. In this paper, we summarize the answers of N = 60 respondents regarding NeuroIS topics and methods.

Keywords Brain · Methods · NeuroIS · Status · Survey · Tools · Topics

1 Introduction

The first NeuroIS Retreat took place in Gmunden, Austria, in 2009. Since then, the NeuroIS community has grown and in 2018 this annual academic conference celebrates the 10 years anniversary in Vienna, Austria. A total of 166 individuals attended this forum for the presentation and discussion of research and development projects in the last decade, and thereby contributed to the prosperous development of the field. Motivated by the fact that the NeuroIS Retreat exists for 10 years now, we developed an online survey to ask all former conference participants about their

T. Fischer (✉) · R. Riedl
University of Applied Sciences Upper Austria, Steyr, Austria
e-mail: thomas.fischer@fh-steyr.at

R. Riedl
e-mail: rene.riedl@fh-steyr.at

F. D. Davis
Texas Tech University, Lubbock, USA
e-mail: fred.davis@ttu.edu

R. Riedl
Johannes Kepler University, Linz, Austria

F. D. Davis et al. (eds.), *Information Systems and Neuroscience*,
Lecture Notes in Information Systems and Organisation 29,
https://doi.org/10.1007/978-3-030-01087-4_1

perspectives on the status of the field. In this paper, we present major results of this survey related to NeuroIS topics and methods. Specifically, we investigated the participants' perspectives on topics and methods, which are currently studied and applied, and what they think about future topics and methods.

2 Survey Characteristics and Sample Demographics

Using the online survey tool SoSci Survey, we conducted a survey amongst a population of all 166 previous participants of the NeuroIS Retreat 2009–2017 in the period 12/04/2017–02/06/2018. The survey contained questions related to impressions of the past developments in the field, but also gave respondents the opportunity to report on their future NeuroIS research and their expectations for the field. Overall, it took about ten minutes to complete the survey. We were able to gather 60 complete responses, amongst 152 individuals who are still involved in academic research (response rate of 39.5%). The remaining 14 individuals are not active researchers anymore and it was not possible to contact them in the context of this study.

Amongst the respondents, 75% were male and a majority of 64% is between 30 and 49 years old (see Fig. 1). We also asked respondents to indicate the country were they are currently employed (see Fig. 2). The results show that most respondents are currently either employed in German-speaking countries (25 individuals are from Austria, Germany, Switzerland, and Liechtenstein) or North America (24 individuals are from the USA and Canada).

We also wanted to know the current academic position of our respondents, which revealed that 39% were full professors, followed by 19% who were Ph.D. candidates and 17% each who were either associate professor or assistant professor. This finding indicates that the field is not only interesting to a selected group of established researchers, but also allows new researchers to enter the arena, such as early-stage

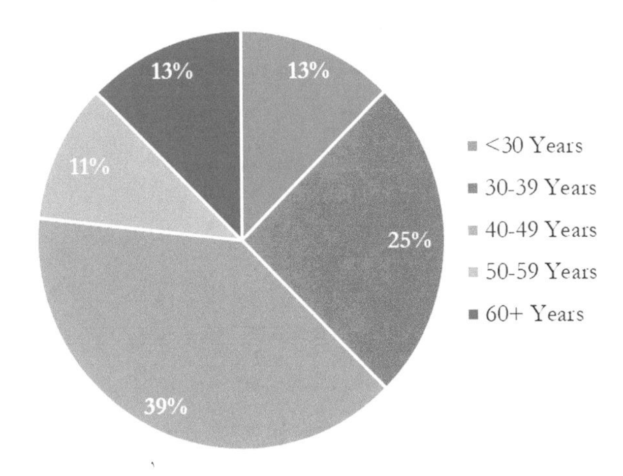

Fig. 1 Share of respondents per age group (N = 60)

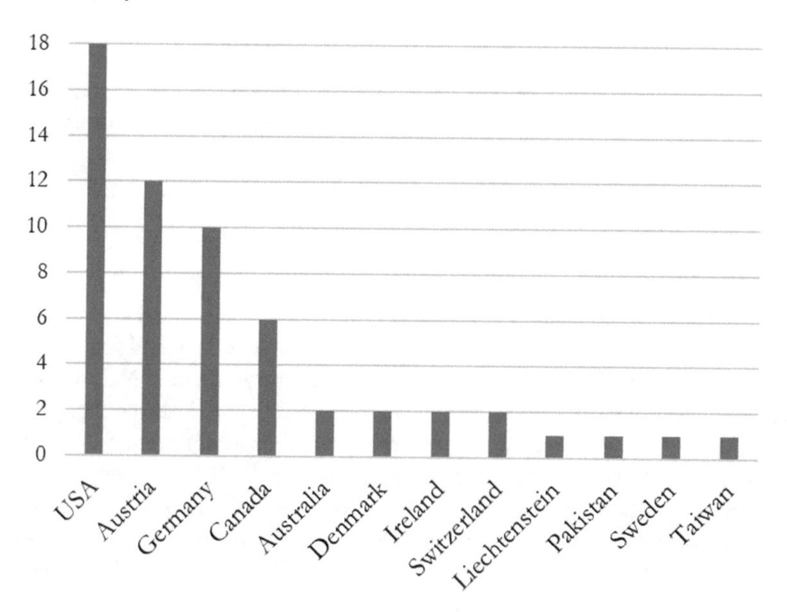

Fig. 2 Number of respondents per country of employment (N = 60)

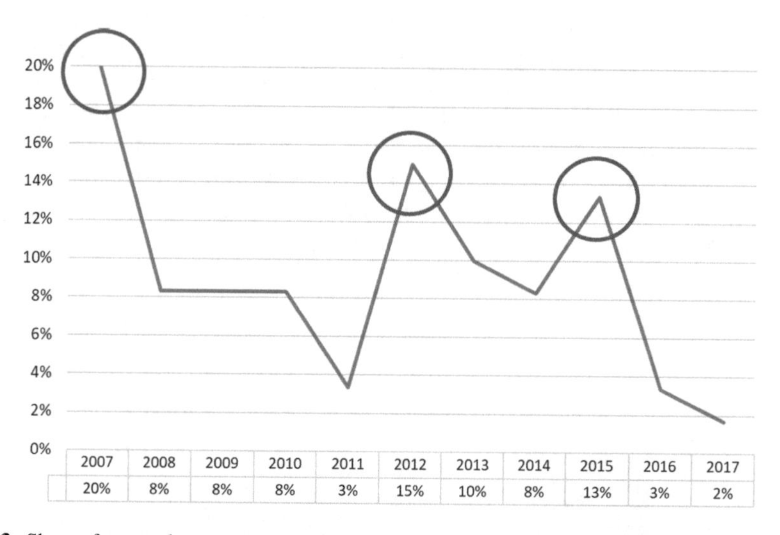

Fig. 3 Share of respondents per year in which they first came into contact with NeuroIS (N = 60)

researchers. This finding is substantiated by the fact that only 20% of our respondents are affiliated with NeuroIS since its establishment in 2007; there is a substantial number of researchers who entered the field later (e.g., 2012 and 2015, see Fig. 3).

Most of these individuals (85%) came into touch with NeuroIS through personal contacts (e.g., Ph.D. students through their professors who had previously attended the NeuroIS Retreat), but also NeuroIS publications were an important source of information (28%). The website www.NeuroIS.org and conference calls were also of some importance (point of contact for 13% of respondents each), but not comparable to word-of-mouth spread throughout the NeuroIS community and related communities such as the more general IS community.

3 Topics

We asked respondents about the NeuroIS topics on which they had focused in their previous research and the topics they think were most important in NeuroIS research in the past decade. As our respondents had the possibility to indicate more than one topic (or construct), we ended up with a list of more than 40 different NeuroIS topics. Here we report the topics which were mentioned by at least 10% of our respondents as a current or future focus in their research or as being amongst the most important NeuroIS topics in the past decade. Through some abstractions (e.g., grouping "emotional responses" and "affective processing" into the category "Emotions"), we ended up with eight main topics (see Fig. 4).

We first looked at the current and future focus in the research of our respondents (see the blue and orange bars in Fig. 4) and found that topics which are established in neuroscience (or related fields such as neuropsychology or neuroergonomics) such as cognitive load, emotions, and stress, are also amongst the most popular topics in NeuroIS research. As shown in Fig. 4, it can be expected that there will be a stronger focus on emotion in future research. In the case of other popular topics (e.g., technology acceptance or trust), our respondents were not so certain whether they will still focus on these topics in their future research. These findings are also in line with a recent review, which showed that cognitive and emotional processes

Fig. 4 NeuroIS topics with share of respondents who currently focus on them (blue bar) and will focus on them in the future (orange bar), importance of the topic in the past decade (green bar) and calls for more attention in future research (yellow bar) (N = 60)

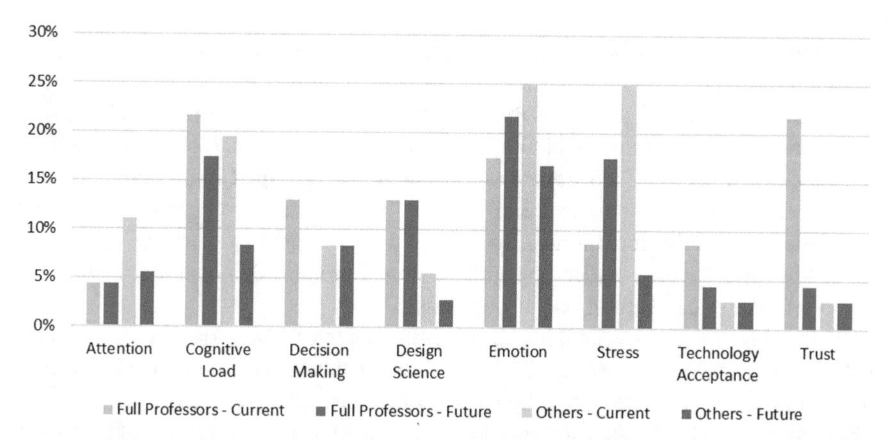

Fig. 5 NeuroIS topics with share of full professors who currently focus on them (light blue bar) and will focus on them in the future (dark blue bar), and researchers with a different tenure status who currently focus on them (light red bar) and will focus on them in the future (dark red bar) (N = 59)

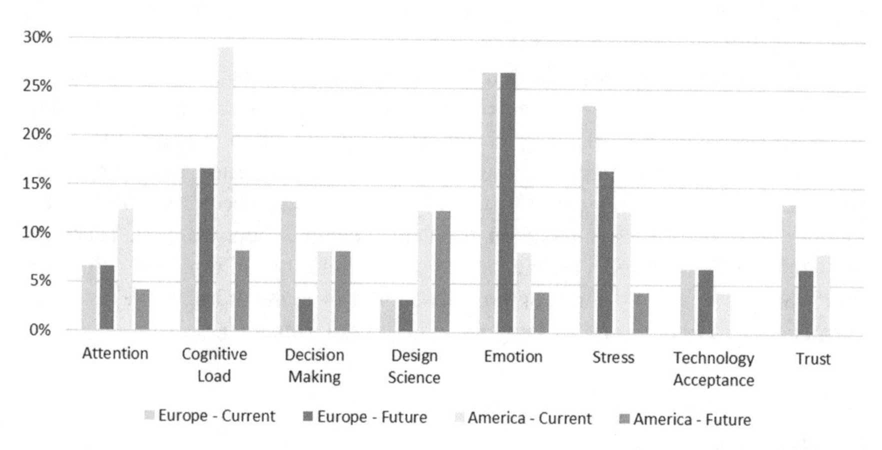

Fig. 6 NeuroIS topics with share of respondents from Europe who currently focus on them (light green bar) and will focus on them in the future (dark green bar), and researchers from North America who currently focus on them (light yellow bar) and will focus on them in the future (dark yellow bar) (N = 54)

have been the main focus in the extant NeuroIS literature, while decision-making processes and social processes were of lower importance [1].

In addition, we asked the respondents to indicate the topics that they felt had been the most important ones in the first decade of NeuroIS research (green bars, Fig. 4) and whether these topics should receive more attention (yellow bars, Fig. 4). Interestingly, emotion is not amongst the top 3 of the most important topics. Instead, most respondents felt that trust was amongst the most important topics, in addition to stress and cognitive load. This result is plausible because early NeuroIS publications in top IS journals had a focus on trust, such as [2]. Still, emotion as a topic received the most votes (i.e., 17%) when it came to the topics that should receive more attention in future research.

We further analyzed the topics that respondents focus on in their current and will focus on in their future research, based on two respondent characteristics, namely their tenure status and their country of employment, grouped into continents. For the tenure status, we looked at the differences between full professors (39% of our respondents) and the remaining respondents. For the country of employment, we looked at differences between researchers from Europe (50% of our respondents) and North America (40% of our respondents).

Regarding the current and future research topics of full professors, we found noteworthy differences (see Fig. 5). In general, most respondents who are currently not full professors are uncertain about the topics on which they will focus in their future research (which can, for example, be explained by the uncertainty of the future funding of their research). Full professors rather than the remaining respondents indicated that they will focus more strongly on emotions, as well as stress, in their future research, while decision-making and trust are topics of lower interest. For most other topics (e.g., attention, cognitive load, or design science) we observe equal interest by full professors in the future.

We also found differences regarding the thematic focus of researchers from Europe and North America (Fig. 6). While emotions and stress are more prevalent topics for European researchers, particularly design science is a topic that is more prevalent in the research of American researchers (note that design science, as defined in our research context, does not necessarily imply systems engineering activities, which are often typical for researchers from German-speaking countries, [3]). There will also likely be some shifts in the thematic focus, with European researchers focusing more strongly on attention and cognitive load research in the future, while American researchers will likely more strongly focus on decision-making.

4 Methods

We also asked the respondents about data collection methods they had previously used in their NeuroIS research, which methods they may use in the future, and whether they thought that certain methods should receive more or less attention in future NeuroIS research. We included a total of 13 data collection methods in our survey (i.e., blood pressure, heart rate related-measures, eyetracking, EMG, EEG, fMRI, NIRS, skin conductance-related measures, hormone measures based on blood, urine, or saliva, neurological patients, and transcranial direct current stimulation). In Figs. 7 and 8, we have summarized the results for each of these methods regarding (1) how many respondents have used them before ("previous use", blue bar), and (2) how those respondents who did not use a method before, intend to use it in the future ("intended use", orange bar; e.g., 20% of the respondents used hormone measures from saliva before and an additional 20% intend to use it in the future). In the case of previously used methods, eyetracking is on top with 58% of respondents

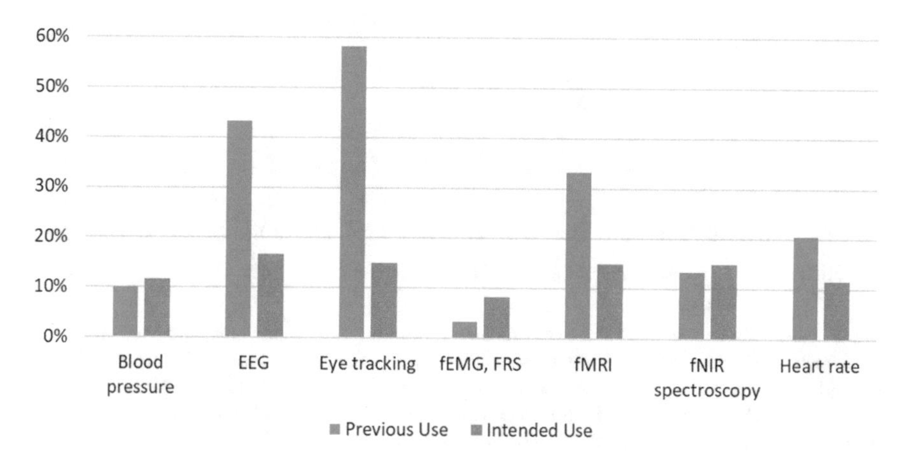

Fig. 7 NeuroIS methods with share of respondents who have previously used them (blue bar) and intent to use them in the future amongst previous non-users (orange bar) (N = 60), Part 1

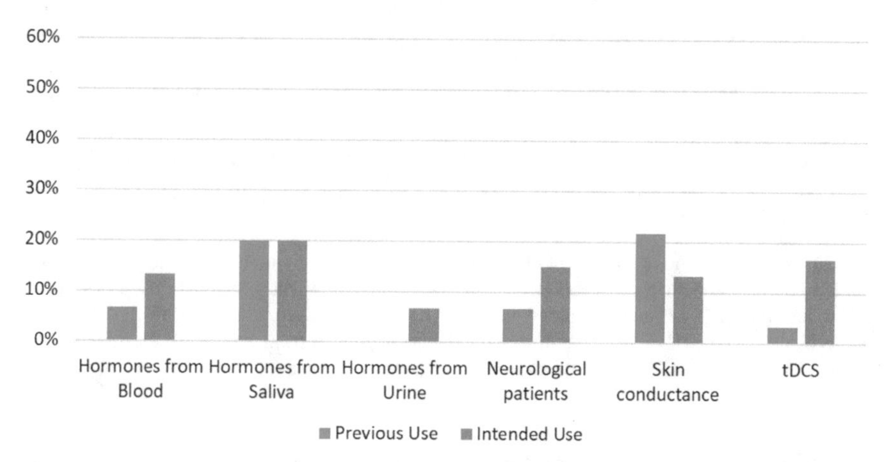

Fig. 8 NeuroIS methods with share of respondents who have previously used them and intent to use them in the future amongst previous non-users (N = 60), Part 2

indicating that they had used this method in their research. For intended use, hormone measurements based on saliva are in the lead, with 20% of respondents indicating that they would like to use this method in their future research (see Fig. 7).

In addition to eyetracking, which is widely employed and will also likely receive further attention in the future, particularly measures that collect data related to processes of the central nervous system (i.e., EEG, fMRI, NIRS, tDCS, and to some extent neurological patients) will be part of the future research of our respondents. It is interesting to see though, that saliva measurements may become more popular in the future as they can, for example, be used to measure physiological stress based on alpha amylase levels as indicator (e.g., [4]). Because the use of saliva samples, if compared to central nervous system measurements, implies less research effort and

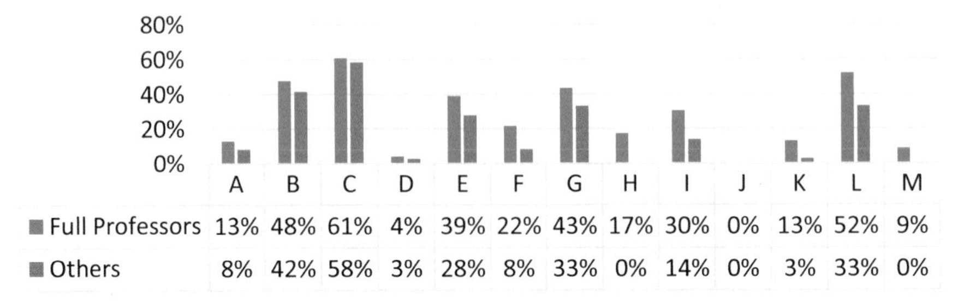

Fig. 9 NeuroIS methods with share of respondents with full professor status who have previously used them (dark blue bar) and share of respondents with other tenure status who previously used them (dark orange bar) (N = 59). Legend: (A) blood pressure, (B) EEG, (C) Eyetracking, (D) fEMG, (E) fMRI, (F) fNIRS, (G) HR, (H) Hormones from Blood, (I) Hormones from Saliva, (J) Hormones from Urine, (K) Neurological Patients, (L) Skin Conductance, (M) tDCS

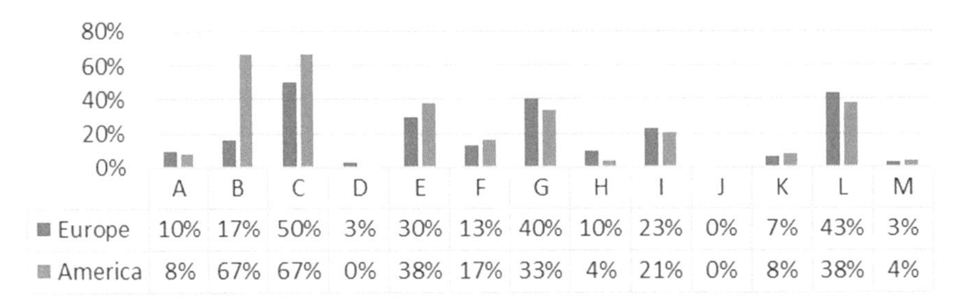

Fig. 10 NeuroIS methods with share of respondents from Europe who have previously used them (dark green bar) and share of respondents from North America who previously used them (dark yellow bar) (N = 54). Legend: (A) blood pressure, (B) EEG, (C) Eyetracking, (D) fEMG, (E) fMRI, (F) fNIRS, (G) HR, (H) Hormones from Blood, (I) Hormones from Saliva, (J) Hormones from Urine, (K) Neurological Patients, (L) Skin Conductance, (M) tDCS

causes lower costs, it seems that many NeuroIS researchers base their research tool selection on pragmatic reasons. Why the intended use of measurements related to autonomic nervous system activity (e.g., heart rate, skin conductance) is rather low in our sample (despite its enormous potential in IS research, see [5]) remains an open question that deserves further investigation.

Some respondents also mentioned additional methods that should be of importance in future NeuroIS research including voxel-based morphometry (VBM), Magnetoencephalography (MEG), genetic measures, measurements made using data from everyday devices (e.g., smartwatches, see [6]), combinations of methods (e.g., eyetracking and fMRI, see [7]) and behavioral measures such as mouse cursor tracking.

In Figs. 9 and 10, we provide an overview of the differences concerning the previous use of NeuroIS methods among our respondents based on tenure status (Fig. 9) and country of employment (Fig. 10).

Based on tenure status, we hardly find differences, though the share of full professors who have used fNIRS (F), hormones from blood (H) or saliva (I), as well as neurological patients (K), in their research is considerably larger than the share of respondents with other tenure status. This could, for example, be explained by the complexity of getting access to the involved materials and data (e.g., in the case of hormones and neurological patients) or the novelty and cost of research methods (e.g., fNIRS), which makes access to these methods harder for individuals with lower tenure status.

We also analyzed differences in previous method use based on the country of employment, again clustered by continent (Fig. 10). We find tendencies for European researchers to more frequently employ methods that can be used to measure the activity of the autonomic nervous system (e.g., (G) heart rate or (L) skin conductance), while North American researchers more frequently employ methods that can be used to measure the activity of the central nervous system (e.g., (B) EEG or (E) fMRI). The largest differences can be found for (B) EEG and (C) Eyetracking, which are more frequently used by North American researchers. Future research must determine the reasons for the observed differences.

5 Conclusion

Based on an online survey among N = 60 former participants of the NeuroIS Retreat, we found that emotional processes will likely be a key topic, eventually *the* key topic, in future NeuroIS research. Methodologically, we found that eyetracking measures and brain-related measures such as EEG or fMRI will be of high relevance in the future. These findings are in line with observations in the NeuroIS literature, as emotional processes have been of major importance in previous NeuroIS research [1], and NeuroIS publications in the most prestigious IS journals have often applied brain-related measures such as fMRI (e.g., [2, 8–12]). Importantly, our survey also revealed the interest of NeuroIS researchers in methods which have not been used frequently thus far, such as EMG and FRS (Face Recognition Software, e.g. to determine user emotion, for details see [13]). It seems that NeuroIS researchers have realized that these and further methods are well suited to reveal insights into the NeuroIS topics of the future (e.g., [14]).

Acknowledgements This research was funded by the Upper Austrian government as part of the Ph.D. program "Digital Business International", a joint initiative between the University of Applied Sciences Upper Austria and the University of Linz.

References

1. Riedl, R., Fischer, T., Léger, P.-M.: A decade of NeuroIS research: status quo, challenges, and future directions. In: Kim, Y.J., Agarwal, R., Lee, J.K. (eds.) Proceedings of the Thirty Eighth International Conference on Information Systems (2017)
2. Dimoka, A.: What does the brain tell us about trust and distrust? Evidence from a functional neuroimaging study. MIS Q. **34**, 373–396 (2010)
3. Heinrich, L.J., Riedl, R.: Understanding the dominance and advocacy of the design-oriented research approach in the business informatics community: a history-based examination. J. Inf. Technol. **28**, 34–49 (2013)
4. Tams, S., Hill, K., Ortiz de Guinea, A., Thatcher, J., Grover, V.: NeuroIS—alternative or complement to existing methods? Illustrating the holistic effects of neuroscience and self-reported data in the context of technostress research. J. Assoc. Inf. Syst. **15**, 723–753 (2014)
5. Riedl, R., Davis, F.D., Hevner, A.: Towards a NeuroIS research methodology: intensifying the discussion on methods, tools, and measurement. J. Assoc. Inf. Syst. **15**, i–xxxv (2014)
6. Fischer, T., Riedl, R.: Lifelogging as a viable data source for NeuroIS researchers: a review of neurophysiological data types collected in the lifelogging literature. In: Davis, F.D., Riedl, R., Vom Brocke, J., Léger, P.-M., Randolph, A.B. (eds.) Information Systems and Neuroscience. Gmunden Retreat on NeuroIS, pp. 165–174. Springer, New York (2016)
7. Vance, A., Jenkins, J.L., Anderson, B.B., Bjornn, D.K., Kirwan, C.B.: Tuning out security warnings: a longitudinal examination of habituation through fMRI, eye tracking, and field experiments (forthcoming) (2018)
8. Riedl, R., Hubert, M., Kenning, P.H.: Are there neural gender differences in online trust? An fMRI study on the perceived trustworthiness of eBay offers. MIS Q. **34**, 397–428 (2010)
9. Anderson, B.B., Vance, A., Kirwan, C.B., Jenkins, J.L., Eargle, D.: From warning to wallpaper: why the brain habituates to security warnings and what can be done about it. J. Manage. Inf. Syst. **33**, 713–743 (2016)
10. Jenkins, J.L., Anderson, B.B., Vance, A., Kirwan, C.B., Eargle, D.: More harm than good? How Messages that interrupt can make us vulnerable. Inf. Syst. Res. **27**, 880–896 (2016)
11. Riedl, R., Mohr, P.N.C., Kenning, P.H., Davis, F.D., Heekeren, H.R.: Trusting humans and avatars: a brain imaging study based on evolution theory. J. Manage. Inf. Syst. **30**, 83–114 (2014)
12. Warkentin, M., Walden, E.A., Johnston, A.C., Straub, D.W.: Neural correlates of protection motivation for secure IT behaviors: an fMRI examination. J. Assoc. Inf. Syst. **17**, 194–215 (2016)
13. Riedl, R., Léger, P.-M.: Fundamentals of NeuroIS: Information Systems and the Brain. Springer, Berlin (2016)
14. Koller, M., Walla, P.: Measuring affective information processing in information systems and consumer research—introducing startle reflex modulation. In: George, J.F. (ed.) Proceedings of the 33rd International Conference on Information Systems (2012)

Improving Security Behavior Through Better Security Message Comprehension: fMRI and Eye-Tracking Insights

Anthony Vance, Jeffrey L. Jenkins, Bonnie Brinton Anderson, C. Brock Kirwan and Daniel Bjornn

Abstract Security warnings are critical to help users make contextual security decisions. Unfortunately, users find these warnings hard to understand, and they routinely expose themselves to unintended risks as a result. Although it is straightforward to determine when users fail to understand a warning, it is more difficult to pinpoint *why* this happens. The goal of this research is to use eye tracking and fMRI to step through the building blocks of comprehension—attention, semantics, syntax, and pragmatics—for SSL and other common security warnings. Through this process, we will identify ways to design security warnings to be more easily understood.

Keywords NeuroIS · Eye-tracking · fMRI · Comprehension
Security messages

1 Introduction

Users routinely disregard protective messages such as software security warnings [2, 3]. One reason for the ineffectiveness of warnings is the mismatch between security concerns and security behavior. For example, individuals' stated security concerns have been found to be inconsistent with their subsequent behavior in response to

A. Vance
Temple University, Philadelphia, PA, USA
e-mail: anthony@vance.name

J. L. Jenkins · B. B. Anderson (✉) · C. Brock Kirwan · D. Bjornn
Brigham Young University, Provo, UT, USA
e-mail: bonnie_anderson@byu.edu

J. L. Jenkins
e-mail: jeffrey_jenkins@byu.edu

C. Brock Kirwan
e-mail: kirwan@byu.edu

D. Bjornn
e-mail: dbjornn@byu.edu

F. D. Davis et al. (eds.), *Information Systems and Neuroscience*,
Lecture Notes in Information Systems and Organisation 29,
https://doi.org/10.1007/978-3-030-01087-4_2

security warnings [11]. These empirical results confirm those of Crossler et al. [5], who called for research that explains the discrepancy between security intentions and behaviors.

One important factor contributing to the disconnect between security concerns and actual behavior is the lack of comprehension. For example, in the case of security warnings, although users may intend to behave securely, they may not comprehend a security warning, which may in turn lead them to make a choice that unintentionally exposes themselves to security risks.

Past research on comprehension of security warnings has highlighted the difficulty users have in understanding security warnings. Felt et al. [6] tested several iterations of text and design for SSL warnings in Google Chrome. They found that users routinely had difficulty determining the threat source and data risk, even after designing interventions to improve comprehension.

However, comprehension is not a binary event, but rather involves interrelated stages that lead to understanding. These stages include [9]:

1. Attention—focused mental resources on a certain object.
2. Semantics—the meaning of individual words and simple phrases.
3. Syntax—the structure of sentences that creates relationships between words.
4. Pragmatics—the application of past experience and knowledge to infer meaning.

The research objectives of this study are to: (1) use eye tracking, fMRI, and users' behavioral responses, through a series of complementary experiments, to determine failures at each of the above stages of comprehension for security warnings. Through this process, we will (2) identify ways to design security warnings to improve comprehension at each stage.

2 Planned Research and Expected Outcomes

2.1 Past Research on Comprehension of Security Warnings

Poor comprehension of security warnings is a common finding in the human–computer interaction literature. For example, researchers found that Android users paid attention to app permissions during installation only 17% of the time, and only 3% of users could correctly answer comprehension questions about permissions they saw [7]. Similarly, in a later study they found that users comprehended the threat source of SSL warnings in Chrome only 37.7% of the time, and comprehended even less what data was at risk. By changing the warning design based on recommendations from warning literature, they improved threat source comprehension nearly 12%. However, the design was not able to improve the comprehension of the risk to data [6].

We build on this past literature by applying behavioral information security to better understand and improve users' security behaviors [1]. Based on our findings, we expect to be able to determine more precisely where and why warning comprehension breaks down both from a neural and behavioral perspective. This will, in turn, allow us to create guidelines to improve comprehension in security warnings.

Previous work on comprehension using eye tracking found that more complex sentence structures result in poorer comprehension. For example, participants who read sentences with confusing ("ambiguous") syntax had poorer ability to answer simple questions about the sentences correctly compared to similar sentences that were changed slightly to be less confusing ("disambiguated"). Specifically, comprehension accuracy decreased by 15–38% when syntax was complex. This impaired comprehension was paired with significantly more re-reading of the ambiguous sentences (27–60% more time spent re-reading). In summary, not only does complex syntax impair comprehension, but re-reading is a reliable indicator of this impairment [4].

2.2 *Description of Project and Expected Outcomes*

To achieve our research objectives, we will record eye tracking data to step through the stages of comprehension (see Fig. 1). Comparable to code debugging, we will work through the different stages of comprehension to determine where comprehension is impeded. We will then improve warning designs to increase attention, ensure clear semantic and syntactic understanding, and promote pragmatic cognition. For example, at the level of attention, use of symbols or animation may help to improve overall attention. Similarly, semantic understanding may be improved through use of more familiar terms, or increased word frequency. By examining each stage individually, we expect to improve comprehension overall.

Eye tracking is an ideal tool for measuring the moment-by-moment allocation of attention. It is also used in psychology and linguistics to explore how people understand written language and to measure comprehension difficulty. For instance, words that are less familiar or unexpected (semantics) are looked at longer, and complex or confusing sentences (syntax) are re-read more often than are simple sentences [10]. In contrast to eye tracking, fMRI can provide information about the underlying neural and cognitive operations in attentional, semantic, and syntactic processing [8].

2.3 *Hypotheses*

We propose an eye-tracking experiment that examines the influence of syntax on users' comprehension of warnings. We will examine whether changing the syntax of the warning, to place the focus on different aspects of the warning, influences the likelihood of a data security breach. In addition to the usual focus on the attacker or

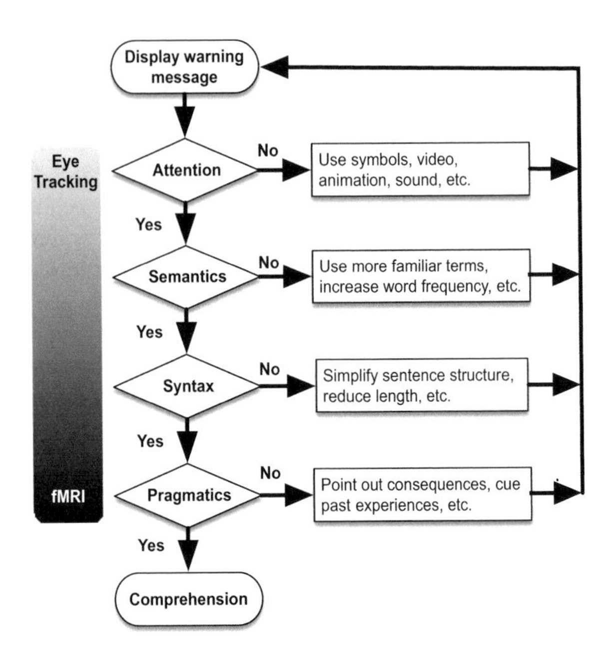

Fig. 1 Evaluating warnings at different stages of comprehension using eye tracking and fMRI

the target website, we will also include a condition where the syntax of the warning shifts the focus to the consequences of ignoring the warning. We hypothesize that:

> Hypothesis 1—changing the focus of the warning will result in significant differences in comprehension as evidenced by a significant difference in the number of regressions (i.e., rereading) across warning focus.

Eye movement regressions are often used as a non-conscious measure of reading comprehension. As such, they may be more sensitive to subtle differences in comprehension between the different warning focus conditions in our experiment. Additionally, syntax changes should result in differences in overt comprehension as measured by performance on post hoc comprehension questions. We hypothesize that:

> Hypothesis 2—eye regressions in turn will significantly predict whether participants correctly understand the warnings as measured by performance on a post hoc comprehension quiz.

3 Eye Tracking Pilot Study

3.1 Participants and Stimuli

A total of 43 college-age individuals (14 male, 29 female) participated in the study. Five participants were not able to fully participate because an accurate calibration was not obtained. Removing these five participants left the sample with 38 individuals (14 male, 24 female). Participants were given course credit for participating in the study.

Warnings were created by sampling four warning types from the Google Chrome browser and the Apple Safari browser, namely malware, phishing, SSL, and unwanted software. The text for the warnings was then manipulated by changing the subject, verb, and object of the statement. For example, warnings from Chrome focus on the attacker as the subject of the statement. An example of this focus can be seen from the SSL warning text, "Attackers might be trying to steal your information from expired.badssl.com (for example, passwords, messages, or credit cards)." Warnings from Safari focus on the website as the subject of the statement. Chrome warning text was manipulated to change the focus to the website and Safari warning text was manipulated to change the focus to the attacker.

Along with the focus on the attacker and the website, a third text condition focused on the potential consequences of ignoring the warning. For example, the chrome SSL warning could be changed to, "Your information from expired.badssl.com (for example, passwords, messages, or credit cards) might be stolen if you visit it." Text from Chrome and Safari warnings were manipulated to fit this design.

The four warning types (i.e., malware, phishing, SSL, and unwanted software) for two browsers (i.e., Chrome and Safari) across three different conditions (i.e., attacker focus, consequence focus, and site focus) provided 24 different warnings. All references to a specific website were changed to "this website" for ease of presentation.

The warning text was overlaid onto a mock warning image for each trial. Warning titles were created from the standard text from the warning type for each browser (e.g., "Your connection is not private" for the Chrome SSL warning and "This Connection Is Not Private" for the Safari SSL warning).

3.2 *Task*

Participants viewed each warning one at a time on the computer screen and then answered a question. Each trial began with a drift check, which required participants to look at a circle on the top left part of the screen and press the spacebar to continue. The warning was then presented and participants read the warning and pressed the spacebar when they were ready to continue. The last part of each trial was the comprehension question which asked, "If this were a real threat and I ignored this warning, an attacker could," and then presented four answer options. Each of the answer options corresponded to a warning type:

- Phishing—"Trick me into installing malicious software or disclose personal information"
- Malware—"Install a dangerous program on my computer that could steal my information or delete my data"
- SSL—"See anything I send or receive from the website"

- Unwanted Software—"Install software that displays ads on my computer or makes changes to my browser".

The answer options were presented in a random order for each trial. The full task consisted of 24 trials. The eye tracker was calibrated before the start of the task and after every 6 trials. Warnings were presented in a random order for each participant.

3.3 *Planned Analysis*

The results of this experiment will be analyzed by examining the behavioral and eye tracking measures of comprehension. For the behavioral analyses, we will calculate the proportions correct for warning type and warning focus separately. Repeated measures ANOVA tests will be run to test these factors individually in order to ensure a large enough number of trials for each bin.

For hypothesis 1, we will test whether warning focus predicts the number of regressions (i.e., rereading the text) by entering the total number of regressions for each trial as a dependent variable into a linear regression model with an independent variable of warning focus. For hypothesis 2, we will test whether the number of regressions predict accuracy on the comprehension test by entering trial accuracy as a dependent variable into a linear regression model with an independent variable of the total number of regressions in the trial. We will also use comprehension of other, non-security, messages and warnings for a comparison. Our post-study survey will contain the standard demographic, education and computer experience, as well as security risk questions, big five personality traits, and general risk propensity profile.

4 Conclusion

Users often respond inappropriately to security warnings. A significant factor in this failure is users' difficulty in comprehending warnings. The insights expected to be gained from this research have the potential to inform the design and evaluation of warnings that more effectively help users to respond to security threats, enhancing the information security of individuals and organizations.

References

1. Anderson, B.B., Kirwan, C.B., Jenkins, J.L., Eargle, D., Howard, S., Vance, A.: How polymorphic warnings reduce habituation in the brain: insights from an fMRI study. In Proceedings of the ACM Conference on Human Factors in Computing Systems (CHI) ACM, Seoul, South Korea (2015)

2. Anderson, B.B., Vance, A., Kirwan, C.B., Eargle, D., Jenkins, J.L.: How users perceive and respond to security messages: a NeuroIS research agenda and empirical study. Eur. J. Inf. Syst. **25**(4), 364–390 (2016)
3. Bravo-Lillo, C., Komanduri, S., Cranor, L.F., Reeder, R.W., Sleeper, M., Downs, J., Schechter, S.: Your attention please: designing security-decision uis to make genuine risks harder to ignore. In Proceedings of the Ninth Symposium on Usable Privacy and Security ACM, Newcastle, United Kingdom, pp. 1–12 (2013)
4. Christianson, K., Luke, S.G., Hussey, E.K., Wochna, K.L.: Why reread? Evidence from garden-path and local coherence structures. Q. J. Exp. Psychol. **70**(7), 1380–1405 (2017)
5. Crossler, R.E., Johnston, A.C., Lowry, P.B., Hu, Q., Warkentin, M., Baskerville, R.: Future directions for behavioral information security research. Comput. Secur. **32**(1), 90–101 (2013)
6. Felt, A.P., Ainslie, A., Reeder, R.W., Consolvo, S., Thyagaraja, S., Bettes, A., Harris, H., Grimes, J.: Improving ssl warnings: comprehension and adherence. In Proceedings of the Conference on Human Factors in Computing Systems (2015)
7. Felt, A.P., Ha, E., Egelman, S., Haney, A., Chin, E., Wagner, D.: Android permissions: user attention, comprehension, and behavior. In Proceedings of the Eighth Symposium on Usable Privacy and Security ACM, pp. 3:1–3:14 (2012)
8. Keller, T.A., Carpenter, P.A., Just, M.A.: The neural bases of sentence comprehension: a fMRI examination of syntactic and lexical processing. Cereb. Cortex **11**(3), 223–237 (2001)
9. Rayner, K.: Eye movements in reading and information processing: 20 years of research. Psychol. Bull. **124**(3), 372–422 (1998)
10. Rayner, K.: Eye movements and attention in reading, scene perception, and visual search. Q. J. Exp. Psychol. **62**(8), 1457–1506 (2009)
11. Vance, A., Anderson, B.B., Kirwan, C.B., Eargle, D.: Using measures of risk perception to predict information security behavior: insights from electroencephalography (eeg). J. Assoc. Inf. Syst. **15**(10), 679–722 (2014)

Neural Activity Related to Information Security Decision Making: Effects of Who Is Rewarded and When the Reward Is Received

Bridget Kirby, Kaitlyn Malley and Robert West

Abstract Breaches of information security resulting from cybercrime represents a significant threat to the security and well-being of individuals, corporations, and governments. Therefore, understanding the neurocognitive processes that lead individuals to violate information security policy represents a fundamental pursuit for NeuroIS researchers. In the current study, we examined the effects of whether an individual or a close associate benefited from a violation of information security, and the temporal delay before the benefit was received on event-related brain potentials (ERPs) related to ethical decision making. The electrophysiological data revealed modulations of the ERPs that were generally sensitive to ethical decision making, or that were specifically sensitive to the recipient or timing of the reward. The components that were sensitive to the two independent variables were observed over the anterior frontal region of the scalp, consistent with the neuroimaging literature demonstrating that several prefrontal structures participate in self-referent processing and intertemporal choice.

Keywords Information security · Ethical decision making
Event-related brain potentials

1 Introduction

As society has become increasingly dependent upon digital information, the impact of cybercrime has increased exponentially. Cybercrime may reduce consumer con-

---

B. Kirby · K. Malley · R. West (✉)
Department of Psychology and Neuroscience, DePauw University, Greencastle, USA
e-mail: robertwest@depauw.edu

B. Kirby
e-mail: bridgetkirby_2020@depauw.edu

K. Malley
e-mail: kaitlynmalley_2019@depauw.edu

F. D. Davis et al. (eds.), *Information Systems and Neuroscience*,
Lecture Notes in Information Systems and Organisation 29,
https://doi.org/10.1007/978-3-030-01087-4_3

fidence [1], create tenuous international relations [2], and is estimated to cost the world economy over three trillion dollars annually [3]. There have been significant advances in the field of computer science leading to enhancements of hardware and software technologies designed to deter cybercrime [4]. However, these advances may not necessarily thwart the actions of individuals within an organization [5, 6], and studies demonstrate that roughly 50% of information security violations result from insider threats [7]. Unfortunately, current deterrence programs focused on information security may not reduce the intention to commit, or the incidence rate of, cybercrime arising from inside an organization [8]. Therefore, the current study builds upon recent work from our laboratory by examining the effects of two independent variables (i.e., the recipient and timing of a benefit) on event-related brain potentials (ERPs) elicited during ethical decision making related to information security. Based upon the extant literature, these two variables are known to influence decision making in various domains [9, 10].

The Information Security Paradigm (ISP) was developed by Hu et al. [11] to measure the neural correlates of ethical decision making related to information security using ERPs. This task was adapted from a survey-based research instrument used in the information systems literature. For the ISP, individuals make decisions as if they are a fictitious IT employee named Josh. In the task, participants read a series of 1–2 sentence scenarios describing violations of information security practices that vary in their degree of severity (i.e., minor or major) or control scenarios that do not involve an ethical violation. Following the scenario, subjects are presented with a decision prompt, and decide whether or not Josh should take the action described in the scenario. Comparing ethical violation scenarios to control scenarios allows one to isolate neural activity that is generally related to ethical decision making; while comparing scenarios related to minor and major violation of information security allows one to isolate neural activity that is sensitive to the severity of the violation.

Hu et al. [11] found that the behavioral and ERP data for the ISP are sensitive to both the presence and severity of ethical violations. The choice data revealed that subjects were less likely to endorse scenarios involving an ethical violation than control scenarios; while the response time data indicated that individuals considered minor violations longer than major violations. The ERP data differentiated control, minor and major ethical violation scenarios over the lateral and medial frontal regions and the right parietal region beginning around 200 ms after the onset of the prompt. In comparison to the control scenarios, ethical violation scenarios elicited an early posterior effect on the N2, that may reflect a limit in the attentional resources available for encoding the prompt. Neural activity was also sensitive to the severity of the ethical violation. Major violations elicited greater activity over the left parietal region between 400 and 600 ms, revealing fairly early processing that was sensitive to the severity of the ethical violations [12]. In addition to the early activity occurring over the parietal region, there was sustained frontal-central-temporal activity that distinguished ethical violation scenarios from control scenarios that persisted for 1.5–2 s after onset of the prompt [11, 12].

In the current study, we utilized an adapted version of the ISP [11] and had two primary goals. First, we sought to provide a conceptual replication of the ERP findings related to our original materials. This goal allowed us to examine the generalizability of the behavioral and ERP data measured in the paradigm with a new set of scenarios and action prompts.

H1 There will be differences in ERP amplitude between the ethical scenarios and control scenarios that emerge beginning at around 200 ms over the occipital-temporal region and then continue over the parietal and frontal regions between 200 and 2000 ms.

Second, we sought to examine the effect of two independent variables (i.e., the benefactor of a reward and the timing of a reward) on the behavioral and ERP data related to decision making in the context information security. Previous research has demonstrated that the perceived benefit of a violation is a significant predictor of the intention to violate IS security policy [13], and here we sought to identify the neural basis of this effect. Additionally, the literature on intertemporal choice demonstrates that individuals are sensitive to the timing of rewards, often discounting a larger distant gain for a immediate smaller gain [9]. The functional neuroimaging literature examining the neural basis of self-referent processing and intertemporal choice has consistently revealed activation of the medial prefrontal cortex related these two constructs [9, 10]. In the ISP for the current study, the benefactor of the reward associated with the violation was either Josh or a friend/relative; and the benefit was received after either 0–3 months or 12–24 months.

H2 Individuals will be more likely to say yes to Josh benefit scenarios than Other benefit scenarios, and to short delay scenarios that to long delay scenarios.

H3 The ERPs will reveal sustained differences in amplitude between Josh versus Other benefit scenarios, and short versus long delay scenarios, over the frontal region of the scalp.

2 Method

Participants. Forty individuals participated in the study, and the demographic information for one individual was lost. The participants were on average 20 years of age; and were 82% female, 79% white, 56% were Democrats and 23% were Republicans, and participants described themselves as being politically moderate (M = 3.11) on a 7-point scale (1 = liberal, 4 = moderate, 7 = conservative).

Materials. The Information Security Paradigm represented a 3 (benefactor: Control, Josh, Other) by 2 (timing of reward: 0–3 month delay or 12–14 month delay) factorial design with eight scenarios presented for each of the six cells of the design. Control scenarios included activities that did not involve an ethical violation; Josh scenarios involved unethical behaviors that he would benefit from; and Other scenarios involved unethical behaviors that another individual would benefit from (e.g., a friend, relative, partner) and explicitly stated the identity of the third party. The 48 scenarios were presented in a different random order for each individual. Scenarios were limited to 300 characters; and prompts were limited to 50 characters and were posed in the form of a question. The prompts did not mention the nature, benefactor, or timing of the reward. Individuals were given an unlimited amount of time to read the scenario and then pressed the spacebar to view the decision prompt. The response time and ERP data were time locked to the onset of the prompt. Individuals responded on a 4 points scale (No, Likely No, Likely Yes, Yes) using the C-V-B-N keys of the keyboard.

Procedure. After arriving at the laboratory for the study, individuals were given a brief overview of the procedure and provided signed informed consent. Individuals then completed a demographic survey and several questionnaires measuring individual differences related to self-control, moral foundations, media exposure, pathological gaming, and grit. After completing the scales, individuals were fitted with a 32 electrode actiCAP and completed the ISP, moral foundations task, and a picture rating task while EEG was recorded. Following this, individuals were debriefed and compensated with either course credit or $15.

EEG recording and analysis. The EEG was recorded from a 32 channel actiCHamp system using the Brain Vision Recorder software and a standard 32 channel actiCAP scalp montage where CP5–CP6 were replaced with active electrodes located below the eyes. During recording the electrodes were grounded to electrode Fpz and referenced to electrode Cz, for data analysis the data were re-referenced to the average reference. The EEG was digitized at 500 Hz and then bandpass filtered between .1 and 30 Hz using an IIR filter implemented in ERPLAB (5.1.1.0) [14] for the analyses. Ocular artifacts associated with blinks and saccades were corrected with ICA implemented in EEGLAB (13.6.5b) [15]. Trials including other artifacts were rejected before averaging using a $\pm 100\ \mu V$ threshold. ERPs were averaged for Control, Josh, and Other scenarios, or Short and Long scenarios from −200 to 2000 ms around onset of the prompt, and mean voltage measurements were made using the measurement tool in ERPLAB. Two to four electrodes were included in the analyses of the mean differences, with most analyses including three electrodes.

3 Results

Behavioral Data. The response choice and response time data were analysed in a set of 3 (scenario: Control, Josh, Other) by 2 (timing: Short or Long delay) ANOVAs (Table 1). The analysis of response choice revealed a significant main effect of scenario, $F(2, 78) = 214.45$, $p < .0.001$, representing a decrease in the likelihood of responding yes from Control to Josh scenarios, $t(39) = 17.68$, $p < .0.001$, and from Josh to Other scenarios, $t(39) = 2.58$, $p = .036$. The difference between Josh and Other scenarios provides support for Hypothesis 2. The main effect of timing was also significant, $F(1, 39) = 11.52$, $p = .002$, revealing that individuals were less likely to respond yes for Short delay scenarios than Long delay scenarios; and a significant interaction, $F(2, 78) = 9.63$, $p < .001$. This finding does not support Hypothesis 2. This interaction resulted from the effect of timing being significant for Josh benefit

Table 1 Descriptive data for choice and response time (in milliseconds) for the ISP

	Cont. short	Cont. long	Josh short	Josh long	Oth. short	Oth. long
Choice M	2.88	2.96	1.60	1.91	1.64	1.62
SD	0.36	0.31	0.48	0.60	0.38	0.50
RT M	2052	2194	1926	2290	2085	1990
SD	778	613	706	1024	776	745

scenarios, $t(39) = 4.93$, $p < .001$, but not for Other, $t(39) = .47$, $p = .64$, or Control, $t(39) = 1.40$, $p = .17$, scenarios. The results of this analysis reveal that individuals were more likely to endorse an unethical behavior that results in a longer term personal gain.

The analysis of response time revealed a nonsignificant main effect of scenario, $F < 1.00$, and a significant main effect of timing, $F(2, 39) = 4.90$, $p = .033$, revealing shorter response times for Short delay scenarios than for Long delay scenarios. The scenario by timing interaction was significant, $F(2, 78) = 4.87$, $p = .01$, and resulted from shorter response times for Short than Long delay scenarios when Josh benefitted, $t(39) = 3.19$, $p = .003$, but not for Other, $t(39) = .99$, $p = .33$, or Control, $t(39) = 1.37$, $p = .18$, scenarios. The results of this analysis reveal that individuals may have thought longer about decisions related to unethical behaviors they were more likely to accept (i.e., the Josh Long delay scenarios).

ERP Data. For the ERP data we examined three comparisons: (1) Differences between Control scenarios and Josh and Other scenarios—collapsed across short and long delay scenarios—were considered to identify neural activity generally related to ethical decision making. (2) Differences between Josh and Other scenarios were considered to identify neural activity related to self-referent processing. (3) Differences between Short and Long delay scenarios—collapsed across Josh and Other scenarios—were considered to identify the effect of temporal delay.

The comparison of Control scenarios versus Josh and Other scenarios provide support for Hypothesis 1, revealing differences in the ERPs between conditions beginning around 200 ms over the occipital region, that were then broadly distributed over the scalp including the parietal, central, and frontal regions until 2000 ms after onset of the prompt (Fig. 1a; Table 2). The comparison of Josh and Other trials revealed sustained ERP activity over the anterior frontal region (electrodes Fp1–Fp2, F3–F4, Fig. 1b) between 300 and 1500 ms after onset of the prompt, $F(1, 39) = 5.06$, $p = .03$, reflecting greater negativity for Other scenarios ($M = -1.12$ μV) that for Josh scenarios ($M = -.47$ μV). The comparison of Short and Long delay scenarios revealed sustained ERP activity over the anterior frontal region (electrodes Fp1–Fp2, Fig. 1c) between 500 and 1500 ms after onset of the prompt that was marginally significant, $F(1, 39) = 4.07$, $p = .051$, and reflected greater negativity for Long delay scenarios ($M = -.83$ μV) than for Short delay scenarios ($M = .20$ μV). Both of these analyses provide support for Hypothesis 3.

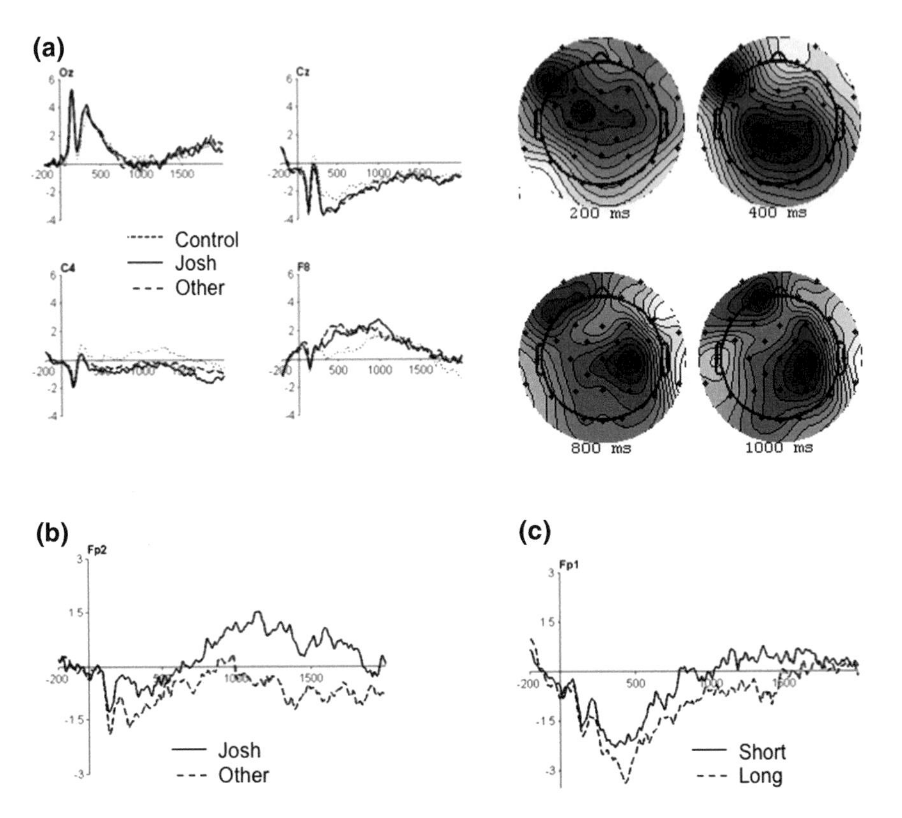

Fig. 1 **a** ERPs and scalp topography maps (Josh—Control) demonstrating the timing and distributions of ERP activity that differed between ethical scenarios and control scenarios. **b** ERPs demonstrating the slow frontal activity that differ for Josh versus other scenarios. **c** ERPs demonstrating the slow frontal activity that differed for Short and Long delay scenarios

Table 2 Mean voltage in microvolts and omnibus F- and p-values for the comparisons of control, josh, and other scenarios

	Occipital 200–300	Central 200–500	Parietal 350–600	RT central 350–1200	RT frontal 300–1000	LT parietal 1000–2000
Control	1.53	−1.73	−.01	.77	.83	−.60
Josh	2.51	−2.49	−.91	.04	1.82	.55
Other	2.84	−2.67	−.60	.07	2.19	.63
F, p	6.56, .002	7.75 < .001	5.42, .006	8.01 < .001	6.04, .004	9.63 < .001

Note Post hoc comparisons revealed the that Control scenarios differed from Josh and Other scenarios, that did not differ from one another

4 Discussion

The first goal of the study was to examine the generalizability of the ISP using a modified set of scenarios and prompts. Supporting Hypothesis 1, the comparison of the ERPs elicited for Control scenarios relative to Josh and Other scenarios revealed differences in ERP amplitude between control scenarios and those that included an ethical violation that were similar in time course and topography to the findings of our previous research [12]. These findings indicate that the ISP provides an assay of a core neural network involving structures within the occipital-temporal, parietal, and lateral and medial frontal cortex that contribute to ethical decision making as related to information security. Additionally, there appears to be considerable overlap between the neural correlates of ethical decision making related to information security and traditional moral reasoning tasks [16, 17]. The greater negativity over the medial frontal region for ethical scenarios relative to control scenarios may reflect conflict processing within the ACC or medial frontal cortex that arises when considering an unethical action. Extensive work using ERPs has associated the ACC with conflict processing [18], and within the ISP there is greater medial frontal negativity when individuals accept rather than reject an unethical action [19]. A finding that is consistent with the moral reasoning literature wherein the ACC is more active for difficult decisions [17]. Slow anterior and lateral frontal ERP activity is consistently observed following more transient medial frontal activity during conflict processing in cognitive control and gambling tasks [20]; in the ISP this slow wave activity may reflect deliberative processing that reflects the weighing of the benefit to be gained against to ethical violation represented in the scenarios.

The second goal of the study was to explore the effects of two independent variables on the neural correlates of ethical decision making in the ISP. The behavioral data provide partial support for Hypothesis 2, revealing an interaction between the independent variables that reflected a greater acceptance, and longer consideration, of long delays trials when Josh benefited relative to the other three types of scenarios involving ethical violations. These data are consistent with previous evidence demonstrating that perceived benefit is a predictor of the intention to violation information security policies [13]. Together with existing evidence, our data indicate that the benefit of a violation of information security may relate to both outcomes that might be mutually positive for the decision maker and organization (e.g., time savings) [13] or be limited to the decision maker (e.g., Josh vs. Other scenarios).

Supporting Hypothesis 3, the benefactor of the reward and the timing of the reward were associated with differences in ERP activity over the anterior frontal region between 300 or 500 and 1500 ms after onset of the prompt. The topography of the effect of these variables on the ERPs was somewhat different from those of the ERPs that distinguished Josh and Other benefit scenarios from the Control scenarios. These findings converge with the neuroimaging literature revealing that self-referent processing and intertemporal choice are consistently related to activity within the anterior frontal cortex [9, 10], and are consistent with the idea that ethical or moral reasoning arises from the recruitment of more general neurocognitive processes rather than proprietary neural circuits that are dedicated to ethical decision making [17].

There are some limitations of the study that should be acknowledged. The sample was predominantly white female undergraduates. There is continued development of the prefrontal cortex into early adulthood that may influence ethical reasoning and decision making, so it may be worthwhile in future studies to examine neural activity in the ISP in a sample in their late 20s or 30s once they entered the workforce. There is also some evidence demonstrating cultural differences in the adoption of information security practices [21], indicating that it could be useful to explore cultural variation in the ISP. Finally, we are in the process of balancing the gender distribution within the sample to examine the possible influence of variables that may differ between males and females and that are related to information security (e.g., video game experience) [22].

In summary, the current findings demonstrate that the ISP provides a sound methodological foundation for probing the activity of a neural system that underpins ethical decision making related to information security, in addition to neural systems associated with other constructs (e.g., perceived benefit and temporal delay) that may influence decision making in this domain. Additionally, other research from our laboratory demonstrates that ERPs elicited during the ISP are sensitive to individual differences in self-control and moral beliefs [11, 12]; variables that predict the occurrence of hacking behavior [22, 23] or the intention to violation information security policy [13]. Finally, we are encouraged by the overlap in the neural systems underpinning decision making related to information security and moral reasoning more generally, and believe that this convergence has the potential to facilitate synergistic collaborations between scholars with interests in information systems, cognitive and decision neuroscience, and moral reasoning.

References

1. Yayla, A.A., Hu, Q.: The impact of information security events on the stock value of firms: the effect of contingency factors. J. Inf. Technol. **26**, 60–77 (2011)
2. Groll, E.: Russian Interference Went Far Beyond DNC Hack, Senate Panel Hears. Foreignpolicy.com (2017, March 30)
3. Cybersecurity Ventures Cybereconomy Infographic: http://cybersecurityventures.com/cybereconomy-infographic/ (2017)
4. Ayuso, P.N., Gasca, R.M., Lefevre, L.: A cluster-based fault-tolerant architecture for stateful firewalls. Comp. Secur. **31**, 524–539 (2013)
5. Posey, C., Bennett, R.J., Roberts, T.L.: Understanding the mindset of the abusive insider: an examination of insiders' causal reasoning following internal security changes. Comp. Secur. **30**, 486–497 (2011)
6. Warkentin, M., Wilson, R.: Behavioral and policy issues in information systems security: the insider threat. Eur. J. Inform. Syst. **18**, 101–105 (2009)
7. Richardson, R.: CSI computer crime and security survey. http://www.GoSCI.com (2011)
8. Hu, Q., Xu, Z., Dinev, T., Ling, H.: Does deterrence work in reducing information security policy abuse by employees? Commun. ACM **54**, 55–60 (2011)

9. Kable, J.W.: Valuation, intertemporal choice, and self-control. In: Glimcher, P.W., Fehr, E. (eds.) Neuroecenomics, 2nd edn, pp. 173–192. Academic Press, London (2014)
10. Mitchell, J.P., Schirmer, J., Ames, D.L., Gilbert, D.T.: Medial prefrontal cortex predicts intertemporal choice. J. Cogn. Neurosci. **23**, 1–10 (2010)
11. Hu, Q., West, R., Smarandescu, L.: The role of self-control in information security violations: insights from a cognitive neuroscience perspective. J. Manage. Inform. Syst. **31**, 6–48 (2015)
12. Budde, E., West, R.: Neural correlates of ethical decision making related to information security. Poster presented at the Midwestern Psychological Association, Chicago, IL (2017)
13. Vance, A., Siponen, M.: Is security policy violations: a rational choice perspective. J. Organ. End User Comput. **24**, 21–41 (2012)
14. Lopez-Calderon, J., Luck, S.J.: ERPLAB: an open-source toolbox for the analysis of event-related potentials. Front. Hum. Neurosci. **8**, 213 (2014)
15. Delorme, A., Makeig, S.: EEGLAB: an open source toolbox for analysis of single-trial EEG dynamics. J. Neurosci. Meth. **143**, 9–21 (2004)
16. de Oliveira-Souza, R., Zohn, R., Moll, J.: Neural correlates of human morality: an overview. In: Decety, J., Wheatley, T. (eds.) The moral brain: a multidisciplinary perspective, pp. 183–196. The MIT Press, Cambridge, MA (2015)
17. Greene, J.D.: The cognitive neuroscience of moral judgment and decision making. In: Decety, J., Wheatley, T. (eds.) The moral brain: a multidisciplinary perspective, pp. 197–220. The MIT Press, Cambridge, MA (2015)
18. Cavanagh, J.F., Frank, M.J.: Frontal theta as a mechanism for cognitive control. Trends Cogn. Sci. **18**, 414–421 (2014)
19. West, R., Budde, E., Malley, K.: Neural correlates of thinking about and making unethical decisions. Talk presented at the meeting of the Psychonomic Society, Vancouver, BC (2017)
20. West, R., Bailey, K., Tiernan, B.N., Boonsuk, W., Gilbert, S.: The temporal dynamics of medial and lateral frontal neural activity related to proactive cognitive control. Neuropsychologia **50**, 3450–3460 (2012)
21. Dinev, T., Goo, J., Hu, Q., Nam, K.: User Behavior towards protective information technologies: the role of national cultural differences. Inform. Syst. J. **19**, 391–412 (2009)
22. Hu, Q., Zhang, C., Xu, Z.: Moral beliefs, self-control, and sports: effective antidotes to the youth computer hacking epidemic. Paper presented at 45th Hawaii international conference on systems science (2012)
23. Xu, Z., Hu, Q., Zhang, C.: Why computer talents become computer hackers. Commun. ACM **56**, 64–74 (2013)

NeuroIS for Decision Support: The Case of Filmmakers and Audience Test Screenings

Sandra Pelzer, Marc T. P. Adam and Simon Weaving

Abstract The application of neuroscience theories, methods, and tools holds great potential for the development of novel decision support systems. In this paper, we develop a theoretical framework for how NeuroIS may support the test screening process of filmmakers where decisions are made about what narrative material is shown to the audience, what sequence it is to be ordered, and what emotional value it must carry. While current methods for audience test screenings commonly rely on standardized questionnaires and focus groups, decision support systems may employ neuroscience tools as built-in functions to provide the filmmaker with novel insights into how their movie is ultimately perceived by the audience. Thereby, a key focus lies on the coherence between the emotional experience intended by the filmmaker and the emotional experience exhibited by the audience. Further, NeuroIS allows an evaluation of how the emotional experience to specific cinematic moments affects overall movie satisfaction.

Keywords Audience testing · Decision support systems · Filmmaker · NeuroIS

1 Introduction

Over the past ten years, the application of neuroscience theories, methods, and tools has provided valuable theoretical and methodological insights for information systems research, particularly in terms of informing the design of IT artifacts and using neuroscience tools for their evaluation. However, only few studies have explored

S. Pelzer
Karlsruhe Institute of Technology, Karlsruhe, Germany
e-mail: sandra.pelzer@web.de

M. T. P. Adam (✉) · S. Weaving
The University of Newcastle, Newcastle, Australia
e-mail: marc.adam@newcastle.edu.au

S. Weaving
e-mail: simon.weaving@newcastle.edu.au

F. D. Davis et al. (eds.), *Information Systems and Neuroscience*,
Lecture Notes in Information Systems and Organisation 29,
https://doi.org/10.1007/978-3-030-01087-4_4

how biosignals can be used as built-in functions of IT artifacts such as decision support systems [1, 2]. In this paper, we explore this promising path of design science research by developing a theoretical framework for how NeuroIS tools may support filmmakers in the process of finishing a film for distribution, when a series of decisions are made in post-production that finally determine what narrative material is shown to the audience, what sequence it is to be ordered, and what emotional value it must carry. The framework enables filmmakers (1) to evaluate the level of coherence between the filmmaker's intentions for the emotional experience at specific moments of visual storytelling and the audience's exhibited emotional experience and (2) to identify how the emotional experiences in response to specific cinematic moments affect overall movie satisfaction.

The global movie box-office for 2017 reached US$40 billion, with movies playing in 125,000 screens in more than 25,000 cinemas across the world [3]. Yet, despite estimates of total global movie production exceeding more than 3,000 films a year [4], the industry is characterised by "high stakes, highly uncertain investments" [5] and high failure rates of individual movies. In addition, industry practitioners rely upon "tradition, conventional wisdom and simple rules of thumb" [6] rather than more scientific approaches to creative and managerial decision-making. In particular, knowledge of the emotional experience of the audience and its link to the success of a movie is scarce, with audience testing typically limited to standardized questionnaires and focus groups to follow up and find more detailed qualitative causes. Feedback is focused primarily upon ascertaining an overall rating (e.g., "Would you recommend this movie to your friends?") and consideration of pre-selected aspects of the film thought to be potentially problematic (e.g., concerns over a main character's likeability). Results from audience testing are often aimed at discovering elements for marketing campaigns, but also provide information so that the filmmaker can make adjustments in the final stages of the post-production process [7].

While the existing approaches of questionnaires and focus groups provide important insights into an audience's overall perception of a movie, they allow for little exploration of the perception of individual cinematic moments at an emotional level. As complex forms of storytelling, movies are developed through screenwriting, brought to life by direction, and finally constructed with editing. Throughout this interconnected process, the intended emotional response of the audience is the primary concern, particularly for the roles of screenwriter, director, and editor. Whatever the genre, movies are designed to give pleasure by provoking emotion: to be successful horror films must scare, thrillers must thrill, and "weepies" must make us cry. For screenwriters, "what we are really after, what we are really concerned about is the emotion […] What is the emotion underpinning the scene, this story?" [7, p. 25]. Directing involves "a passion for the human condition and characters and their emotional state of mind from moment to moment" [8, p. 3], whilst the art of editing places the highest value on being "true to the emotion of the moment" [9, p. 18].

In this paper, we elaborate on how utilizing biosignals may support the decision making of filmmakers in the post-production process. Thereby, we build on the recent advances in NeuroIS research in employing neurophysiological measurements such as electroencephalography (EEG), heart rate variability, skin conductance, and startle reflex modulation as measures for human affective processing [10–12], and the integration of these measures as built-in functions of IT artifacts [1, 2, 13].

2 Theoretical Framework

Traditional approaches for gathering feedback from audiences prior to the release of a movie generally focus on the audience's perception of the movie as a whole. However, the making of a movie involves a multitude of decisions around how the narrative is to be delivered by means of a complex, and yet sequential, set of audio-visual sensory stimuli. Hence, an audience's overall perception of a movie is a function of how they experience this sequence of audio-visual sensory stimuli, and the way this experience leads to a re-construction of the story world in the mind of the audience. The development of our theoretical framework starts from the rationale that neurophysiological measurements may provide important insights for the filmmaker into how their selection of audio-visual stimuli is ultimately experienced by the audience, and how the experience of individual segments is reflected in their overall movie satisfaction.

Importantly, our framework particularly focuses on those segments of a movie that the filmmaker believes to play a critical role in the perception of a movie, referred to in the following as *cinematic moments*. While a cinematic moment may last anywhere between only a few seconds to several minutes, it draws its significance from the meaning that the filmmaker intends against the backdrop of the plot and the story's characters. In most instances, the meaning inherent to a cinematic moment is carried at least partially by an emotional experience (e.g., anger, relief) intended by the filmmaker. Building on this rationale, our proposed theoretical framework sees the concept of cinematic moments and the emotional experience created through them as antecedents of overall movie satisfaction (see Fig. 1).

The framework establishes the notion that a filmmaker may induce emotions through two groups of interrelated components. First, the *narrative function* refers to the meaning that the cinematic moment carries for plot and/or the characters of the

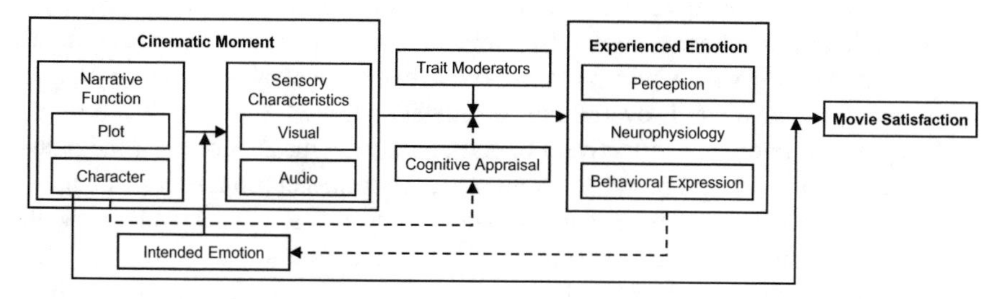

Fig. 1 Theoretical framework

film. For instance, the moment when Mrs. Bates (Norman's mother) is revealed to be a skeleton at the end of the movie *Psycho* (1960), both solves the plot's mystery of who the killer is, and makes us re-evaluate the character of Norman Bates (all accomplished while provoking shock and terror in the audience). Second, *sensory characteristics* refer to the specific auditory and visual elements that are chosen by the filmmaker to fulfil the narrative function of the cinematic moment. Filmmakers make choices about what we see and hear, and these choices are not only made to maximise narrative comprehension but have an intended emotional response in mind. For instance, in the skeleton reveal scene, director Alfred Hitchcock chooses to cut to a close up just as the skeletal face of Mrs. Bates is revealed, and punctuates the moment with a scream and the dramatic violin-dominated theme music.

Further, we conceptualize that the emotional experience that the filmmaker intends to invoke in the audience (intended emotion) ultimately leads to an actual emotional experience in the audience (experienced emotion) as conveyed by the sensory characteristics of the cinematic moment. The experience of emotions depends on the cognitive appraisal of the narrative function of the cinematic moment (e.g., confusion might be a desirable emotion for the climax of a psychological thriller, but not for establishing the story world of a drama) as well as individual characteristics of the audience (e.g., movie preferences, expectations of genre). Thereby, each experienced emotion may comprise perception, neurophysiological activation patterns, and behavioural expressions. The set of intended emotions is revealed in the sensory elements a filmmaker chooses when delivering the narrative function of a cinematic moment. We therefore conceptualize that the way narrative functions are delivered by audio-visual elements is moderated by a filmmaker's intended emotions. Further, even though a cinematic moment may trigger a range of emotions in the audience, this emotional experience may not necessarily be in tune with what the filmmaker intended.

Based on the conceptualization of cinematic moments, and their interplay with the audience's emotional experience and overall movie satisfaction, the framework enables us to examine a range of relationships between the investigated constructs (e.g., intended emotion, experience emotion, movie satisfaction) that may provide important insights for the movie production process. For instance, it enables the filmmaker to see (1) the coherence of emotions experienced by an audience and the emotions intended by the filmmaker for each cinematic moment, (2) the degree of divergence in emotional experience across different members of the audience (e.g., if some audience members are bored by a long action sequence, whilst others find it thrilling), and (3) the way in which the emotional experience to individual cinematic moments contribute towards overall movie satisfaction.

We posit that in order to effectively operationalize the measurement of the pathways expressed in the proposed theoretical framework, neurophysiological measurements provide a promising avenue for audience test screenings. Combined with self-report data on the nature of the emotional state and the audience's overall satisfaction with the movie, neurophysiological measurements allow for the collection of information on how an audience experiences individual cinematic moments without interrupting the overall movie experience. Previous research has involved the analysis of movie preferences using neurophysiological signals (e.g., EEG [14], heart rate [15]), but these approaches were not intended to provide decision support for filmmakers, whose active involvement in the selection of the cinematic moments, and the definition of the intended emotional experience for each, form the criteria around which measurement occurs. Operationalizing the measurements this way can provide detailed decision-support for filmmakers prior to distribution, with the aim of reducing risks of failure in a highly competitive global market.

In particular, over the past decade NeuroIS scholars have explored a range of neurophysiological measures to investigate human affective processing in human-computer interaction. Thereby, a common approach is to follow a dimensional perspective of emotion as expressed in Russel's circumplex model of affect [16], which considers the dimensions of hedonic *valence* (from unpleasant to pleasant) and *arousal* (from unaroused to aroused) as key aspects of a person's emotional state. As for the valence dimension, scholars have employed measures such as EEG (e.g., to predict e-loyalty in websites [10]) and startle reflex modulation (e.g., to predict attitudes towards brands and virtual reality [12]). Complementarily, measures such as skin conductance response and certain aspects in heart rate variability (e.g., in electronic auctions [11, 13]) have been employed to assess the arousal dimensions of users' emotional experience. Further, eye tracking has been used to synchronize neurophysiological activity with the time at which a user perceives a certain stimulus on the screen (e.g., users' responses to email pop-up notifications [17]). Finally, several studies have explored how such biosignals may be used as real-time system input for information systems (e.g., for stress management [2] and emotion regulation [13]). Applied to the case of decision supports systems for filmmakers, such measures could be used in audience test screenings to provide insight into the emotions an audience experiences during cinematic moments identified by the filmmaker without interrupting the film. The system may then contrast these data with the emotional experience intended by the filmmaker, and compare them with data collected using follow up questionnaires on the audience's perceptions of specific cinematic moments after the end of the film (e.g., identification of emotional states).

3 Discussion and Future Work

The theoretical framework presented in this paper may provide a first step towards employing NeuroIS tools for decision support in audience test screenings. At this stage, filmmakers have little information as to how the emotional experiences they intend to create through audio-visual stimuli lead to overall satisfaction with the movie. The theoretical framework allows us to identify key constructs in the perception of a movie that drive movie satisfaction, which in turn enable us to devise operationalizations with neurophysiological measurements for the design of decision support systems. Conceptually, such a system will enable a feedback loop between a filmmaker's intentions in the making of a movie, and the audience's actual experience—information that may turn out critical for decision making in post-production.

Building on the proposed framework, a proof-of-concept has been implemented using the software platform *Brownie* [18, 19]. Its main purpose was to conduct an initial test of the framework using a nearly completed short film where a number of cinematic moments and their intended emotions were identified by the filmmaker, and compared to the emotional experience of a small group of viewers. Because an audience's satisfaction with a movie is also subject to their expectation towards that movie [20], the proof-of-concept deliberately avoided setting specific expectations about the film experience. The results provide support for our rationale that the emotional experience and satisfaction for individual cinematic moments contributes to the audience's overall movie satisfaction. Further, a higher degree of convergence between a filmmaker's intention and the audience's actual emotional experience is also associated with a higher degree of movie satisfaction, which supports our hypothesis that if an audience member does not feel the intended emotions with the intended intensity, the satisfaction with a moment or the movie as a whole drops. Finally, a higher degree of divergence in emotional experience across the audience is related to a lower degree of overall movie satisfaction.

We suggest further research in a number of areas. The first is in identifying a set of particular emotional states that are most relevant for the movie watching experience and the link between these and the overall satisfaction of a movie experience. This set of emotional states is essential to support the filmmaker in selecting cinematic moments and defining the intended emotional experience. Secondly, as "overall satisfaction" is not a discrete emotion but presumably formed as part of the emotional journey experienced, further research would be useful in establishing the links between the extent to which specific emotions are experienced as cinematic moments unfold, and the overall sensation of being "satisfied" with the whole movie. Thirdly, we suggest research into how neurophysiological data can be used in decision support systems to forecast satisfaction with a more diverse range of genres (e.g., comedy, drama) and with creative content beyond traditional feature films (e.g., documentaries, music videos). Such research would be a highly positive step for filmmakers, who are in search of meaningful tools to replace the rules of thumb and long-established processes that dominate decision-making in their highly competitive and risky business environment.

References

1. vom Brocke, J., Riedl, R., Léger, P.-M.: Application strategies for neuroscience in information systems design science research. J. Comput. Inf. Syst. **53**, 1–13 (2013)
2. Adam, M.T.P., Gimpel, H., Maedche, A., Riedl, R.: Design blueprint for stress-sensitive adaptive enterprise systems. Bus. Inf. Syst. Eng. **59**, 277–291 (2017)
3. Comscore: Comscore reports worldwide box-office hits a new all-time record. https://ir.comscore.com/news-releases/news-release-details/comscore-reports-worldwide-box-office-hits-new-all-time-record
4. The numbers: movies released in 2017. https://www.the-numbers.com/movies/year/2017 (2017)

5. Chisholm, D., Fernandez-Blanco, V., Ravid, S., Walls, W.: Economics of motion pictures: the state of the art. J. Cult. Econ. **39**, 1–13 (2015)
6. Eliashberg, J., Elberse, A., Leenders, M.: The motion picture industry: critical issues in practice, current research & new research directions. Mark. Sci. **25**, 638–661 (2005)
7. Bauer, I.: Screenwriting Fundamentals: The Art and Craft of Visual Writing. Routledge, New York/London (2017)
8. Schreibman, M.: The Film Director Prepares: A Practical Guide to Directing for Film and TV. Lone Eagle Publishing, New York (2006)
9. Murch, W.: In the Blink of an Eye. A Perspective on Film Editing. Silman-James Press, Hollywood (1992)
10. Gregor, S., Lin, A.C.H., Gedeon, T., Riaz, A.: Neuroscience and a nomological network for the understanding and assessment of emotions in information systems research. J. Manage. Inf. Syst. **30**, 13–48 (2014)
11. Teubner, T., Adam, M.T.P., Riordan, R.: The impact of computerized agents on immediate emotions, overall arousal and bidding behavior in electronic auctions. J. Assoc. Inf. Syst. **16**, 838–879 (2015)
12. Koller, M., Walla, P.: Measuring affective information processing in information systems and consumer research: introducing startle reflex modulation. In: ICIS 2012 Proceedings, pp. 1–16, Orlando, USA (2012)
13. Astor, P.J., Adam, M.T.P., Jerčić, P., Schaaff, K., Weinhardt, C.: Integrating biosignals into information systems: a NeuroIS tool for improving emotion regulation. J. Manage. Inf. Syst. **30**, 247–278 (2013)
14. Dmochowski, J.P., Bezdek, M.A., Abelson, B.P., Johnson, J.S., Schumacher, E.H., Parra, L.C.: Audience preferences are predicted by temporal reliability of neural processing. Nat. Commun. **5**, 1–9 (2014)
15. Christoforou, C., Christou-Champi, S., Constantinidou, F., Theodorou, M.: From the eyes and the heart: a novel eye-gaze metric that predicts video preferences of a large audience. Front. Psychol. **6**, 1–11 (2015)
16. Russel, J.A.: A circumplex model of affect. J. Pers. Soc. Psyc. **39**(6), 1161–1178 (1980)
17. Léger, P.-M., Sénecal, S., Courtemanche, F., Ortiz de Guinea, A., Titah, R., Fredette, M., Labonte-LeMoyne, É.: Precision is in the eye of the beholder: application of eye fixation-related potentials to information systems research. J. Assoc. Inf. Syst. **15**, 651–678 (2014)
18. Hariharan, A., Adam, M.T.P., Dorner, V., Lux, E., Müller, M.B., Pfeiffer, J., Weinhardt, C.: Brownie: A platform for conducting NeuroIS experiments. J. Assoc. Inf. Syst. **18**, 264–296 (2016)
19. Jung, D., Adam, M.T.P., Dorner, V., Hariharan, A.: A practical guide for human lab experiments in information systems research: a tutorial with Brownie. J. Syst. Inf. Technol. **19**, 228–256 (2017)
20. Aurier, P., Guintcheva, G.: Using affect-expectations theory to explain the direction of the impacts of experiential emotions on satisfaction. Psyc. Mark. **31**, 900–913 (2014)

Measuring the Impact of Mind Wandering in Real Time Using an Auditory Evoked Potential

Colin Conrad and Aaron Newman

Abstract In this research-in-progress paper, we propose an experiment to investigate the neurophysiological correlates of mind wandering using electroencephalography (EEG). Auditory oddball event related potentials have been observed to be sensitive to the mind wandering state and can be used as a real-time passive measure. This has advantages over standard survey techniques because it is an objective, non-disruptive real time measure. We describe an experiment to observe the neurophysiological correlates of mind wandering in online learning environments using an auditory oddball. In doing so, we introduce a new experimental paradigm to the IS literature which could be used to extend other attention-related research.

Keywords Auditory oddball · Mind wandering · Online learning · EEG

1 Introduction

Mind wandering refers to processes commonly described as "daydreaming," or "spontaneous thoughts" [1]. More precisely, mind wandering represents a phenomenon where sustained attention is brought away from a stimulus and toward self-generated experiences [2]. It is commonly thought that mind wandering occurs in the higher education environment, and though it varies from student to student, it is often perceived to have an overall negative impact on student performance [3, 4]. In the case of common online learning systems such as Massive Open Online Courses (MOOCs), it is tempting to make similar inferences, as they often likewise follow a lecture format. One key difference between the classroom and the online lecture format however, is that good classroom teachers can often observe behaviors characteristic of mind wandering and improve their teaching to increase

C. Conrad (✉) · A. Newman
Dalhousie University, Halifax, Canada
e-mail: colin.conarad@dal.ca

A. Newman
e-mail: aaron.newman@dal.ca

F. D. Davis et al. (eds.), *Information Systems and Neuroscience*,
Lecture Notes in Information Systems and Organisation 29,
https://doi.org/10.1007/978-3-030-01087-4_5

engagement. Detecting mind wandering in an online learning environment would be useful to improving e-learning systems and identifying improved methods for objectively measuring mind wandering would be instrumental to the improvement of such systems.

In order to measure the impact of mind wandering on education, we explore using two electroencephalography (EEG) measures. The first measure is commonly referred to as the P1-N1-P2 auditory event related potential (ERP), which consists of a sequence of three peaks that consistently appear in response to the onset of auditory stimuli, with characteristic timing and scalp distributions [5]. Studies in mind wandering have found an effect where the amplitude of the P2 elicited by auditory oddball stimuli is reduced in individuals who have attention directed away from task-relevant stimuli and toward self-generated information [2, 6]. The second measure consists of oscillatory patterns in specific frequency bands, commonly referred to as delta, theta and alpha activity, which have been found to be correlated with mind wandering [6]. In this research-in-progress paper, we describe an experiment to identify the differences in these two patterns and their correlation with self-reported mind wandering. We propose employing these methods to conduct research in real-time changes in covert attention, which are relevant to predicting performance in online learning.

2 Hypothesis Development

Mind wandering is a common phenomenon that plays a significant role in general thought processes, even taking up to 50% of our waking time [7]. Mind wandering is also understood to be associated self-generated thoughts and with the default mode network, which is the series of mental functions active in the absence of an explicit task. The activation of self-generated thought processes carry both costs and benefits from the perspective of cognition, depending on the context in which they are active. Self-generated thoughts have been observed to contribute to absentmindedness and unhappiness, but also have the benefit of facilitating creativity and planning [8].

Though self-generated thoughts seem to play an essential role in common human experience, the role they play in learning is inconclusive. In the context of information technology, Sullivan, Davis and Koh performed exploratory work on this subject and found that not all types of mind wandering are detrimental to learning and some forms might in fact be beneficial [9]. However, other studies affirm its overall negative impact on learning. In a study of 463 undergraduate students, Lindquist and McLean found that students who experienced frequent task-unrelated images and thoughts performed poorer in course examinations and that experiencing task-unrelated thoughts was negatively correlated with degree of course interest [4]. Mind wandering has also been found to be correlated with the activation of brain regions associated with cognitive control and executive networks, and may even compete for resources with learning stimuli [10]. Though it far from conclusive, we can hypothesize that mind wandering is generally detrimental to knowledge acquisition, at least when it comes to the sorts of knowledge acquired with executive networks, such as rote learning.

H1 Reported mind wandering will be negatively correlated with rote learning.

2.1 Measuring Mind Wandering Using Neurophysiological Indicators

Though mind wandering can be effectively measured using ex post questionnaires, these methods do not offer insight into the temporal impact of mind wandering. It is desirable to develop measures that can offer insight on the changes in mind wandering patterns over time, as temporal data can help identify which portions of an online learning system account for changes in mind wandering patterns. One method for doing this is experience sampling, a series of very short self-reports designed to capture the temporal experience of participants. Studies using these methods often employ a simple yes/no measure in order to determine the occurrence of mind wandering [11, 12]. This comes with the advantage of measuring mind wandering in real time, but with the disadvantage of disrupting the person's current cognitive processes, be they task-related or mind wandering.

Neuroimaging can be used to mitigate this problem. Oddball protocols can be used to elicit event-related potential responses from a given stimulus during a sustained task such as an e-learning session and have already been demonstrated in the IS context [13]. The P1-N1-P2 complex is a series of event related potentials triggered by an auditory or visual stimulus and can be adapted to this task. Established by Hillyard, Vogel and Luck, this complex is a mandatory response triggered by early attention control mechanisms in the occipital regions [5]. The P1-N-P2 complex has been found to be a significant indicator of the switch of general selective attention from one stimulus to another, most notably by differences in amplitude between attended and ignored stimuli. The amplitude of the P2 component was also observed to be sensitive to mind wandering by Braboszcz and Delorme [6]. Using an passive auditory oddball protocol, they demonstrated significant differences between the P2 amplitudes between participants reporting to be in a mind-wandering state versus on task.

In addition to the P2 response, correlations between oscillatory activity and mind wandering have been found at the delta, theta, alpha and beta bands [6]. Neural oscillations are caused by neural activity in the central nervous system and underline at least two modes if cerebral activity: fast-frequency waves reflective of high degrees of task-related attention (beta activity at 12–30 Hz) and a low-frequency waves reflective of low task-related attention (theta activity at 3–7 Hz). Braboszcz and Delorme also observed the impact of oscillatory activity on mind wandering ultimately found theta and beta to be significant correlates of mind wandering, while noting that delta and alpha activity was suggestive. We are led to the following hypotheses:

H2a Mean P2 amplitude will be positively correlated with reported mind wandering.

H2b Delta power will be positively correlated with reported mind wandering.
H2c Theta power will be positively correlated with reported mind wandering.
H2d Alpha power will be positively correlated with reported mind wandering.
H2e Beta power will be negatively correlated with reported mind wandering.

3 Experiment Design

Participants will be asked to attend to a 51-min English language video on Machine Learning as auditory stimuli are presented [14]. The subject matter and video were chosen because the subject matter is not commonly taught in the standard business curriculum, had some utility to the participants, and was observed triggering variations in mind wandering during pilot studies. The video consists of a standard lecture along with a visual aid created in Microsoft PowerPoint. Participants are asked to pay attention to the video, while being presented with one of two audio stimuli every 1–1.5 s (mean 1.25). Participants are asked to report when they experience mind wondering by pushing a button on the computer keyboard, which is recorded on the parallel port. Following the video, participants complete a multiple-choice quiz to measure retention. Participants also complete a short multiple-choice quiz before and after the video. The differences in results are used as a measure of rote learning.

3.1 Participants

Twenty-four healthy students between the ages of 19 and 29 will be recruited from Dalhousie University to participate in the study. Power analysis on the oddball response suggest that this number would be for 99% confidence. Participants will be screened for neurophysiological, emotional, medical, hearing and vision conditions that could lead to abnormal EEG. Participants will also be excluded if they are majoring in computer science, have taken a course related to machine learning or are not fluent in English. Participants will be compensated CAD $25 for their time.

3.2 Experimental Stimuli

All stimuli will be presented in a controlled computer environment in a small, quiet testing room. Audio stimuli consist of 100 ms tones delivered every 1–3 s (randomly distributed with mean of 2 s). Task standard stimuli (80% of trials) consist of 500 Hz tones while the oddball (20% of trials, pseudo-randomly distributed) stimuli are 1000 Hz. Exactly 2448 tones are presented in the course of the experiment. The PsychoPy Python library is used to present the audio stimuli and record manual responses [15]. The onset of each audio tone is communicated to the EEG amplifier via TTL codes sent from PsychoPy via the parallel port.

3.3 Procedure

After completing the informed consent procedure, participants are fitted with the EEG cap (see next section) and brought to the controlled environment. Participants watch the 51 min machine learning video, and are asked to press a button on the computer keyboard every time they become aware that their mind is wandering. Following the study, participants complete a multiple-choice quiz to test their retention of the material presented in the video.

3.4 EEG Data Acquisition

Participants are fitted with 32-channel scalp electrodes (ActiCap, BrainProducts GmbH, Munich, Germany) positioned at standard locations according to the International 10-10 system, and referenced during recording to the midline frontal (FCz) location. Bipolar recordings are made between the outer canthi of the two eyes, and above and below one eye, to monitor for eye movements and blinks. Electrode impedances are kept below 15 kΩ throughout the experiment. EEG data are sampled at 512 Hz using a Refa8 amplifier (ANT, Enschede, The Netherlands), bandpass filtered between 0.01 and 170 Hz, and saved digitally using ASAlab software (ANT).

3.5 Artifacts Correction and Data Processing

The MNE-Python library [16] is used for data preprocessing. A 0.1–40 Hz bandpass filter is applied to the data, followed by manual identification and removal of electrodes and epochs with excessive noise. The data are then segmented into epochs spanning 200 ms prior to the onset of each auditory tone, to 1 s after. Independent Components Analysis is then used to identify and remove artifacts such as eye blinks and movements [17]. The epochs that occur in the 10 s before the reported mind wandering (excluding the 1 s window before the report) are assigned a "mind wandering" label, while epochs that occur in the 10 s after the reported mind wandering (excluding the 1 s window after the report) are assigned an "on-task" label. Figure 1 illustrates how the data are prepared for analysis.

Planned comparisons are between standard and oddball stimuli, within and between mind wandering and on-task conditions. Pilot results ($n = 11$; see below) suggest a high variance in mind wandering reports among participants, ranging from 1 to 60 responses. Following the recommendations of Braboszcz and Delorme [6],

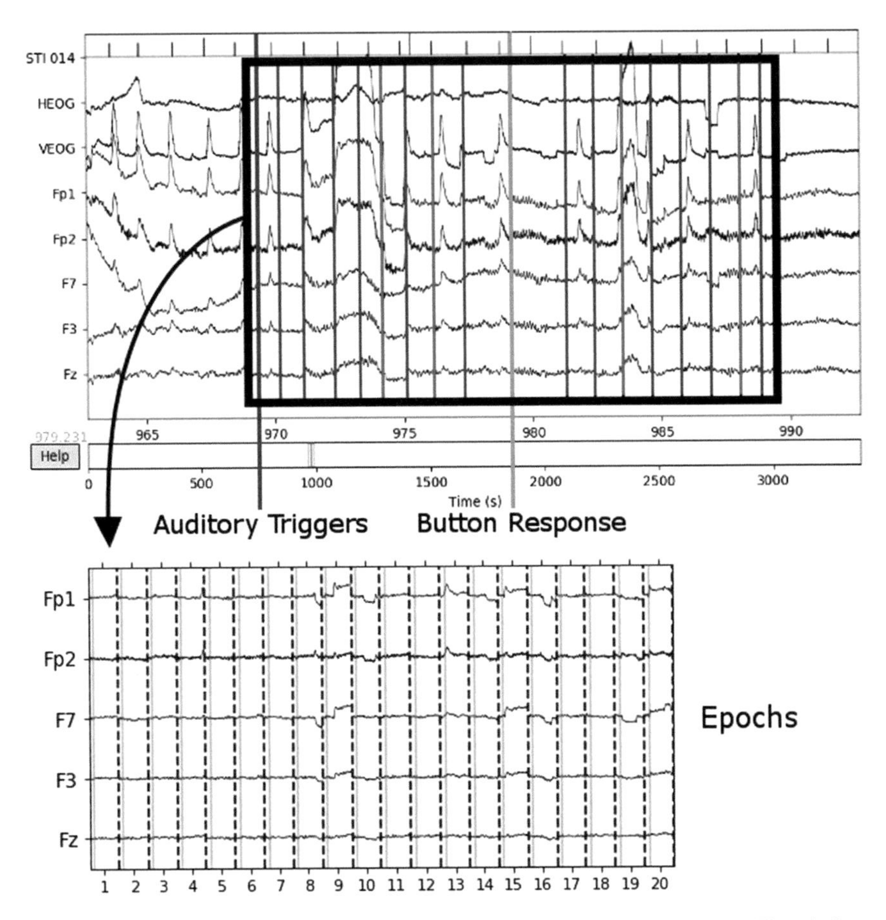

Fig. 1 Auditory events are triggered in PsychoPy and recorded in the parallel port. Though thousands of events are recorded, only the 1.2 s epochs from the auditory events in the 10 s before the button response ('mind wandering condition') and the 10 s following the response ('on task condition') are compared

participants with fewer than 20 oddball responses will be excluded. Each participant is expected to yield between 20 and 140 mind wandering or on-task oddball events. In addition to temporal domain (ERP) analyses of the P1-N1-P2 components, time-frequency analysis will be investigated in the 10 s pre- and post-report. These longer epochs will assessed for power spectral density ($\mu V^2/Hz$) in each standard EEG frequency band.

4 Pilot Study and Future Work

We conducted a pilot study of this paradigm with 11 participants. Of the 11 participants recruited, 3 had to be excluded due to technical errors or lack of mind wandering measures. After data processing there were 2251 standard and 474 oddball epochs with the "on task" label, and 1887 standard and 417 oddball usable epochs with the "mind wandering" label. Figure 2 visualizes the differences in the grand average between the standard and oddball ERP and the two conditions.

In both mind wandering and on-task conditions, clear differences were observed between standard and oddball stimuli over midline frontal electrodes at two times: at approximately 200 ms—with a greater negativity for oddballs—and from approximately 300–400 ms—with oddball stimuli showing a more positive amplitude over midline frontal electrodes. These correspond to the typically described N1 and P3 components, respectively. Though the N1 effect appears similar to that observed by Braboszcz and Delorme [6], the enhanced positivity occurs on the P3 component, rather than on the P2 as reported by Braboszcz and Delorme. The P3 is commonly elicited by oddball stimuli in paradigms such as this, however it is more commonly associated with task-relevant stimuli—whereas here the stimuli were to be ignored. Interestingly however, the P3 appears larger in the present data during the mind wandering thank on-task periods. We speculate that this could be caused by participants' paying greater attention to the auditory stimuli when their attention was less focused on the video (i.e., during mind wandering) the auditory stimuli drawing attention away from the video to a greater degree in the mind wandering state. As these were pilot data no statistical analyses were performed, but linear mixed effects analysis will be used once the full sample has been collected.

These preliminary results provide encouraging support for the proposal that this paradigm represents an automatic, covert, and temporally sensitive measure of mind wandering that can be applied in a range of task settings. If the auditory oddball correlates of mind-wandering are successfully established for online learning research, we

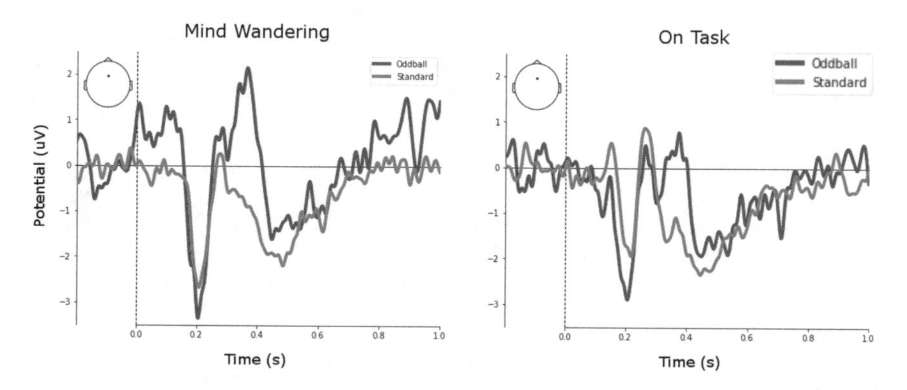

Fig. 2 Grand average ERP observed during mind wandering and on task conditions for channel Fz

can envision extending this measure to answer questions about the role of mind wandering in other technology environments. This could complement other psychophysiological measures such as eye movements or electrodermal activity, which could in turn be used to investigate the role of mind wandering outside of human-computer interaction, such as in-group dynamics or conversation [18, 19]. Additionally, a robust understanding of these correlates open up the potential of attention-adaptive interfaces, which have applications to information technology generally.

Acknowledgements This research is supported by the Killam and NSERC Doctoral scholarships to Colin Conrad, and an NSERC Discovery Grant to Aaron Newman. We also thank the participants of the 2017 NeuroIS training course for their feedback.

References

1. Christoff, K., Gordon, A., Smith, R.: The role of spontaneous thought in human cognition. In: Vartanian, O., Mandel, D. (eds.) Neuroscience of Decision Making, pp. 259–284. Psychology Press, New York (2011)
2. Smallwood, J.: Mind wandering and attention. In: Fawcett, J., Risk, E., Kingstone, A. (eds.) The Handbook of Attention, pp. 233–255. MIT Press, Cambridge (2015)
3. Wilson, K., Korn, J.H.: Attention during lectures: beyond ten minutes. Teach. Psychol. **34**(2), 85–89 (2007)
4. Lindquest, S.I., McLean, J.P.: Daydreaming and its correlates in an educational environment. Learn. Individ. Differ. **21**(2), 158–167 (2011)
5. Hillyard, S.A., Vogel, E.K., Luck, S.J.: Sensory gain control (amplification) as a mechanism of selective attention: electrophysiological and neuroimaging evidence. Philos. Trans. R. Soc. Lond. B Biol. Sci. **353**(1373), 1257–1270 (1998)
6. Braboszcz, C., Delorme, A.: Lost in thoughts: neural markers of low alertness during mind wandering. Neuroimage **54**(3), 3040–3047 (2011)
7. Schooler, J.W., Smallwood, J., Christoff, K., Handy, T.C., Reichle, E.D., Sayette, M.A.: Meta-awareness, perceptual decoupling and the wandering mind. Trends Cogn. Sci. **15**(7), 319–326 (2011)
8. Smallwood, J., Andrews-Hanna, J.: Not all minds that wander are lost: the importance of a balanced perspective on the mind wandering state. Front. Psychol. **4**(441) (2013)
9. Sullivan, Y., Davis, F., Koh, C.: Exploring mind wandering in a technological setting. In: Proceedings of the 36th International Conference on Information Systems, Fort Worth (2015)
10. Moss, J., Schunn, C.D., Schneider, W., McNamara, D.S.: The nature of mind wandering during reading varies with the cognitive control demands of the reading strategy. Brain Res. **1539**, 48–60 (2013)
11. Smallwood, J., Schooler, J.W.: The restless mind. Psychol. Bull. **132**(6), 946–958 (2006)
12. Schooler, J.W., Reichle, E.D., Halpern, D.V.: Zoning out while reading: evidence for dissociations between experience and metaconsciousness. In: Levin, D.T. (ed.) Thinking and Seeing: Visual Metacognition in Adults and Children, pp. 204–226. MIT Press, Cambridge (2004)
13. Léger, P.M., Sénecal, S., Courtemanche, F., Oritz de Guinea, A., Titah, R., Fredette, M., Labonte-LeMoyne, E.: Precision is in the eye of the beholder: application of eye fixation-related potentials to information systems research. J. Assoc. Inf. Syst. **15**(10), 651–678 (2014)
14. Grimson, E.: Introduction to Machine Learning. YouTube (2017)
15. Peirce, J.W.: Generating stimuli for neuroscience using PsychoPy. Front. Neuroinformatics **2**(10) (2009)

16. Gramfort, A., Luessi, M., Larson, E., Engemann, D.A., Strohmeier, D., Brodbeck, C., Parkkonen, L., Hämäläinen, M.S.: MNE software for processing MEG and EEG data. Neuroimage **86**, 446–460 (2014)
17. Delorme, A., Makeig, S.: EEGLAB: an open source toolbox for analysis of single-trial EEG dynamics including independent component analysis. J. Neurosci. Methods **134**(1), 9–21 (2004)
18. Douglas, K., Tremblay, A., Newman, A.: Measuring the N400 during naturalistic conversation: an EEG hyperscanning study. Poster to be presented at the neurobiology of language meeting, Quebec City, 18–22 Aug (2018)
19. Tremblay, A., Flick, G., Blanchard, A., Cochingyan, E., Dickie, K., Macgillivray, K., Ashraf Mahmoud-Ahmed, H., Schwartz, S., Sher, C., Asp, E., Newman, A.J.: Free, unscripted conversation with MEG recordings is a viable experimental paradigm for research into language processing. Paper presented at the 40th Atlantic Provinces Linguistic Association (APLA-40), 28–29 Oct 2016, Mount Saint Vincent University, Halifax, Canada (2016)

Exploring Eye-Tracking Data for Detection of Mind-Wandering on Web Tasks

Jacek Gwizdka

Abstract Mind-wandering (MW) is a phenomenon that affects most of us; it affects our interactions with information systems. Yet the literature on its effects on human-computer interaction is only scant. This research aims to contribute to establishing eye-tracking measures that could be used to detect periods of MW while a user is engaged in interaction with online information. We conducted a lab study (N = 30) and present an exploratory analysis of eye-tracking data with a focus on finding differences between periods of MW and not-MW. We found 12 eye tracking measures that were significantly different between periods of MW and not-MW. We also show promising classification results of the same variables. Our results indicate plausibility of using eye-tracking data to infer periods of MW.

Keywords Mind-wandering · Mindless reading · Eye-tracking · Pupillometry

1 Introduction

Most people have experienced mental state when their mind has wandered. This phenomenon is quite common and many people can remember when, for example, their reading did not result in any meaningful understanding of the text. In this case, a wandering mind can be a harmful thing as it makes us less efficient, prone to errors and to making incorrect decisions. If an information system were able to detect when a person's mind is wandering, it could offer an intervention. For example, if it detects that an e-commerce website user has spent a significant time MW while reviewing purchase options, the system could ask for additional verification before the purchase is made.

The goal of this project is to establish eye-tracking based measures of MW that could be used to detect periods of MW while a user is engaged in interaction with online information. We present an exploratory analysis of eye-tracking data (includ-

J. Gwizdka (✉)
School of Information, University of Texas at Austin, Austin, TX, USA
e-mail: neurois2018@gwizdka.com

F. D. Davis et al. (eds.), *Information Systems and Neuroscience*,
Lecture Notes in Information Systems and Organisation 29,
https://doi.org/10.1007/978-3-030-01087-4_6

ing pupillometry) with a focus on finding differences between periods of task-related and task-unrelated thoughts.

2 Related Work

While it is known that MW can have positive effects on human thought processes and, in particular, on creativity [1], MW is typically detrimental to the tasks that require focused attention. MW has been shown to negatively affect reading [2, 3], and the ability to resolve conflicts in displayed information [4], because the executive function is impaired. These are just two examples of negative influence of MW on user interaction with IS. MW has received increased attention from cognitive scientists and psychologists in the last decade [1, 3, 5], but we are only beginning to understand these processes.

Interestingly, it has been demonstrated that MW while reading is related to changes in eye-fixation patterns. For example, fixation durations were found to be longer and less affected by lexical and linguistic variables, while eye movements were found more erratic during MW periods than during reading [2]. Pupil dilation was found to change more spontaneously during MW periods [6, 7]. Results from these studies suggest that eye movements during MW are controlled by different cognitive processes than during normal reading. People may engage in internally focused cognitive tasks, which, presumably, are associated with different cognitive processes than externally focused tasks and unfocused MW. Recent work compared eye movements during goal-directed internally focused cognitive task and reading task. Eye behavior during the former was different and characterized, among others, by more and longer blinks, fewer microsaccades, more and shorter fixations, more saccades and saccades of higher amplitude [8]. These results suggest that some aspects of eye behavior may be coupled with internally generated information and related internal cognitive processes. We will come back to this in the Discussion section.

While MW is an important phenomenon with potentially significant explanatory power for human interaction with IS, research in this area outside psychology is still rather scant. Notable exceptions include, a theoretical model of MW in a technological setting proposed in a doctoral dissertation, with a specific goal to better understand costs and benefits of this phenomenon on technology users [9, 10]; a person-independent detection of MW based on eye-tracking data proposed by a group of computer scientists [11]; and the use of eye-tracking data and web cams in a large scale detection of MW in an education online setting [12].

Encouraged by previous research that showed relationship between episodes of MW and eye-tracking data, we aimed to (1) use more realistic stimuli than in psychology research (e.g., [6, 7]), (2) perform analysis without considering text characteristics (in contrast to [2] and to local features reported in [11]), and (3) examine differences in eye-tracking variables between periods of MW and nMW (not reported in [11], but reported in their second paper [13]).

Our research questions are as follows:

RQ1. Which eye-tracking measures differ significantly between periods of MW and nMW?

RQ2. Can eye-tracking measures be used to classify periods of MW and nMW?

3 Method

We conducted an eye-tracking lab experiment (N = 30, 20 females) in Information eXperience (IX) lab in the School of Information at University of Texas at Austin. Eye tracking data was collected using remote eye-tracker Tobii TX-300. The experiment was approved by IRB. Each lab session typically lasted 30 min. At the completion of a session, each participant received $12.

3.1 Procedure and Materials

Participants were pre-screened for their native or near-native level of English, and for normal to corrected-to-normal eyesight. Each participant filled out background questionnaire, performed a training task, and three online reading tasks shown in randomized order. After each task participants answered comprehension questions. These questions were included to provide motivation for attentive reading. The task design followed a simulated work-task approach [14], where tasks are presented with reasons for their performance. In our study, participants were informed that they needed to read three articles in order to prepare for a course "Technology and Society" they were taking. The articles were taken from the UBC-Hampton Reading Comprehension Test Suite [15]; their sources are listed in Table 1. Each text was presented on several web pages and was displayed in black Arial font on white background. The pages were designed to show about the same number of text lines on the page and thus each screen had about the same luminescence. Text lines height was uniform at 27px, which corresponds to 0.45° of visual angle and is approximately equal to the eye-tracker's accuracy reported by the manufacturer as 0.4°–0.5°.

To capture incidents of MW, participants were periodically probed [16] and asked to indicate whether they were reading or MW [17]. We used a visual probe [18]—a

Table 1 Articles for online reading tasks

1—A quick overview of digital activism. A blog post by Curiouscatherine (2011)
2—Taking the slack out of slacktivism. A popular press article published by Radio Free Europe, 2011.02.17
3—'Free the spectrum!' Activist encounters with old and new media technology. A journal article by Dunbar-Hester, C., published in New Media & Society

Fig. 1 Pop-up window with the mind-wandering probe

pop-up window with two response buttons (Fig. 1), which was displayed at random times controlled to be between 40 and 60 s and shown for 12–16 s at several pseudo-randomly selected locations on screen.

3.2 Variables

Independent variable was MW state with two levels: MW or nMW (i.e. reading). Dependent variables were obtained from eye-tracking data for two 5-s-long epochs that started respectively 5 and 10 s before the response to the probe. Eye-tracking data was cleaned by removing bad quality fixations (as marked by Tobii). Data from epochs with very few fixations (<3 standard deviations (std) below the mean) was discarded (<5% of epoch data) as it indicated epochs with many missing data points. We used eight types of variables: fixation count, regressions count, fixation duration, saccade duration, length, velocity, angle (four categories), and relative pupil change. For fixation duration and numerical saccade measures we calculated: mean, std, total, min and max values. For relative pupil change, which we calculated separately for each user and then denoised it with cubic interpolation of missing data (e.g., blinks), we calculated mean and std. Saccade angle was categorized to indicate (approximately) eye movement (1) forward (−10°; 10°), (2) backward (170°; 180°) in the same text line. Angles outside these ranges were categorized as (3) forward or (4) backward above or below the text line. This process yielded twenty-seven (27) variables.

4 Data Analysis and Results

Two participants have reported no periods of MW. Following prior work [6], we treated them as "outliers" and removed their data. Thus, we report data from 28 participants. The obtained proportion of MW responses (27.3%) matches the expectations [11, 13]. Response times to the probe were significantly longer for MW as compared with nMW periods (Table 2). This supports participants' correct self-classification of their internal thought processes.

We performed inferential statistics and classification. In inferential statistics, due to not-normal distribution of variables and lack or homogeneity of variance, we used non-parametric Mann-Whitney U test (M-WU). Given the exploratory nature of our research, we conducted individual M-WU tests on each variable (Table 3).

Table 2 Responses to the probe

Mind-wandering	Count	%	Response time [ms] mean (sd)
YES	115	27.3	3159 (1192)
NO (reading)	306	72.7	2879 (933)
Total	421	100	M-WU: $z = 2.2$, $p = 0.03$

Table 3 Descriptive statistics and Mann-Whitney U tests for significantly different variables

Epoch	Variable	# data samples (read/MW)	MW mean (sd)	nMW mean (sd)	M-W U
5 s	total_fixation_duration	306/115	3792 (761)	3915 (764)	$z = 2.44$; $p = 0.015$
10 s	avg_fixation_duration	288/110	254 (87)	238 (36)	$z = -2.02$; $p = 0.043$
5 s	fixation_count	306/115	16 (3)	16.5 (3.5)	$z = 2.02$; $p = 0.043$
10 s	fixation_count	288/110	16 (3)	16.8 (3.2)	$z = 2.46$; $p = 0.014$
10 s	tot_saccade_len	288/110	2072 (909)	2220 (847)	$z = 2.2$; $p = 0.027$
10 s	max_saccade_len	288/110	710 (355)	779 (342)	$z = 1.97$; $p = 0.049$
5 s	max_saccade_dur	306/115	362 (307)	301 (357)	$z = -3.18$; $p = 0.0015$
10 s	max_saccade_dur	288/110	324 (311)	298 (344)	$z = -2.78$; $p = 0.0055$
5 s	avg_saccade_dur	306/115	76 (68)	71 (90)	$z = -2.36$; $p = 0.019$
10 s	avg_saccade_dur	288/110	68 (63)	64 (67)	$z = -2.74$; $p = 0.0061$
5 s	std_saccade_dur	306/115	105 (109)	91 (154)	$z = -3.1$; $p = 0.002$
10 s	std_saccade_dur	288/110	98 (132)	85 (130)	$z = -2.98$; $p = 0.0028$
10 s	min_saccade_vel	288/110	0.6 (0.68)	0.63 (2.3)	$z = 3.02$; $p = 0.0025$
10 s	angle_cat_bck_count	288/110	3.2 (1.6)	3.7 (1.9)	$z = 2.17$; $p = 0.03$
5 s	avg_pupil_change	306/115	−0.015 (0.06)	−0.0042 (0.05)	$z = 1.93$; $p = 0.0535$
10 s	avg_pupil_change	288/110	−0.015 (0.06)	−0.0019 (0.05)	$z = 2.55$; $p = 0.011$
5 s	std_pupil_change	302/109	0.034 (0.015)	0.027 (0.014)	$z = -4.3$; $p < 0.0001$

Table 4 Classification results—best features

Epoch	Data sampling	Best features (in the order of weights from information gain ranking filter)
5 s	Intact	std_pupil_change, std_saccade_dur
	SMOTE (synthetic)	angle_cat_fwd_ud_count, regression_count, std_pupil_change, std_saccade_dur, max_saccade_dur, min_saccade_vel
	Random with replacement	std_pupil_change, avg_saccade_dur, std_saccade_dur, max_saccade_dur, min_saccade_vel, angle_cat_fwd_count
10 s	Intact	max_saccade_dur, std_saccade_dur, avg_saccade_dur
	SMOTE (synthetic)	std_saccade_dur, max_saccade_dur, avg_saccade_dur, avg_pupil_change
	Random with replacement	std_saccade_dur, avg_pupil_change, avg_saccade_dur, angle_cat_bck_count, max_saccade_dur, angle_cat_bck_ud_count, std_pupil_change

We run classifications using Weka 3.8 [19]. Due to the imbalanced number of samples between MW and nMW classes, we used two data sampling methods to improve the balance, (1) synthetic sample generation SMOTE [20] and (2) random sample generation with replacement. We present best classification results obtained by applying random forest classifier with 10-fold cross-validation [21] (Table 5) and the best features for each classifier (Table 4). These features overlap with significant results in (Table 3).

5 Discussion

Responding to *RQ1*, we found that 12 eye-tracking measures (44% of all measures considered) significantly differed between periods of MW and nMW—seven in 5 s epoch and ten in 10 s epoch (in that, five measures overlapped in both epochs). These results, taken together with the confirmatory answer to *RQ2*, i.e. the reasonably promising classification results, indicate plausibility of using eye-tracking data to infer periods of MW, at least on the tasks similar to ours.

Our results generally match the previous research. For example, similarly as in [2], we found that average fixation duration (avg_fixation_duration) tended to be longer in MW periods (in 10 s epochs). We also found a higher variability of changes in pupil dilation (higher avg_pupil_change and std_pupil_change) in MW periods. This is similar to results presented in [6, 7], where the authors reported more spontaneous changes in pupil dilation during MW periods. Bixler and D'Mello [13] found a longer minimum saccade duration in MW periods, while we found mean and maximum

Table 5 Classification results (Random Forest with 10-fold cross validation)

Epoch	Data sampling	Samples nMW/MW	Accuracy [%]	ROC [%]	F-measure [%]	F-measure: for MW class [%]
5 s	Intact (no sampling)	305/115	73.2	62.7	68.3	28.0
	SMOTE (synthetic)	306/230	77.4	84.0	77.3	72.8
	Random with replace-ment	210/210	87.6	96.2	87.6	87.9
10 s	Intact (no sampling)	288/110	70.9	59.3	64.3	17.1
	SMOTE (synthetic)	288/220	76.0	83.2	75.8	71.1
	Random with replace-ment	199/199	89.0	96.3	88.9	89.1

saccade duration to be longer during MW, as well as a higher saccade duration variability (std_saccade_dur) during MW.

Several findings from our study are in some contrast to [8]. We found fewer fixations (in both 5 and 10 s epochs) and longer average fixation duration (in 10 s epochs) in MW periods, while [8] reported more and shorter fixations on their goal-directed internally focused cognitive task. This suggests that internal cognitive processes during MW periods caught in our study are different from cognitive processes during internally focused cognition described in [8]. It further suggests that the differences in cognitive processes associated with internally focused cognition, MW and reading are reflected in eye behavior and thus can be discerned from eye-tracking measures.

Compared with [10, 12], accuracy of our classifications is higher (73–89% in our work vs. 59–72% and 52–74% in their first and second work, respectively). This difference may be due to our use of different classification algorithm and a somewhat different set of features. However, our low values of F-measures for MW class (Table 5) indicate poor classification performance for this class when no resampling was used. This points to the need for more data samples from each user.

Contrary to prior work [2, 10], we did not find a significant difference in regression counts. Although it was marginally significant at $p = 0.098$ for 5 s epoch, the difference was in the direction opposite to the expected, that is, we found fewer regressions in MW than in nWM periods. The same unexpected relationship was found in the related measure angle_cat_bck_count, which was significantly different in 10 s epoch. We also have not found a higher number of line crossing saccades during MW (reported in [13]). We don't have yet a good explanation for these findings.

5.1 Limitations

Limitations of our work include unbalanced number of data samples from MW and nMW segments. This is expected and is due to typical frequency of MW occurrences and suggests the need to collect more data before classification algorithms are trained. We also plan to use a wider variety of epoch lengths and in the future studies, use different tasks.

6 Conclusion

We believe that MW is a phenomenon that will grow in importance and will be more widely studied in the context of human interaction with information systems. A broader impact of implicit detection of MW lies in its potential applicability to e-commerce and e-learning systems, where upon detection of MW an intervention could be offered to a customer or learner.

Acknowledgements This project was supported, in part, by the Temple Teaching Fellowship 2016–2017. We thank Ms. Xueshu Chen, who was a Graduate Research Assistant, for her contributions to this project.

References

1. Schooler, J.W., Mrazek, M.D., Franklin, M.S., Baird, B., Mooneyham, B.W., Zedelius, C., Broadway, J.M.: The middle way: finding the balance between mindfulness and mind-wandering. In: Ross, B.H. (ed.) Psychology of Learning and Motivation, pp. 1–33. Academic Press, Cambridge (2014)
2. Reichle, E.D., Reineberg, A.E., Schooler, J.W.: Eye movements during mindless reading. Psychol. Sci. **21**, 1300–1310 (2010)
3. Smallwood, J.: Mind-wandering while reading: attentional decoupling, mindless reading and the cascade model of inattention. Lang. Linguist. Compass. **5**, 63–77 (2011)
4. Choi, H., Nam, C.S., Feng, J.: A wandering mind cannot resolve conflicts in displayed information. In: Proceedings of the Human Factors and Ergonomics Society Annual Meeting, Vol. 59, pp. 1397–1401 (2015)
5. Smallwood, J.: Distinguishing how from why the mind wanders: a process–occurrence framework for self-generated mental activity. Psychol. Bull. **139**, 519–535 (2013)
6. Franklin, M.S., Broadway, J.M., Mrazek, M.D., Smallwood, J., Schooler, J.W.: Window to the wandering mind: pupillometry of spontaneous thought while reading. Q. J. Exp. Psychol. **66**, 2289–2294 (2013)
7. Smallwood, J., Brown, K.S., Tipper, C., Giesbrecht, B., Franklin, M.S., Mrazek, M.D., Carlson, J.M., Schooler, J.W.: Pupillometric evidence for the decoupling of attention from perceptual input during offline thought. PLoS ONE **6**, e18298 (2011)
8. Walcher, S., Körner, C., Benedek, M.: Looking for ideas: eye behavior during goal-directed internally focused cognition. Conscious. Cogn. **53**, 165–175 (2017)
9. Sullivan, Y.: Costs and benefits of mind wandering in a technological setting: findings and implications. https://digital.library.unt.edu/ark:/67531/metadc862836/

10. Sullivan, Y., Davis, F., Koh, C.: Exploring mind wandering in a technological setting. In: ICIS 2015 Proceedings (2015)
11. Bixler, R., D'Mello, S.: Toward fully automated person-independent detection of mind wandering. In: Dimitrova, V., Kuflik, T., Chin, D., Ricci, F., Dolog, P., Houben, G.-J. (eds.) User Modeling, Adaptation, and Personalization, pp. 37–48. Springer International Publishing, Switzerland (2014)
12. Zhao, Y., Lofi, C., Hauff, C.: Scalable mind-wandering detection for MOOCs: a webcam-based approach. In: Data Driven Approaches in Digital Education, pp. 330–344. Springer, Cham (2017)
13. Bixler, R., D'Mello, S.: Automatic gaze-based detection of mind wandering with metacognitive awareness. In: Ricci, F., Bontcheva, K., Conlan, O., Lawless, S. (eds.) User Modeling, Adaptation and Personalization, pp. 31–43. Springer International Publishing, Switzerland (2015)
14. Borlund, P.: The concept of relevance in IR. J. Am. Soc. Inf. Sci. Technol. **54**, 913–925 (2003)
15. Freund, L., Kopak, R., O'Brien, H.: The effects of textual environment on reading comprehension: implications for searching as learning. J. Inf. Sci. **42**, 79–93 (2016)
16. Smallwood, J., Schooler, J.W.: The restless mind. Psychol. Bull. **132**, 946–958 (2006)
17. Smallwood, J., Schooler, J.W., Turk, D.J., Cunningham, S.J., Burns, P., Macrae, C.N.: Self-reflection and the temporal focus of the wandering mind. Conscious. Cogn. **20**, 1120–1126 (2011)
18. Feng, S., D'Mello, S., Graesser, A.C.: Mind wandering while reading easy and difficult texts. Psychon. Bull. Rev. **20**, 586–592 (2013)
19. Witten, I.H., Frank, E., Hall, M.A., Pal, C.J.: Data Mining, Fourth Edition: Practical Machine Learning Tools and Techniques. Morgan Kaufmann Publishers Inc., San Francisco, CA, USA (2016)
20. Chawla, N.V., Bowyer, K.W., Hall, L.O., Kegelmeyer, W.P.: SMOTE: synthetic minority over-sampling technique. ArXiv11061813 Cs (2011)
21. Breiman, L.: Random forests. Mach. Learn. **45**, 5–32 (2001)

Attentional Characteristics of Anomaly Detection in Conceptual Modeling

Karl-David Boutin, Pierre-Majorique Léger, Christopher J. Davis, Alan R. Hevner and Élise Labonté-LeMoyne

Abstract We use eye tracking to better understand the attentional characteristics specific to successful error detection in conceptual models. This phase of our multi-step research project describes the visual comportments associated with successful semantic and syntactic error identification and diagnosis. We test our predictions, based on prior studies on visual attention in an error detection task, or studies comparing experts and non-experts in diverse tasks, in a controlled experiment where participants are tasked with detecting and diagnosing errors in 75 BPMN® models. The results suggest that successful error diagnostics are linked with shorter total view time and shorter fixation duration, with a significant difference between semantic and syntactic errors. By identifying the visual attention differences and tendencies associated with successful detection tasks and the investigation of semantic and syntactic errors, we highlight the non-polarity of the 'scale' of expertise and allow clear recommendations for curriculum development and training methods.

Keywords Eye tracking · Conceptual modeling · Attentional characteristics

K.-D. Boutin (✉) · P.-M. Léger · É. Labonté-LeMoyne
HEC Montréal, Montréal, Québec, Canada
e-mail: karl-david.boutin@hec.ca

P.-M. Léger
e-mail: pierre-majorique.leger@hec.ca

É. Labonté-LeMoyne
e-mail: elise.labonte-lemoyne@hec.ca

C. J. Davis
University of South Florida, Saint Petersburg, FL, USA
e-mail: davisc@mail.usf.edu

A. R. Hevner
University of South Florida, Tampa, FL, USA
e-mail: ahevner@usf.edu

F. D. Davis et al. (eds.), *Information Systems and Neuroscience*,
Lecture Notes in Information Systems and Organisation 29,
https://doi.org/10.1007/978-3-030-01087-4_7

1 Introduction

Business process modeling has become a central activity in IS practice [1]. Conceptual models facilitate communication about business domains and their processes [2–4]. Such models have become a primary medium used in design activities. This phase of our research strives to deepen understanding of visual attention during error detection tasks [5]. While researchers have explored the variations between novice and experienced modelers [6], the differences in the visual attention between successful and unsuccessful error detection tasks in conceptual modeling are yet to be deeply investigated.

For this research, we employ the Business Process Modeling Notation (BPMN^{R}), an international standard for business processes. Its popularity in commercial settings prompted its selection for this phase of our work. Visual notations such as BPMN are composed of visual syntax—symbolic vocabulary and grammar—and visual semantics—elements that give meaning to each symbol and symbol relationship [7, 8]. However, while evaluating notations or their usage, the syntactic component is rarely discussed [8]. This presents an opportunity to contribute to the literature by comparing both the semantic and syntactic error identification process. The main objective of this study is to explore the differences in the attentional characteristics between successful and unsuccessful diagnostics in a detection task. We use eye tracking to monitor the visual attention of subjects as they search for and diagnose semantic and syntactic errors in a controlled experiment.

2 Prior Research and Hypotheses Development

Studies regarding the difference in the visual attention in an error detection task conclude that, on average, errors are fixated more often and longer than irrelevant information [9–11], and that a high number of fixations on the stimulus is correlated with an ineffective search [11, 12]. Furthermore, the longer the participant spends looking for an error, the lower the chances of success [9], possibly due to too much cognitive resource being drawn away for the encoding of the stimulus. Studies that compare novices and experienced modelers point to attentional characteristics that might be associated with expertise, and thus, generally, with better performance [1, 13].

Several meta-analyses that use eye tracking to compare experts and novices in a range of domains conclude that those classified as experts spend less time looking at stimuli before fixating relevant areas or anomalies [5, 14–16]. More efficient scan patterns [13–15] or more detailed and completed schemata [17, 18] are offered as explanations. Experts also tend to have fewer fixations, suggesting less cognitive effort to decipher and understand the stimuli [13, 15], and shorter fixation durations [5], which are also associated with lower cognitive processing effort [11]. Therefore, we propose three study hypotheses:

H1 *Successful error detections in conceptual modeling will require less time spent looking at the stimulus than unsuccessful error detections.*

H2 *Successful error detections in conceptual modeling will require, in total, fewer fixations than unsuccessful error detections, but with a higher proportion of fixations on the error.*

H3 *Successful error detections in conceptual modeling will require, on average, shorter fixation duration than unsuccessful error detections, but with longer fixation duration on the error.*

3 Research Method

Our experiment was conducted on a sample of 18 participants (7 males, 11 females) with different ages and experience. All our participants were offered a $20 gift card as a compensation for their participation. The research was approved by our institution's Research Ethics Board (REB), and each participant signed a consent form.

3.1 Research Design and Protocol

In order to confirm our hypotheses, we tasked our participants with identifying and diagnosing errors in conceptual models written in BPMN. Each participant had to inspect five (5) distinct sets of 15 models (for a total of 75 models), where each set represented a business process scenario (e.g. airport check-in process). An example can be seen in Fig. 1. Five (5) versions of each scenario were presented as BPMN 'sentences'. These were further manipulated to present three (3) versions: one with no known errors; one with a known semantic error, and one with a known syntactic error [7].

To mitigate the effect of prior knowledge of the business domains of the models [19, 20], the stimuli were created using simple scenarios, well-known to all potential participants. Furthermore, we limited the range of symbols used and the scope of the 'sentences' to 10–12 elements, favoring numerous accessible models rather than more complex and domain knowledge-dependent stimuli. To train the participants and mitigate the effect of learning through the trial, the experiment started with a short presentation on BPMN [21]. The symbols used in the study, as well as the two different types of errors, were shown and explained. The training was concluded with a practice task where participants were shown three (3) models, each one with a different type of error. Just like in the real task, the participant had to pinpoint the error and to diagnose the error type, both by clicking in the corresponding area on the stimuli. The correct location of the error, as well as the right error type, was revealed after each practice model. To avoid bias, the models used in the practice task were not related to the sets of models used later in the experiment, and the conditions were the same as in the experiment [21].

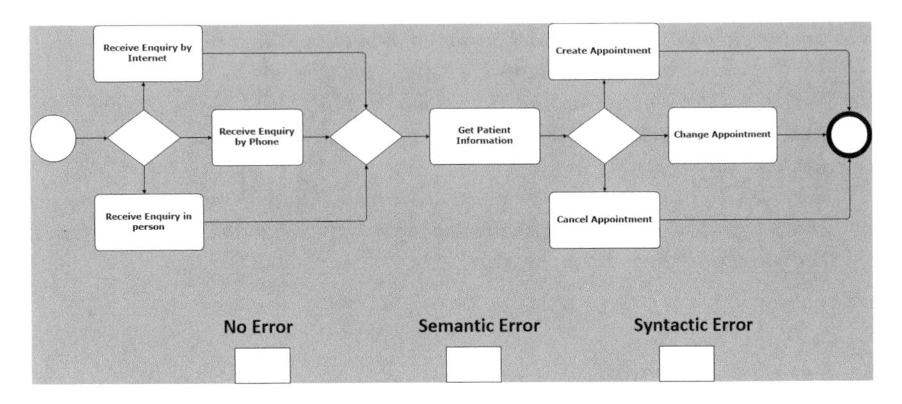

Fig. 1 Example of a model with boxes to indicate type of error detected

The participants then started the first task with their first set of models. The fifteen models included in the set were shown in random order and without any time limit. After identifying and diagnosing the error in a model, participants had to manually advance to the next model, using the space bar on their keyboard. They then proceed to the next set of models until they completed the five (5) sets.

3.2 *Apparatus and Measures*

Eye tracking (Red 250, SensoMotoric Instruments GmbH, Teltow, Germany) was used to gather the behavioral measures throughout the experiment, at a sampling frequency of 60 Hz. The number of fixations, which is the stabilization of the eye on an object [13], and their duration were gathered for each area of interest (AOI), as the literature tends to agree that fixation is related to the cognitive processing of visual information [13, 22]. The fixation duration threshold was set at 200 ms [11, 23]. The time before the first fixation in an AOI and the total view time of a stimulus were also collected. One (1) to three (3) AOIs were placed on the correct choice of error type and on the actual location of the error(s). For each participant, the eye tracker was calibrated to a maximum average deviation of 0.5°, using a 9-points predefined calibration grid.

4 Preliminary Results

We briefly present several preliminary results from our study. Hypothesis 1 states that successful identification and diagnosis of errors in conceptual models will take less time than unsuccessful answers. A linear regression with mixed model and a two-tailed level of significance is performed to compare the Total View Time for each value of the variable Answer (i.e. if the participant successfully diagnosed the error, Answer = 1, if not, Answer = 0). Results suggest that successful detection of error, including models without an error, is linked with lower total time spent on each stimulus ($B = -0.3934$, $p < .0001$). Furthermore, successful detection of semantic errors shows a faster time to first fixation on the area of interest (i.e. the zone containing the error) ($B = -0.3333$, $p = .0027$). However, there are no significant results linking the detection of syntactic or no errors with the time to first fixation.

Hypothesis 2, which stipulates that successful error detection will be linked with fewer fixations, but with a higher proportion of fixations on the error, is tested using a Poisson regression with mixed model and a two-tailed level of significance of Fixation Count on Answer. A significant relation is found between successfully detecting an error in a model and lower fixation count ($B = -0.4402$, $p < .0001$). Moreover, greater proportions of fixation are allocated to the zone containing semantic errors ($B = 0.5448, p < .0001$) and syntactic errors ($B = 0.9379, p < .0001$). However, while correct diagnosis of semantic errors are linked with a decrease in the number of fixation in the areas of interest ($B = -0.04956, p = .0897$), the opposite is found for the successful detection of syntactic errors, where more fixations on the AOIs are required ($B = 0.3654, p < .0001$).

In order to test Hypothesis 3, we apply a linear regression with mixed model and a two-tailed level of significance of Fixation Durations on Answer, allowing us to identify a significant correlation between successful diagnostics and shorter fixation duration ($B = -0.373$, $p < .0001$). However, fixations in the AOIs are longer for successful diagnosed semantic errors ($B = 0.1654$, $p = .0061$) and syntactic errors ($B = 0.4436$, $p < .0001$), which indicate that the participants, when successfully identifying the errors, spend more time on the errors, but less time on the rest of the stimuli.

5 Discussion and Conclusion

Our preliminary results suggest that H1, H2 and H3 are partially supported. Significant links between successfully detecting an error and a lesser amount of time spent on a stimulus (H1), and between an accurate diagnostic and shorter fixation duration are found (H3). Furthermore, H2, which proposed fewer fixations and a greater proportion of fixations on the errors when successfully detecting an error, is supported. However, successful detection of syntactic errors is significantly associated with a greater number of fixations in AOIs, which suggests that the error was detected, but the correction response is inhibited [9], possibly due to a higher level of complexity in syntactic errors. No significant link between syntactic errors and the number of fixations in the entire stimulus is found. Thus, this study presents evidence that the characteristics of visual attention of experienced modelers, such as lower number and duration of fixations, are generally related with successful error detection. On the other hand, attributes normally associated with novices, such as higher time spent on a stimulus or higher fixation duration, lead to unsuccessful error detection.

Our research so far offers both theoretical and practical contributions. The differences between the process and repertoire of attentional characteristics in the detection of semantic errors and syntactic errors reinforce Moody's [8] propositions about their complex interdependence. Syntactic errors require more attentional fixation than semantic errors. This finding runs contrary to our expectations and highlights the need for further studies to more fully articulate the differences and the metrics that might be used to measure them. The next phase of our research will address this challenge. From a practical standpoint, deeper insights into differences between the attentional characteristics will offer guidance to the evolution of BPMN and other notations and recommendations for curriculum development and training methods.

Limitations of this exploratory phase of our study center on the models used as stimuli for our experiment. While we tried to minimize the impact of domain-specific knowledge by using processes well-known to all potential participants, it is virtually impossible to negate the influence of the variation of domain familiarity between participants. Another limitation can be found in our sample. A bigger and more equally distributed sample will allow more complex statistical analyses and provide more significant findings. Finally, no measure was taken to evaluate the 'stopping rule', which is when a subject decides to terminate his information search because he judges that he has enough information to complete his task [24, 25]. The next step of our research should evaluate this concept in order to better understand our eye tracking data, especially the measures linked to the view time.

References

1. Recker, J.C., Dreiling, A.: Does it matter which process modelling language we teach or use? An experimental study on understanding process modelling languages without formal education. In: Toleman, M., Cater-Steel, A., Roberts, D. (eds.) 18th Australasian Conference on Information Systems, Toowoomba, Australia, pp. 356–366 (2007)
2. Parsons, J., Cole, L.: What do the pictures mean? Guidelines for experimental evaluation of representation fidelity in diagrammatical conceptual modeling techniques. Data Knowl. Eng. **55**(3), 327–342 (2005)
3. Gemino, A., Wand, Y.: Comparing mandatory and optional properties in conceptual data modeling. In: Proceedings of the Tenth Annual Workshop on Information Technologies and Systems, Brisbane, pp. 97–102 (2000)
4. Gemino, A., Wand, Y.: Evaluating modeling techniques based on models of learning. Commun. ACM **46**(10), 79–84 (2003)
5. Gegenfurtner, A., Lehtinen, E., Säljö, R.: Expertise differences in the comprehension of visualizations: a meta-analysis of eye-tracking research in professional domains. Educ. Psychol. Rev. **23**(4), 523–552 (2011)
6. Shanks, G.: Conceptual data modelling: an empirical study of expert and novice data modellers. Australas. J. Inf. Syst. **4**(2) (1997)
7. Davis, C.J., Hevner, A.R., Labonte-LeMoyne, É., Léger, P.-M.: Expertise as a mediating factor in conceptual modeling. In: Information Systems and Neuroscience, pp. 85–92. Springer, Berlin (2018)
8. Moody, D.: The "physics" of notations: toward a scientific basis for constructing visual notations in software engineering. IEEE Trans. Softw. Eng. **35**(6), 756–779 (2009)

9. Van Waes, L., Leijten, M., Quinlan, T.: Reading during sentence composing and error correction: a multilevel analysis of the influences of task complexity. Read. Writ. **23**(7), 803–834 (2009)
10. Henderson, J.M., Hollingworth, A.: The role of fixation position in detecting scene changes across saccades. Psychol. Sci. **10**(5), 438–443 (1999)
11. Holmqvist, K., Nyström, M., Andersson, R., Dewhurst, R., Jarodzka, H., Van de Weijer, J.: Eye Tracking: A Comprehensive Guide to Methods and Measures. OUP, Oxford (2011)
12. Goldberg, J.H., Kotval, X.P.: Computer interface evaluation using eye movements: methods and constructs. Int. J. Ind. Ergon. **24**(6), 631–645 (1999)
13. Yusuf, S., Kagdi, H., Maletic, J.I.: Assessing the comprehension of UML class diagrams via eye tracking. In: Program Comprehension, 2007. ICPC'07, 15th IEEE International Conference, pp. 113–122 (2007)
14. Krupinski, E.A.: The importance of perception research in medical imaging. Radiat. Med. **18**(6), 329–334 (2000)
15. Reingold, E.M., Sheridan, H.: Eye movements and visual expertise in chess and medicine. Oxf. Handb. Eye Movements, 528–550 (2011)
16. Sheridan, H., Reingold, E.M.: Expert vs. novice differences in the detection of relevant information during a chess game: evidence from eye movements. Front Psychol. **5**, 941 (2014)
17. Glaser, R.: Education and thinking: the role of knowledge. Am. Psychol. **39**(2), 93 (1984)
18. Lurigio, A.J., Carroll, J.S.: Probation officers' schemata of offenders: content, development, and impact on treatment decisions. J. Pers. Soc. Psychol. **48**(5), 1112 (1985)
19. Gemino, A., Wand, Y.: A framework for empirical evaluation of conceptual modeling techniques. Requirements Eng. **9**(4), 248–260 (2004)
20. Birkmeier, D., Kloeckner, S., Overhage, S.: An empirical comparison of the usability of BPMN and UML activity diagrams for business users. In: ECIS, p. 2 (2010)
21. Bavota, G., Gravino, C., Oliveto, R., De Lucia, A., Tortora, G., Genero, M., Cruz-Lemus, J.A.: Identifying the weaknesses of UML class diagrams during data model comprehension. In: International Conference on Model Driven Engineering Languages and Systems, pp. 168–182. Springer, Berlin (2011)
22. Just, M.A., Carpenter, P.A.: Eye fixations and cognitive processes. Cogn. Psychol. **8**(4), 441–480 (1976)
23. Rayner, K.: Eye movements in reading and information processing: 20 years of research. Psychol. Bull. **124**(3), 372 (1998)
24. Browne, G.J., Pitts, M.G.: Stopping rule use during information search in design problems. Organ. Behav. Hum. Decis. Process. **95**(2), 208–224 (2004)
25. Nickles, K.R., Curley, S.P., Benson, P.G.: Judgment-Based and Reasoning-Based Stopping Rules in Decision Making Under Uncertainty, vol. 7285. University of Minnesota (1995)

Paying Attention Doesn't Always Pay off: The Effects of High Attention Load on Evaluations of Ideas

Goran Calic, Nour El Shamy, Khaled Hassanein and Scott Watter

Abstract Creativity is a key driver of success for organizations in the digital age. Managers engaged in evaluating the creativity of new ideas are often subject to a myriad of technology-mediated distractors that compete for their attention. In this work in progress paper, we investigate whether attentional overload results in an upward bias for IT-mediated creativity evaluations. We report on promising early results that examines this phenomenon and set out to study its implications on IT design complexity.

Keywords Creativity · Cognitive load · Cognition · Electroencephalography
Eye tracking · Pupil dilation

1 Introduction

Creative ideas—ideas that are original and useful—are considered by most CEO's to be a crucial factor for success and sustainability in an ever turbulent environment [1]. Conversely, conventional ideas are discarded because they provide no competitive edge, as profits from these ideas have already been competed away [2]. Information Technology (IT) platforms are frequently utilized to facilitate creative idea generation and exchange between individuals. To this end, managers are investing in technology

G. Calic · N. E. Shamy · K. Hassanein (✉)
DeGroote School of Business, McMaster University, Hamilton, Canada
e-mail: hassank@mcmaster.ca

G. Calic
e-mail: calicg@mcmaster.ca

N. E. Shamy
e-mail: elshamyn@mcmaster.ca

S. Watter
Department of Psychology, Neuroscience and Behaviour, McMaster University, Hamilton, Canada
e-mail: watter@mcmaster.ca

F. D. Davis et al. (eds.), *Information Systems and Neuroscience*,
Lecture Notes in Information Systems and Organisation 29,
https://doi.org/10.1007/978-3-030-01087-4_8

that fosters employee creativity [3, 4]. The objective of such investments is to get an edge in an increasingly competitive marketplace.

Previous Information Systems (IS) research has investigated how IT could be designed to foster creativity through priming—the presentation of a stimulus designed to subconsciously implant a concept in working memory that alters subsequent behavior [5]. Other studies have investigated the influence of IT design artefacts on users' cognitive demands during idea generation [6–8]. Much less is known about the idea evaluation process – the process of distinguishing creative ideas from conventional ones. Particularly, it would be valuable to understand whether and how attentional load imposed by IT design artefacts influence users' ability to accurately evaluate the creativity of ideas.

When resources required to implement ideas are scarce, trial-and-error strategies of testing whether ideas are creative, rather than conventional, become prohibitively costly. Under such circumstances, a manager's capacity to make unbiased (i.e., accurate) evaluations about the creativity of an idea is highly valuable. Such evaluations require cognitive work. According to cognitive load theory [9], more attentional resources will result in a deeper understanding of the problem and its solution. In creative problems solving, this process begins with the retrieval of information from long-term memory and the subsequent analysis of that information in working memory. Attentional work that competes with this process may disrupt idea evaluation. Managers are increasingly engaged in IT-mediated work and online communication that competes for their attention. When presented with an idea online or through an Electronic Brainstorming (EBS) platform, manager evaluation may be biased due to the cognitive demands imposed by the system (e.g., system complexity) [10] or the attention-grabbing nature of other technology-mediated distractors (e.g., a new e-mail arrives, tweets, or even their cell phone) [11].

In this work in progress, we investigate whether attention overload biases IT-mediated creativity evaluations. We set out to test a broad conjecture that attention load will result in upward biased creativity evaluations. This is because attention load will attenuate an individual's capacity to make accurate judgments about an idea's future utility making the idea appear more surprising. A feeling of surprise has been shown to be highly related to the feeling of creativity and may result in the individual evaluating an idea as creative, even when it is conventional [12]. The implication is that in cognitive states of attention load, individuals will incorrectly evaluate conventional ideas as being creative. Evaluations of ideas are biased because they differ from evaluations the same individual would make in a state free of attention load. Thus, we coined the term "surprise bias" to describe this phenomenon.

2 Theoretical Background

In the next subsections we describe the theoretical lens used to develop our hypothesis about the link between attention overload and idea evaluation. Following this, we develop the concept of creativity and its relation to attention.

2.1 Attention Load

Cognitive load is the mental effort exerted by an individual to solve a problem or accomplish a task, during which information is retrieved from long term memory and temporarily stored in working memory for processing [13, 14]. While knowledge stored in long-term memory can be virtually unlimited, working memory capacity is limited to around seven elements or schemas of information at a time [14, 15]. In order to further integrate, compare, and process information in working memory, additional working memory capacity is required, which further constrains the amount of information elements that can be simultaneously processed at a given moment [14].

Cognitive Load Theory (CLT) argues that task performance can be enhanced by optimizing cognitive resources through effective instructional design [9, 14]. CLT's design principles have been influential in understanding and enhancing individuals' IT-mediated creative ideation in Group Decision Systems (GSS) and Electronic Brainstorming (EBS). Potter and Balthazard [16] provide evidence that information load, in the form attending to others' feedback, reduced the number and quality of ideas generated by subjects in an EBS context. Heninger et al. [11] found that subjects who exchanged information simultaneously (i.e., contributing information while processing others' input) in GSS performed worse than subjects who performed these tasks separately. This difference was attributed to individual cognition and increased demands on working memory during dual-task performance rather than social processes [11]. Numerous studies building on CLT extend IS design principles that foster productivity by reducing cognitive demands [5, 8, 10, 17].

From a neurophysiological perspective, high attention load is associated with several electroencephalographic (EEG) and event-related potential (ERP) correlates as well as the dilation of the pupils. Attention load is associated with Event-Related Desynchronization (ERD) of neural oscillations in the Alpha band (8–12 Hz) [18, 19]. This effect can be observed in the higher Alpha band (10–12 Hz), particularly in the frontal region of the brain responsible for executive functions and working memory encoding [19, 20]. Additionally, under high attention load conditions, the amplitude of the P300 component during the presentation of stimuli is significantly reduced, indicating that cognitive resources are partly diverted away from the processing stimuli information [21, 22]. Further, attention load is associated with an involuntary dilation of the pupils, one of the sympathetic reactions of the Autonomous Nervous System (ANS) that prepare the organism to deal with stressful tasks [23].

2.2 Creativity

Creative ideas are defined as ideas that are both useful and unique in a particular context [24]. They are synthesized by reconstructing distant knowledge in novel ways to achieve a particular objective. Several EEG parameters have been found to correlate with creative ideation. Increases in the Alpha band power have been consistently observed during the generation of creative ideas [24–26]. Fink and Benedek [24] conclude that this relationship reflects an internal focused attention during the retrieval and recombination of distant knowledge from long-term memory. Additionally, the novelty and suddenness of a creative idea has been linked to increased activity in the Gamma band (38–44 Hz), particularly in the parieto-occipital region [27]. This has been interpreted as the result of successfully retrieving distant knowledge from long-term memory and a feeling of surprise, or "Aha!", by the creative idea [27].

Creativity is particularly valuable in situations characterized by attention load—an occurrence that has already pushed information processing of individuals to its limiting point. For example, a CEO might be burdened while evaluating the creativity of EBS-mediated responses to a rapidly escalating public relations crisis that can cost millions of dollars. Research suggests that individuals suffering from attention load generate fewer and less creative ideas [16, 27–29]. Attention load will focus information processing to the task at hand and block the retrieval of remote and distant knowledge from long-term memory and knowledge which, when combined with task relevant easy to retrieve knowledge, results in the synthesis of creative ideas [30]. West [31] argues that competition, severity of challenge, time constraints imposed by the organization or environment, and other states which call on focused attention will inhibit creativity. Together, these literatures suggest that attention load inhibits the generation of creative ideas.

We suggest that attentional load also reduces evaluative capability, and in an unexpected way: attention load will bias managers to evaluate conventional ideas as creative. The implication is that such errors result in decision making errors—managers may invest in conventional ideas because they believe these ideas to be creative. During creative cognition, individuals rely on one or more heuristics (e.g., hypothesis testing [32], trial-and-error [33], or experiential search [34]) to form subjective beliefs about outcomes of implementing an idea. By reducing the information processing capacity available to form unbiased beliefs about the future utility of an idea, attention load increases an individual's subjective feeling of creativity, which we refer to as "surprise bias". Thus:

Hypothesis: Higher attention load will result in higher creativity evaluations.

3 Methodology and Preliminary Results

3.1 Experimental Design

An initial pilot study was conducted to test our hypothesis using EEG and eye tracking methods. We used Cognionics Quick-20 Dry EEG headsets to collect neural activity signals at 500 Hz. Tobii Pro X2-60 was used to collect eye tracking data including fixations and pupil diameter at 60 Hz. Noldus Syncbox was used to send markers to both Tobii Studio and Cognionics EEG Acquisition software to ensure proper synchronization between the systems.

The study followed a modified version of the Alternative Uses (AU) creativity experiment. AU tests are commonly utilized in the study of creative ideation using EEG, during which participants are asked to generate as many creative AU's for common objects [25, 26]. We modified this design by providing participants a pre-generated set of five AU's to each of three common day objects (i.e., brick, tin can, ping pong ball). We asked participants to evaluate each of the AU's provided on a scale of 1–5 (1-strongly disagree 5-strongly agree) on the two creativity dimensions of novelty and usefulness. E-Prime v 2.0 was utilized for the presentation of instructions and stimuli, and Noldus E-Prime Server was used to transfer E-Prime events and subject responses to the Observer XT software for analysis.

Participants were randomly assigned to either a baseline or an attention load condition. In the attention load treatment, participants were asked to simultaneously perform an auditory oddball task during the evaluation of AU's. Auditory stimuli, alternating between a frequent low-pitched tone and a rare high-pitched tone, were played every second for both conditions during the AU's tasks. Participants in the baseline condition were asked to disregard the noise, while those in the attention load condition were asked to count the number of times the target (i.e., high-pitched) tone was played and disregard the distractor tone. After completing the five AU's evaluation tasks for each object, participants in the attention load condition were asked to indicate the number of times the target tone was played. At the end of the experiment, participants in both conditions were asked to rate how distracting they felt the auditory stimuli were on a scale of 1–5 (1-not distracting at all 5-very distracting).

We received clearance from McMaster University's research ethics board, and 12 MBA students (i.e., 1 female, 11 male) were recruited through McMaster Digital Transformation Research Centre (MDTRC) and participated in the pilot study so far. Participants were screened for eye health related problems, and two participants were excluded from the EEG analysis due to handedness (i.e., left handed). The experiment lasted 20 min on average and participants received monetary compensation for their time.

3.2 Analysis and Preliminary Results

This research is still in its nascent stage, more participants are currently being recruited and data analysis is still underway. However, preliminary results seem to be promising. Participants in the attention load condition were more distracted (M_{load} $= 3.3$) by the auditory stimuli than participants in the control group ($M_{control} = 2.4$, t $= 1.8$, $p = 0.055$) indicating that the attention load manipulation is successful. Initial ERP analysis (Fig. 1) provides information on differences in cognitive processing between the two groups. Participants in the control group exhibited sustained voltage negativity over frontal, and especially prefrontal, electrode sites compared to relatively smaller negative values for the attention load condition. Extended negativity in the frontal region has been associated with greater executive control demands and

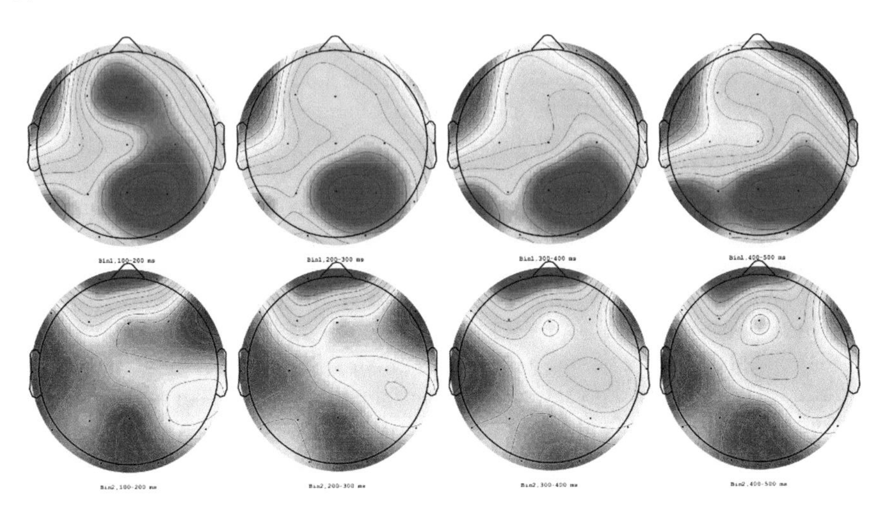

Fig. 1 Mean scalp voltage topography from 100 to 500 ms after the onset of AU's in 100 ms intervals for the attention overload (top row) and control (bottom row) conditions

focus on active task processing [35, 36]. In response to the onset of the AU trial, participants in the attention load condition are partly directing their cognition to process the auditory stimuli, and thus exhibit lower prefrontal negativity relative to the AU task [35, 36]. As hypothesized, creativity ratings are higher in the attention load than the baseline condition for 11 of the 15 AU's. While the differences are not statistically significant, the sample size at the time of submitting this work is too small to be conclusive. A larger sample to be tested in the near future will furnish a more rigorous test of the central hypothesis of this study.

4 Discussion and Next Steps

Our initial results are promising, yet analysis of the pupil dilation differences and EEG data is still ongoing, and more participants are required for these results to be conclusive. Prefrontal Alpha and parieto-occipital Gamma differences may particularly reveal interesting variations in executive control and evaluations of creative ideas under attention load. Our next steps will involve testing our hypothesis for IT-inflicted cognitive load in creativity evaluations. Particularly, we plan to study the impact of system complexity on managers' evaluation of EBS-mediated ideas, as well as on consumers' evaluation of crowdfunding ideas (e.g., Kickstarter). Studies have shown that ideas presented with more information and content had higher chances of success in attracting crowdfunding [37]. However, this goes against basic CLT design principles of parsimony in presenting information to avoid attention load [10]. It will be extremely rewarding to theory and practice to examine this phenomenon in our future work.

References

1. IBM: IBM 2010 Global CEO Study: Creativity Selected as Most Crucial Factor for Future Success. News releases (2010)
2. Gavetti, G.: Toward a behavioral theory of strategy. Organ. Sci. **23**, 267–285 (2012)
3. Bhagwatwar, A., Massey, A., Dennis, A.R.: Creative virtual environments: effect of supraliminal priming on team brainstorming. Proceedings of the Annual Hawaii International Conference on System Sciences, pp. 215–224. (2013). https://doi.org/10.1109/hicss.2013.152
4. Di Gangi, P.M, Wasko, M.M., Hooker, R.E.: Getting customers' ideas to work for you: learning from dell how to succeed with online user innovation communities. MIS Q. Executive **9** (2010)
5. Dennis, A.R., Minas, R.K., Bhagwatwar, A.P.: Sparking creativity: improving electronic brainstorming with individual cognitive priming. J. Manage. Inf. Syst. **29**. IEEE: 195–216. (2013). https://doi.org/10.2753/mis0742-1222290407
6. Davern, M., Shaft, T., Te'eni, D.: Cognition matters: enduring questions in cognitive IS research. J. Assoc. Inf. Syst. **13**, 273–314 (2012)
7. Antunes, P., Ferreira, A.: Developing collaboration awareness support from a cognitive perspective. Proceedings of the Annual Hawaii International Conference on System Sciences. (2011). https://doi.org/10.1109/hicss.2011.161
8. Kolfschoten, G.L.: Cognitive load in collaboration—brainstorming. Proceedings of the Annual Hawaii International Conference on System Sciences. (2011). https://doi.org/10.1109/hicss.2011.107
9. Sweller, J.: Cognitive load theory, learning difficulty, and instructional design. Learn. Instr. **4**, 295–312 (1994). https://doi.org/10.1016/0959-4752(94)90003-5
10. Wang, Q., Yang, S., Liu, M., Cao, Z., Ma, Q.: An eye-tracking study of website complexity from cognitive load perspective. Decis. Support Syst. **62**, 1–10, Elsevier B.V. (2014). https://doi.org/10.1016/j.dss.2014.02.007
11. Heninger, W.G., Dennis, A.R., Hilmer, K.M.: Research note: individual cognition and dual-task inference in group support systems. Inf. Syst. Res. **17**, 415–424 (2006). https://doi.org/10.1287/isre.l060.0102
12. Simonton, D.K.: Creative thought as blind variation and selective retention: why creativity is inversely related to sightedness. J. Theor. Philos. Psychol. **33**, 253–266 (2012). https://doi.org/10.1037/a0030705
13. Sweller, J.: Cognitive load during problem solving: effects on learning. Cogn. Sci. **12**, 257–285 (1988)
14. Sweller, J., van Merrienboer, J.J.G., Paas, F.G.W.C.: Cognitive architecture and instructional design. Educ. Psychol. Rev. **10**, 251–296 (1998)
15. Miller, G.: The magical number seven, plus or minus two: some limits on our capacity for processing information. Psychol. Rev. **63**, 81–97 (1956)
16. Potter, R.E., Balthazard, P.: The role of individual memory and attention processes during electronic brainstorming. MIS Q. **28**, 621–643 (2004)
17. Williams, M.L., Dennis, A.R., Stam, A., Aronson, J.E.: The impact of DSS use and information load on errors and decision quality. Eur. J. Oper. Res. **176**, 468–481 (2007). https://doi.org/10.1016/j.ejor.2005.06.064
18. Stipacek, A., Grabner, R.H., Neuper, C., Fink, A., Neubauer, A.C.: Sensitivity of human EEG alpha band desynchronization to different working memory components and increasing levels of memory load. Neurosci. Lett. **353**, 193–196 (2003). https://doi.org/10.1016/j.neulet.2003.09.044
19. Krause, C.M., Sillanmäki, L., Koivisto, M., Saarela, C., Häggqvist, A., Laine, M., Hämäläinen, H.: The effects of memory load on event-related EEG desynchronization and synchronization. Clin. Neurophysiol. **111**, 2071–2078 (2000). https://doi.org/10.1016/S1388-2457(00)00429-6
20. Klimesch, W., Schimke, H., Pfurtscheller, G.: Alpha frequency, cognitive load and memory performance. Brain Topogr. **5**, 241–251 (1993)
21. Parasuraman, R.: Neuroergonomics: brain, cognition, and performance at work. Curr. Dir. Psychol. Sci. **20**, 181–186 (2011)

22. Lavric, A., Forstmeier, S., Rippon, G.: Differences in working memory involvement in analytical and creative tasks: an ERP study. Cogn. Neurosci. **11**, 1613–1618 (2000). https://doi.org/10.1097/00001756-200006050-00004
23. Riedl, R., Léger, P.-L.: Fundamentals of NeuroIS. In: Studies in Neuroscience, Psychology and Behavioral Economics. Berlin, Heidelberg: Springer
24. Fink, A., Benedek, M.: EEG alpha power and creative ideation. Neurosci. Biobehav. Rev. **44**, 111–123, Elsevier Ltd. (2014). https://doi.org/10.1016/j.neubiorev.2012.12.002
25. Fink, A., Benedek, M., Grabner, R.H., Staudt, B., Neubauer, A.C.: Creativity meets neuroscience: experimental tasks for the neuroscientific study of creative thinking. Methods **42**, 68–76 (2007). https://doi.org/10.1016/j.ymeth.2006.12.001
26. Dietrich, A., Kanso, R.: A review of EEG, ERP, and neuroimaging studies of creativity and insight. Psychol. Bull. **136**, 822–848 (2010). https://doi.org/10.1037/a0019749
27. Sandkühler, S., Bhattacharya, J.: Deconstructing insight: EEG correlates of insightful problem solving. PLoS ONE **3**, e1459 (2008). https://doi.org/10.1371/journal.pone.0001459. (Edited by Paul Zak)
28. de Dreu, C.K.W., Nijstad, B.A., Baas, M., Wolsink, I., Roskes, M.: Working memory benefits creative insight, musical improvisation, and original ideation through maintained task-focused attention. Pers. Soc. Psychol. Bull. **38**, 656–669 (2012). https://doi.org/10.1177/0146167211435795
29. Smith, S.M. Getting into and Out of Mental Ruts: A Theory of Fixation, Incubation, and Insight. The MIT Press (1995)
30. Kasof, J.: Creativity and breadth of attention. Creativity Res. J. **10**, 303–315 (1997). https://doi.org/10.1207/s15326934crj1004_2
31. West, M.A.: Sparkling fountains or stagnant ponds: an integrative model of creativity and innovation implementation in work groups. Appl. Psychol. **51**, 355–387 (2002)
32. Finke, R.A., Smith, S.M., Ward, T.B.: Creative Cognition Theory, Research, and Applications. MIT Press, Cambridge, Massachusetts (1996)
33. Campbell, D.T.: Blind variation and selective retentions in creative thought as in other knowledge processes. Psychol. Rev. **67**, 380–400 (1960)
34. Getzels, J.W. Creativity, intelligence, and problem finding: retrospect and prospect. Front. Creativity Res., 88–102 (1987)
35. Watter, S., Heisz, J.J., Karle, J.W., Shedden, J.M., Kiss, I.: Modality-specific control processes in verbal versus spatial working memory. Brain Res. **1347**, 90–103 (2010). https://doi.org/10.1016/j.brainres.2010.05.085
36. Kiss, I., Watter, S., Heisz, J.J., Shedden, J.M.: Control processes in verbal working memory: an event-related potential study. Brain Res. **1172**, 67–81 (2007). https://doi.org/10.1016/j.brainres.2007.06.083
37. Du, Q., Qiao, Z., Fan, W., Zhou, M., Zhang, X., Wang, A.G.: Money talks: a predictive model on crowdfunding success using project description. In ACIS **2015**, 1–8 (2015)

Using Gaze Behavior to Measure Cognitive Load

Lisa Perkhofer and Othmar Lehner

Abstract Measuring and influencing cognitive load during information processing can be seen as a promising instrument to mitigate the risk of information overload while increasing processing capabilities. In this study, we demonstrate how to use cross-sectional time-series data generated with an eye tracking device to indicate cognitive load levels. Thereby we combine multiple measures related to fixations, saccades and blinks and calculate one comprehensive and robust measure. Applicability is demonstrated by conducting two separate experiments in a decision-making scenario in the context of information visualization.

Keywords Eye tracking · Cognitive load · Structural equation modelling

1 Introduction

Cognitive load is defined as "the amount of working memory resources required in cognitive task execution" [1] p. 381. The monitoring of cognitive load levels during information processing and learning in situations in which a user operates near his or her working memory capacity limits can be identified as a major application [2]. If these capacity limits are breached the user enters the state of information overload in which processing capabilities are seriously impaired, further learning is prohibited, and thus biased decisions may often be the consequence [3, 4]. To avoid these disadvantageous effects research in the field of human-computer interaction, with topics such as interface design and visual computing, has begun to focus on the construct of cognitive load already at design and evaluation stages [1, 5–13].

L. Perkhofer (✉)
Management Accounting, University of Applied Sciences Upper Austria, Steyr, Austria
e-mail: lisa.perkhofer@fh-steyr.at

O. Lehner
Said Business School, University of Oxford, Oxford, UK
e-mail: othmar.lehner@sbs.ox.ac.uk

F. D. Davis et al. (eds.), *Information Systems and Neuroscience*,
Lecture Notes in Information Systems and Organisation 29,
https://doi.org/10.1007/978-3-030-01087-4_9

Consequently, multiple models for the assessment of cognitive load have been suggested so far [1, 7, 14]. Proposed measurement possibilities range from simple self-assessment to a more high-end analysis based on fMRI [14]. Unfortunately, when it comes to the actual application in empirical studies, physiological measures are still scarce especially when it comes to information visualization [9]. Based on the foundational paper by Zagermann et al. [12], this study therefore contributes to closing this gap by empirically assessing various complementary eye tracking measures as proxies for a single comprehensive and robust cognitive load measure. Two separate experiments in the context of InfoVis were conducted and analyzed to demonstrate reliability of the construct. Furthermore, its applicability is tested in a real-case decision-making scenario and evaluated based on a structural equation model.

2 Cognitive Load Measurement

Cognitive load is a construct, which is not directly observable, but can be assessed using indicators [14]. Brücken et al. [14] summarize different measurement models for those indicators, split in subjective measurement via self-reports and objective measurements via physiological, behavioral, outcome-oriented or brain activity measures. Up to now however, the majority of studies conducted in InfoVis measure cognitive load either by analyzing results on decision-making outcome (task accuracy or task completion time), or by subjective assessment via self-reported data (e.g. NASA TLX, SWAT) [9, 15, 16]. This contemporary practice misses robustness and is therefore susceptible to multiple measurement biases, possibly even leading to spurious correlations [16]. Further results cannot be used reliably for prediction or real time assessment given an optimal and user specific support. More robust methods from NeuroIS could help in triangulating early results from literature and further produce more robust empirical evidence which can also be used real time [9].

One physiological measurement method, which is increasingly gaining attention in this context, is eye tracking [13, 16–19]. By relying on eye tracking related measures it should be possible to distinguish between the mental demand of different interface or system designs [7, 8] while predicting it's influence on decision-making outcome during cognitive task execution [6, 13]. In a context of static visual stimuli, eye movements measures on fixations and saccades (mostly voluntary), but also measures on pupil dilation and blink related data (mostly involuntarily movement), have been associated with an increase in mental demand [18]. In the following, we are going to individually and shortly describe possible eye movement indicators and summarize their respective implications in the context of cognitive load.

- **Fixations**. Fixations are cognitively controlled short dwells where the eye stops, and one processes information. An increase in fixation duration is said to indicate higher demands on working memory, whereas a high fixation rate signals great visual and/or cognitive complexity (task dependent searching behavior) [17–19].

- **Saccades**. Saccades refer to the shifts between fixation locations. During the actual movement of the eye, no information acquisition can take place because one is blind. However, it has been shown that a long saccade length indicates higher cognitive load [18] and also a high saccade count is associated with higher visual complexity [13].
- **Pupil**. Most studies on cognitive load and eye tracking focus on pupilometry [13]. In states of high mental load, the pupil diameter enlarges proportionally [1, 13, 20]. However, in order to account for individual and situational differences it is necessary to evaluate pupil diameter with respect to an adaptive baseline [13].
- **Blinks**. Blinking is said to indicate information processing and they occur before and after high states of cognitive load. High blink rates as well as long blink durations are associated with higher mental demand and thinking [18, 20].

3 Development of One Comprehensive Construct

Looking at the data generated by eye tracking devices, we are confronted with cross-sectional time-series data [21]. Each task results in a scan-path, which is a series of fixations, saccades and blinks stringed together. More precisely, a scan-path starts with the onset of the stimuli and ends with its offset and it further includes all periods of task dependent visual search behavior as well as all periods of active mental processing [17, 18]. What we can observe is therefore not a linear trend but a tendency to mean reversion for all eye tracking measures (an example for mean reversion of saccade duration, and pupil diameter is presented in Fig. 1). This means in case of high mental load, which according to previous research indicates mental processing, high states of saccade duration, pupil diameter, blink duration etc. can be observed [16, 20].

Summing up the findings as presented in the previous section and shown in Fig. 1, we can see an increase in value in all measures in stages of actual mental processing, while during visual search periods values should normalize and return to average. To

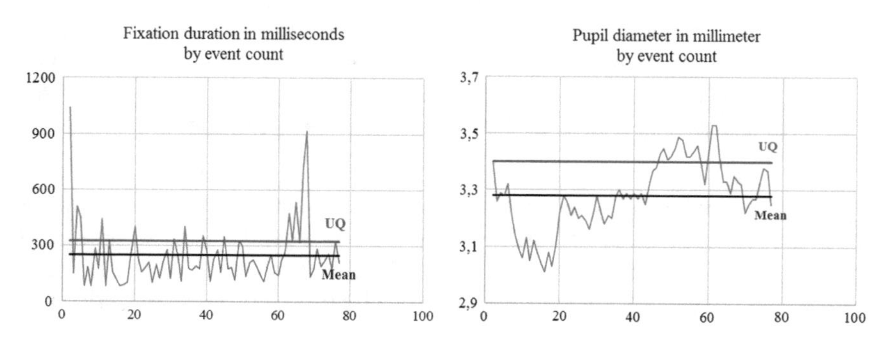

Fig. 1 Mean reversion tendency of fixation duration and pupil diameter (example one participant, one stimulus and one task)

compare the mental demand of one stimulus with another, comparing their respective peaks seems logical, as these should represent actual processing periods. To mitigate the effect of outliers but still focus on phases of high mental demand, we only focus on the upper quintile (red line in Fig. 1) of each eye tracking related measure (saccade duration, fixation duration, blink duration, pupil diameter). For aggregation of the resulting time-series a quantile-regression form (weighted average) is applied [21].

This procedure allows for a calculation of individual values per task, stimulus and participant for all relevant eye tracking measures, which can be used for further assessment. In contrast to most studies, we are trying to create one comprehensive measure (including information on fixations, saccades, pupil dilation, and blinks) following the example of Siegle et al. [20] instead of focusing on just one. Eye tracking indicators provide complementary information but their combination into one comprehensive construct necessitates the use of partial least squares structural equation modelling.

4 Experiments

As indicated in Sect. 2, eye tracking related measures used in this experiment were saccade duration, saccade count, fixation duration, fixation count, pupil diameter, pupil diameter difference, blink duration, and blink count. While count related measures were used as is, duration related measures were calculated according to the procedure explained in Sect. 3. With respect to the baseline introduced for pupil dilation we used the median of the first five fixations per task as after and before task execution pupil dilation is normalized and this allowed us to account for situational and personal differences [13, 22]. Furthermore, constant lightning conditions and an undisturbed and quiet environment were ensured throughout the experiment.

- In **study I** 118 students participated in the experiment. Participants were randomly assigned to one out of four groups. They had to answer questions with varying complexity levels (4 in total) based on different visualization types (16 in total). 1888 observations were recorded for analysis.[1]
- In **study II** 60 students participated in the experiment. Participants were again randomly assigned to one out of three groups. They had to answer questions with varying complexity levels (3 in total) based on visualization dashboards with three levels of data density (3 in total). 720 observations were recorded for analysis (see Footnote 1).

Data was recorded using an SMI stationary eye tracking system with nine-point calibration and 120 Hz sampling frequency. Eye tracking data was recorded and analyzed with Experiment Center as well as BeGaze (Version 3.7). The stimulus material was presented in randomized order in each experiment. For the experimental

[1]Task types and the stimulus material can be downloaded from the author's homepage.

tasks in both studies, participants slipped into the role of the CEO of a fictitious company.

Eye tracking measures were only used if the quality of recordings was high: The tracking ratio per stimulus, which is the time being recorded by the eye tracking system divided by the time of the stimulus, needed to be above 95%. In the case of missing or excluded values mean replacement was used. For each first order construct (saccade, fixations, pupil, and blinks), two indicators were measured. For CL a repeated measurement approach was used, meaning that each indicator used for the first order construct is reused for the second order construct representing CL. This approach is the most frequently used method for estimating higher-order constructs and allows to assess the impact of each eye tracking related event in order to gain a more compressive understanding of cognitive load. As a result predictability on decision-making outcome per user can be evaluated. For the generation of one comprehensive measure we used PLS-SEM (SmartPLS version 3.2.7) [23]. Results are presented in Table 1.

5 Conclusion

As discussed in the beginning, measuring cognitive load is of high relevance for practice because system and interface designers need to avoid creating situations of information overload for the respective users [24, 25]. Therefore, the field is in need of a comprehensive and reliable measure beyond the old proxies of time and error [26]. In this study, we propose to include physiological measures based on eye tracking data to determine cognitive load levels, which then can be used to compare different tasks as well as complete system designs. Some early form of reliability of the proposed measurement is demonstrated as both experiments show similar results although different tasks and different visual stimuli were used. Further, its applicability in InfoVis is shown, because both studies were embedded in decision-making scenarios which is representative for the field and the explanatory power of cognitive load on decision making outcome was high in both (R^2 is roughly 60%). As potential future research endeavor, a more robust regression method may be applied to further improve accuracy. The authors suggest a sampling lasso quantile regression for this.

As potential implications, the introduced measure could further help discern differences between tools and design options, and enhance our understanding of individual differences [5]. By doing so, it could help researchers in quantifying personalization and adaption needs [1] and can even be used in situ to change properties or layout features according to the respective user's cognitive state [18].

Table 1 Results on Study I and Study II

Study I	Study II
Step 1: Evaluation of the second order formative construct	

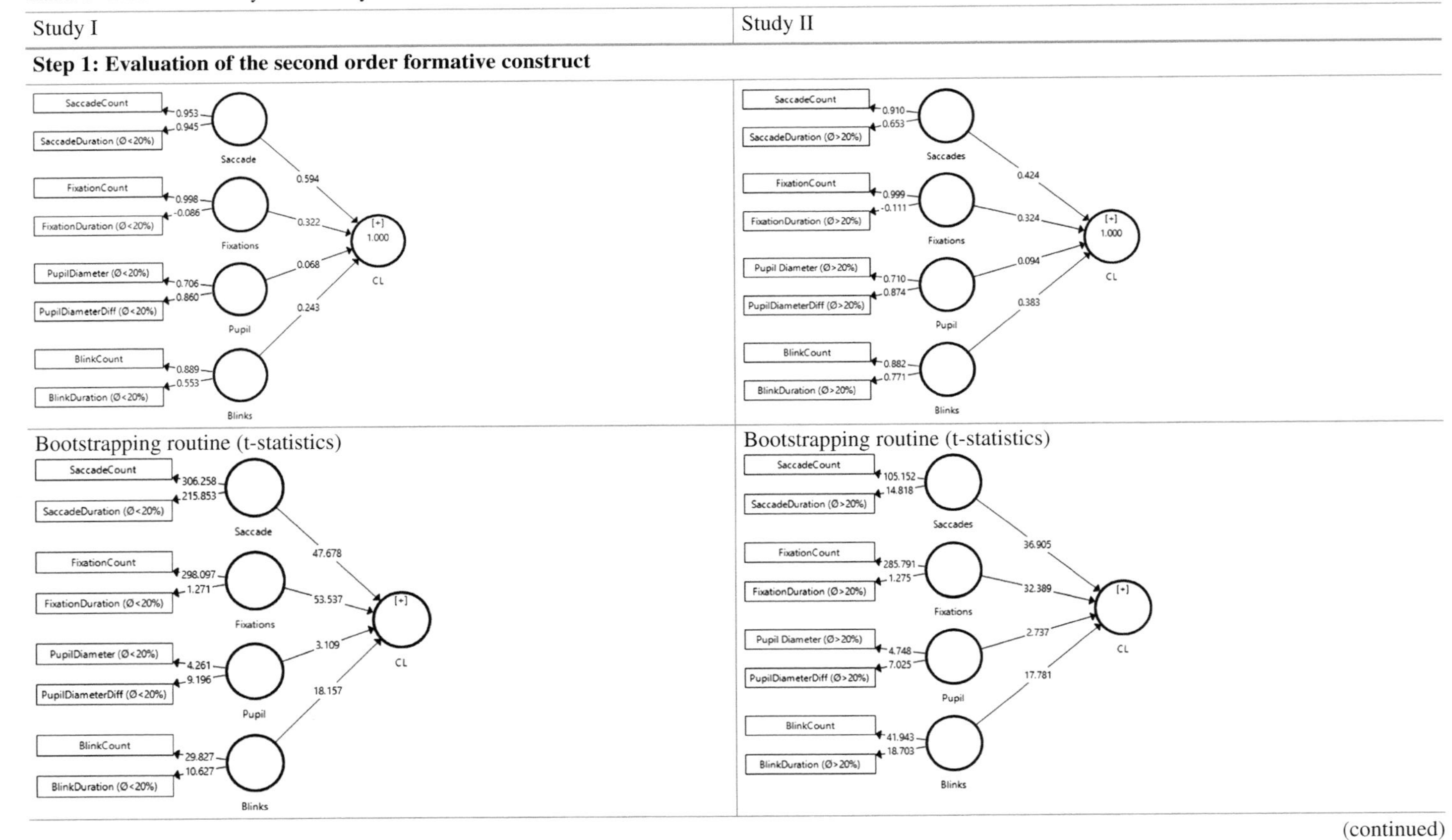

(continued)

Table 1 (continued)

Study I	Study II

Study I — Cross loadings:

	Blinks	Fixations	Pupil	Saccade	CL
BlinkCount	**0.889**	0.543	0.009	0.299	0.569
BlinkDuration (Ø<20%)	**0.553**	0.181	-0.004	0.202	0.313
FixationCount	0.545	**0.998**	0.068	0.740	0.898
FixationDuration (Ø<20%)	0.062	**-0.086**	-0.078	-0.074	-0.062
PupilDiameter (Ø<20%)	-0.025	0.043	**0.706**	0.074	0.099
PupilDiameterDiff (Ø<20%)	0.026	0.069	**0.860**	0.086	0.138
SaccadeCount	0.355	0.776	0.091	**0.953**	0.909
SaccadeDuration (Ø<20%)	0.294	0.628	0.103	**0.945**	0.842

Study II — Cross loadings

	Blinks	Fixations	Pupil	Saccades	CL
BlinkCount	**0.882**	0.438	0.035	0.569	0.724
BlinkDuration (Ø>20%)	**0.771**	0.289	0.035	0.337	0.535
FixationCount	0.455	0.999	0.103	0.891	0.885
FixationDuration (Ø>20%)	0.102	-0.111	-0.059	-0.085	-0.038
Pupil Diameter (Ø>20%)	-0.039	0.086	**0.710**	0.120	0.131
PupilDiameterDiff (Ø>20%)	0.084	0.084	**0.874**	0.112	0.189
SaccadeCount	0.524	0.989	0.116	**0.910**	0.918
SaccadeDuration (Ø>20%)	0.346	0.255	0.118	**0.653**	0.503

Construct reliability is good for all measures (above the 0.5 threshold for AVE), however, loadings on the construct fixation indicate a problem in both experimental studies. Fixation duration is not significant and has a negative impact on cognitive load. Additionally fixation count shows colliniarity with saccade count, blink count and saccade duration (highlighted in red). Based on these two observations, we decided to exclude the first oder construct fixation from the second order construct cognitive load

Step 2: Evaluation of the formative construct without fixations

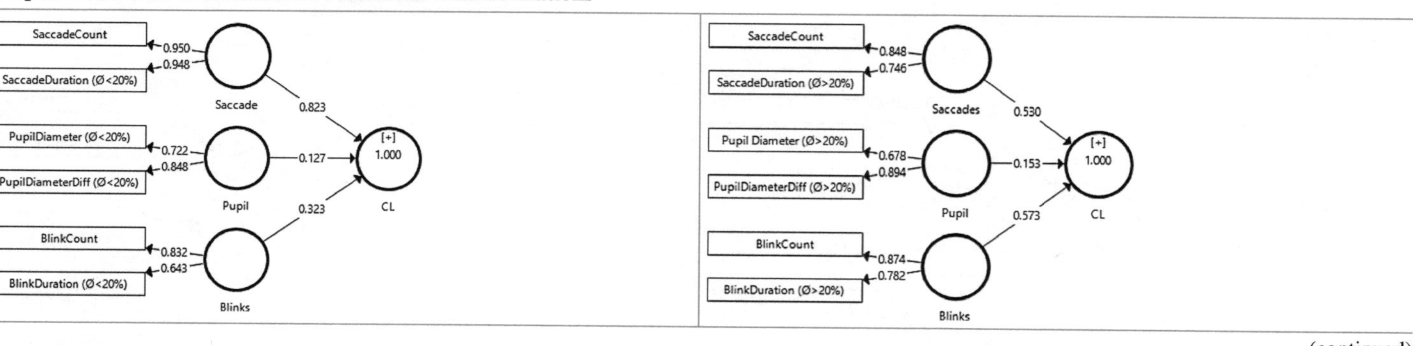

(continued)

Table 1 (continued)

Study I

Bootstrapping routine (t-statistics)

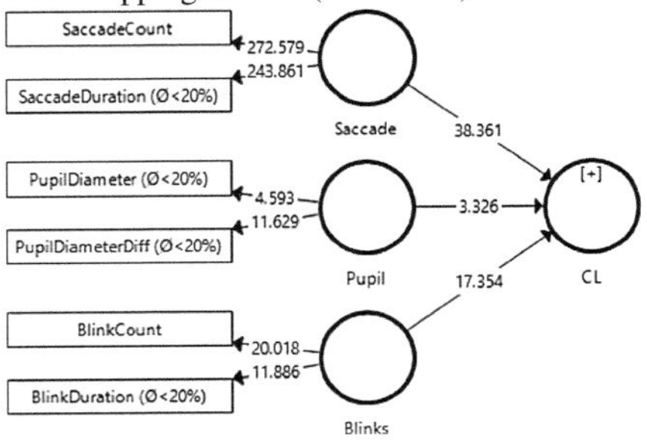

SaccadeCount ← 272.579 — Saccade
SaccadeDuration (Ø<20%) ← 243.861 — Saccade
PupilDiameter (Ø<20%) ← 4.593 — Pupil
PupilDiameterDiff (Ø<20%) ← 11.629 — Pupil
BlinkCount ← 20.018 — Blinks
BlinkDuration (Ø<20%) ← 11.886 — Blinks
Saccade — 38.361 → CL [+]
Pupil — 3.326 → CL [+]
Blinks — 17.354 → CL [+]

Cross loadings:

	Blinks	Pupil	Saccade	CL
BlinkCount	**0.832**	0.009	0.298	0.515
BlinkDuration (Ø<20%)	**0.643**	-0.005	0.202	0.373
PupilDiameter (Ø<20%)	-0.030	**0.722**	0.075	0.143
PupilDiameterDiff (Ø<20%)	0.028	**0.848**	0.086	0.187
SaccadeCount	0.356	0.091	**0.950**	0.908
SaccadeDuration (Ø<20%)	0.292	0.103	**0.948**	0.887

Study II

Bootstrapping routine (t-statistics)

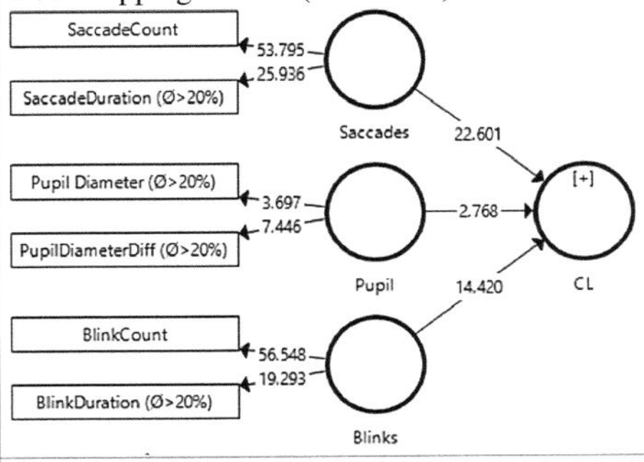

SaccadeCount ← 53.795 — Saccades
SaccadeDuration (Ø>20%) ← 25.936 — Saccades
Pupil Diameter (Ø>20%) ← 3.697 — Pupil
PupilDiameterDiff (Ø>20%) ← 7.446 — Pupil
BlinkCount ← 56.548 — Blinks
BlinkDuration (Ø>20%) ← 19.293 — Blinks
Saccades — 22.601 → CL [+]
Pupil — 2.768 → CL [+]
Blinks — 14.420 → CL [+]

Cross loadings:

	Blinks	Pupil	Saccades	CL
BlinkCount	**0.874**	0.039	0.564	0.806
BlinkDuration (Ø>20%)	**0.782**	0.039	0.327	0.627
Pupil Diameter (Ø>20%)	-0.039	**0.678**	0.127	0.149
PupilDiameterDiff (Ø>20%)	0.085	**0.894**	0.111	0.245
SaccadeCount	0.522	0.116	**0.848**	0.767
SaccadeDuration (Ø>20%)	0.343	0.115	**0.746**	0.610

Excluding fixation as a construct does also mean excluding the respective indicators form the CL measure. All three remaining first order constructs are reliable (AVE is above 0.5) and no colliniarity issues or cross loadings (except a small exceedance in the second experiment with respect too blink count and saccade count) in the outer model can be observed. By running a bootstrapping routine also significance for all measures ($p<0.01$) can be obtained, which is indicated by T-statistics (values above 1.96 indicate significance below a 0.05 level). With respect to the weights of the first order constructs we can see that saccadic and blink related data have a stronger explanatory power than pupil related information

(continued)

Table 1 (continued)

Study I	Study II
Step 3: Using the formative construct to predict decision making-outcome	
In this last step, we show applicability of the introduced cognitive load measure in a decision-making context. The new construct has a high explanatory power on decision making outcome measured by task accuracy and task time (R^2 is roughly 60%) in both conducted studies. It can be shown, that the higher cognitive load, the lower decision making outcome	
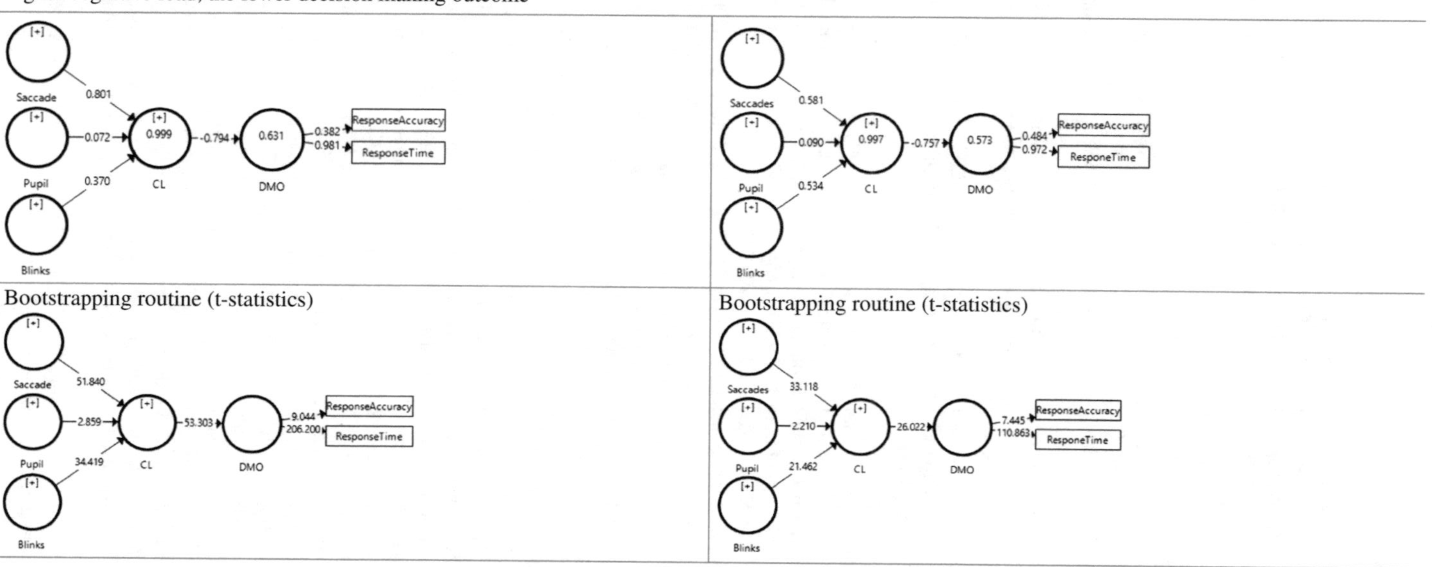	

References

1. Hossain, G., Yeasin, M.: Understanding effects of cognitive load from pupillary responses using hilbert analytic phase. In: 2014 IEEE Conference on Computer Vision and Pattern Recognition Workshops, Columbus, OH, USA, pp. 381–386 (2014)
2. Sweller, J.: Cognitive load theory, learning difficulty, and instructional design. Learn. Instr. **4**(4), 295–312 (1994)
3. Sweller, J.: Element interactivity and intrinsic, extraneous, and germane cognitive load. Educ. Psychol. Rev. **22**(2), 123–138 (2010)
4. Bawden, D., Robinson, L.: The dark side of information: overload, anxiety and other paradoxes and pathologies. J. Inf. Sci. **35**(2), 180–191 (2009)
5. Huang, W., Eades, P., Hong, S.H.: Measuring effectiveness of graph visualizations: a cognitive load perspective. Inf. Vis. **8**(3), 139–152 (2009)
6. Falschlunger, L., Lehner, O., Treiblmaier, H.: InfoVis: the impact of information overload on decision making outcome in high complexity settings. In: Proceedings of the 2016 SIG HCI, Dublin, pp. 1–5 (2016)
7. Anderson, E.W., Potter, K.C., Matzen, L.E., Shepherd, J.F., Preston, G.A., Silvia, C.T.: A user study of visualization effectiveness using EEG and cognitive load. Comput. Graph. Forum **30**(3), 791–800 (2011)
8. Cole, M.J., Gwizdka, J., Liu, C., Belkin, N.J., Zhang, X.: Inferring user knowledge level from eye movement patterns. Inf. Process. Manage. **49**(5), 1075–1091 (2013)
9. Dimoka, A., Pavlou, P.A., Davis, F.D.: Research commentary—NeuroIS: the potential of cognitive neuroscience for information systems research. Inf. Syst. Res. **22**(4), 687–702 (2011)
10. Kalyuga, S.: Effects of learner prior knowledge and working memory limitations on multimedia learning. Proc. Soc. Behav. Sci. **83**, 25–29 (2013)
11. Falschlunger, L., Treiblmaier, H., Lehner, O.: Cognitive differences and their impact on information processing: an empirical study comparing survey and eye tracking data. In: Information Systems and Neuroscience and Organizations, Gmunden, pp. 137–144 (2015)
12. Weber, B., Neurauter, M., Pinggera, J., Zugal, S., Furtner, M., Martini, M., Sachse, P.: Measuring cognitive load during process model creation. In: Information Systems and Neuroscience and Organizations, Gmunden, pp. 129–136 (2015)
13. Lallé, S., Conati, C., Carenini, G.: Prediction of individual learning curves across information visualizations. User Model. User-Adap. Inter. **26**(4), 307–345 (2016)
14. Brücken, R., Plass, J.L., Leutner, D.: Direct measurement of cognitive load in multimedia learning. Educ. Psychol. **38**(1), 53–61 (2003)
15. Paas, F., Tuovienen, J.E., Tabbers, H., Van Gerven, P.W.M.: Cognitive load measurement as a means to advance cognitive load theory. Educ. Psychol. **38**(1), 63–71 (2003)
16. Paas, F., van Merriënboer, J.J.G., Adam, J.J.: Measurement of cognitive load in instructional research. Percept. Motor Skills **79**(1 Pt 2), 419–430 (1994)
17. Wang, Q., Yang, S., Liu, M., Cao, Z., Ma, Q.: An eye-tracking study of website complexity from cognitive load perspective. Decis. Support Syst. **62**, 1–10 (2014)
18. Zagermann, J., Pfeil, U., Reiterer, H.: Measuring cognitive load using eye tracking technology in visual computing. In: Proceedings of the Beyond Time and Errors on Novel Evaluation Methods for Visualization—BELIV '16, Baltimore, MD, USA, pp. 78–85 (2016)
19. Goldberg, J.H., Helfman, J.I.: Comparing information graphics: a critical look at eye tracking. In: Proceedings of the 3rd BELIV'10 Workshop BEyond time and errors novel evaLuation methods for Information Visualization, pp. 71–79 (2010)
20. Siegle, G.J., Ichikawa, N., Steinhauer, S.: Blink before and after you think: blinks occur prior to and following cognitive load indexed by pupillary responses. Psychophysiology **45**(5), 679–687 (2008)
21. Marcus, A., Harding, M., Lamarche, C.: Quantile regression for time-series-cross-section data. Int. J. Stat. Manage. Syst. **6**(1–2), 47–72 (2011)

22. Iqbal, S.I., Adamczyk, P.D., Zheng, X.S., Bailey, B.P.: Towards an index of opportunity: understanding changes in mental workload during task execution. In: Proceedings of the CHI 2005, Portland, Oregon, USA, April 2–7, pp. 311–320 (2005)
23. Hair, J., Hollingsworth, C.L., Randolph, A.B., Chong, A.Y.L.: An updated and expanded assessment of PLS-SEM in information systems research. Industr. Mngmnt. Data Syst. **117**(3), 442–458 (2017)
24. Hair, J., Hollingsworth, C.L., Randolph, A.B., Chong, A.Y.L.: An updated and expanded assessment of PLS-SEM in information systems research. Industr. Mngmnt. Data Syst. **117**(3), 442–458 (2017)
25. Baddeley, A.D., Hitch, G.J.: Development of working memory: should the Pascual-Leone and the Baddeley and Hitch models be merged? J. Exp. Child Psychol. **77**(2), 128–137 (2000)
26. Elmqvist, N., Yi, J.S.: Patterns for visualization evaluation. In: Proceedings of the 2012 BELIV Workshop on Beyond Time and Errors—Novel Evaluation Methods for Visualization—BELIV '12, Seattle, Washington, pp. 1–8 (2012)

The Impact of Gestures on Formal Language Learning and Its Neural Correlates: A Study Proposal

Selina Christin Wriessnegger, Christopher Hacker, Manuela Macedonia and Gernot R. Müller-Putz

Abstract This pilot study reports about the impact of gestures on learning a formal language like Python. The aim of this research-in-progress is to find out if memory performance will benefit from the coupling of gestures and words in the learning phase. Previous research has demonstrated that gestures accompanying speech have an impact on memory for verbal information. This is the first study applying the body-mind concept to formal language learning. We introduce the study design and the results of one person.

Keywords Formal language · Memory · Embodiment · Gestures · EEG

1 Introduction

Previous studies on language acquisition have demonstrated that performing representative gestures during encoding (enactment) enhances memory for concrete words. Furthermore, gestures support memory for verbal information in native [1] and foreign language [2]. There is neuroscientific evidence for the fact that words are not abstract symbols [3], moreover, they are at the basis of the body [4–8].

Macedonia and Klimesch [9] showed that learning novel words with gestures enhances memory compared to learning novel words audio-visually in the long-term range. They suggested that words learned audio-visually are shallow and decay fast.

S. C. Wriessnegger (✉) · C. Hacker · G. R. Müller-Putz
Graz University of Technology, Graz, Austria
e-mail: s.wriessnegger@tugraz.at

C. Hacker
e-mail: Christopher.hacker@students.tugraz.at

G. R. Müller-Putz
e-mail: gernot.mueller@tugraz.at

M. Macedonia
Johannes Kepler University Linz, Linz, Austria
e-mail: manuela@macedonia.at

F. D. Davis et al. (eds.), *Information Systems and Neuroscience*,
Lecture Notes in Information Systems and Organisation 29,
https://doi.org/10.1007/978-3-030-01087-4_10

In contrast, complex sensorimotor codes created by pairing a novel word with a gesture are deep, that is retaining information better and decay slower. For instance, if the word is connected to a motor act that occurred during learning, the word's network comprises a motor component. Another study showed that hearing "pick" activates the cortical region that controls hand movements, "kick" activates foot movements, and "lick" activates tongue movements [10].

The prediction that learners have better memory for words encoded with gestures has further been supported by [11, 12]. They investigated the impact of enactment on abstract word learning in a foreign language. In a transfer test, participants produced new sentences with the words they had acquired. Items encoded through gestures were used more frequently, demonstrating their enhanced accessibility in memory. The results are interpreted in terms of embodied cognition. The results of a recently performed EEG study with 30 participants, based on the work of Macedonia and colleagues [2, 11, 12], gives further evidence for the beneficial impact of learning with gestures on memory performance. Based on these findings, we were interested if the body-mind concept could also have a beneficial impact on formal language acquisition. Furthermore we will investigate the neural coupling of cognition and action by means of EEG. Since a formal language is quite similar to a natural language in terms of syntax and semantics, we hypothesized that the learning process will also be enhanced if it goes along with gestures. In this paper we introduce the preliminary experimental procedure, the study design and the results of one test session.

2 Materials and Methods

2.1 Participants

One female student (21 years) participated in this pilot study. She had normal vision and no experience in programming. She gave written consent participating in the study.

2.2 Experimental Paradigm

Before the experiment started the experimenter carefully instructed the participant about the purpose and the procedure of the following pilot study. The introduction included the studying of a document which contains a detailed description of the programming code of a calculator (Fig. 1). Figure 1 shows an excerpt of the document (first 3 out of 13 functions) with the different functions and their explanations. The other functions include the choice of operation, the numbers the participants selected and the final output (see Fig. 4, left).

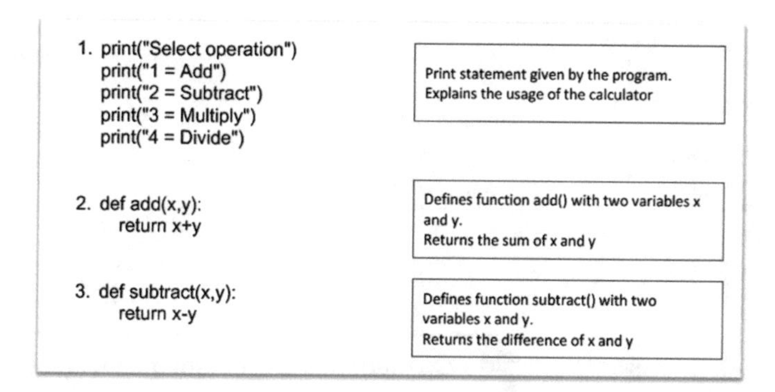

Fig. 1 The first three functions and definitions of the calculator

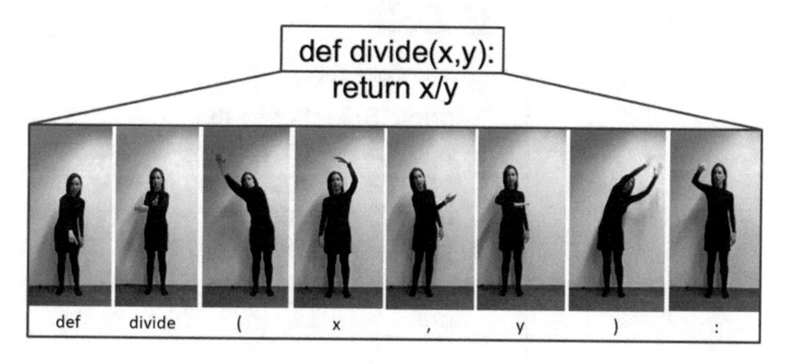

Fig. 2 Example of a function with corresponding gestures

In total 13 code parts with a detailed description of the functions were shown. The subject was additionally advised to the importance of the sequence of code lines. The task was to learn the different functions of the formal language (Python) resulting in a small calculator program. The functions were shown by short videos. Each variable was assigned to one specific gesture. The gestures were symbolically meaningful to the variables and operations they present. For example the first picture in Fig. 2 shows a part of the movie representing the variable "`def`". The subject had to learn the functions with the according gestures 3 times on 3 different days (Fig. 2).

Figure 2 shows the sequence of the gestures which are assigned to the function "`def divide(x,y):`", defining the function divide() with two variables x and y. After each learning phase the participant had to fill out a google questionnaire to control the learning progress. The whole program consisted of 13 functions including 42 gestures which have been imitated by the subject. The training schedule included 3 training days with a repetition of the 13 functions 2 times on each day (Table 1). Participants learned the different variables, resulting in 13 corresponding functions, by watching and imitating the gestures they saw in the video (Fig. 2).

On the third day (session 3) the recall test consisted of the programming of the calculator, that is, the participants have to remember all 13 functions in the correct order

Table 1 Training Schedule Pilot study

Day 1	Day 2	Day 3
Training 1	Training 2	Training 3
Session 1 (20 min) Session 2 (20 min)	Session 1 (20 min) Session 2 (20 min)	Session 1 (20 min) Session 2 (20 min)
Break (5 min)	*Break (5 min)*	*Break (5 min)*
Recall (Google form)	Recall (Google form)	Coding Test

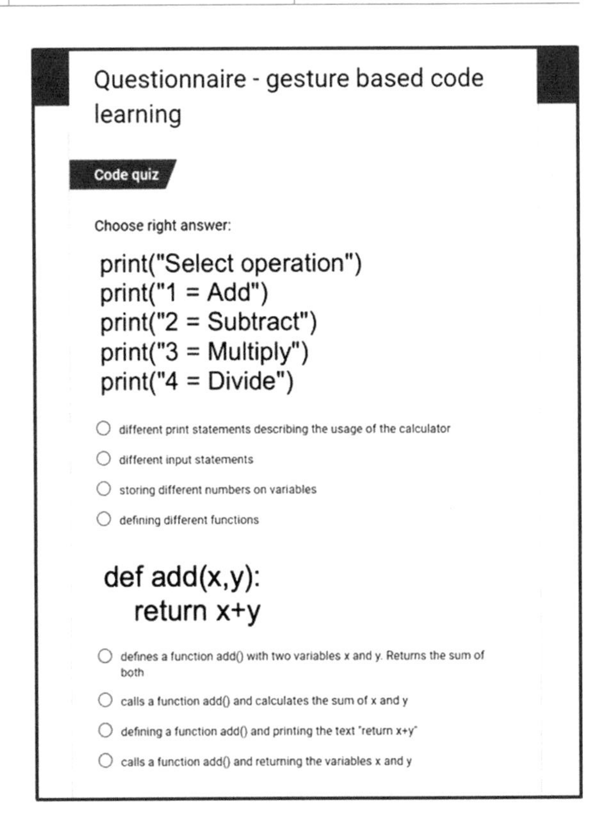
Questionnaire - gesture based code learning

Code quiz

Choose right answer:

```
print("Select operation")
print("1 = Add")
print("2 = Subtract")
print("3 = Multiply")
print("4 = Divide")
```

- different print statements describing the usage of the calculator
- different input statements
- storing different numbers on variables
- defining different functions

```
def add(x,y):
   return x+y
```

- defines a function add() with two variables x and y. Returns the sum of both
- calls a function add() and calculates the sum of x and y
- defining a function add() and printing the text "return x+y"
- calls a function add() and returning the variables x and y

Fig. 3 Screenshot of the used questionnaire in the recall test

(Fig. 3). For this purpose, an editor was used where participants will get immediately feedback about the success of the recall phase (Fig. 4).

2.3 *EEG Recording*

In the main experiment we will additionally record the EEG. For this purpose we will use a mobile and wireless 32 channel EEG system (LiveAmp; Brain Products). Beside the behavioral recall performance we are also interested in the brain activity

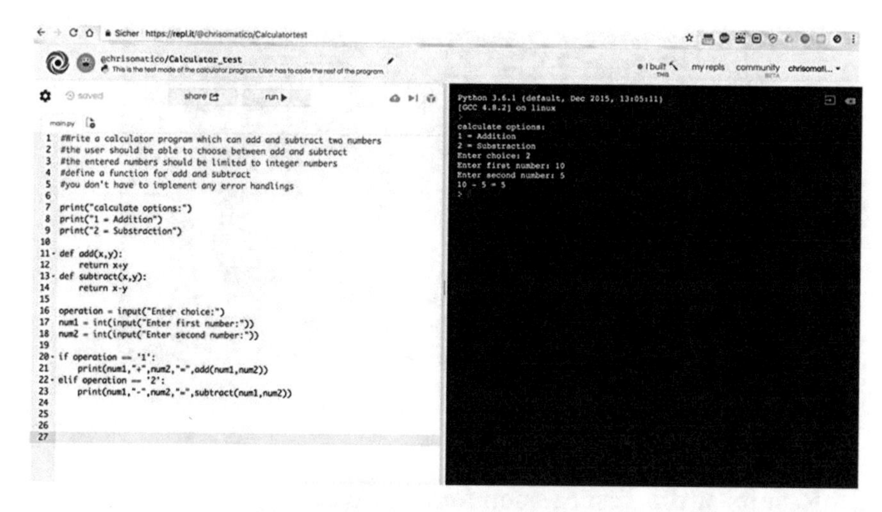

Fig. 4 Screenshot of the editor in the final recall phase "programming the calculator"

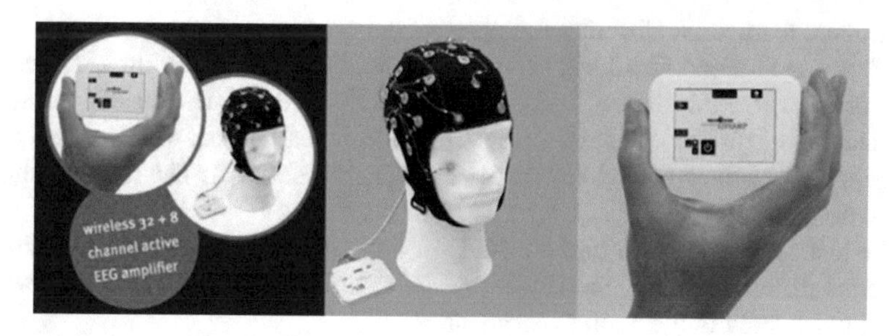

Fig. 5 Mobile EEG System (LiveAmp Brain Products)

during the recall phase (memory processes) and especially what does the brain do when the subject is moving, that is performing the gestures in the learning phase (Fig. 5).

For example [12] found in their fMRI study, activity in the premotor cortices for words encoded with iconic gestures. In contrast, words encoded with meaningless gestures elicited a network associated with cognitive control. These findings suggest that memory performance for newly learned words is not driven by the motor component as such, but by the motor image that matches an underlying representation of the word's semantics. We were interested in finding comparable results by means of EEG.

Moreover, by using a sophisticated connectivity analysis we will further investigate correlations between activity in motor related and frontal areas due to the body-mind concept.

To test the usability of the program and the study design we made a pilot study with one subject. In the following the results of this test session will be presented.

```
Python 3.6.1 (default, Dec 2015, 13:05:11)
[GCC 4.8.2] on linux
>
calculate options:
1 = Addition
2 = Substraction
Enter choice: 2
Enter first number: 10
Enter second number: 5
10 - 5 = 5
>
```

Fig. 6 Screenshot of the subject's final recall phase (programming of the calculator)

3 Results of the Test Session and Discussion

The student learned the different functions of the calculator on three different days. In the first and second recall test she made some errors like forgetting a semicolon, colon or bracket. Nevertheless, surprisingly she answered all questions regarding the functions of the different codes correctly. In the last recall phase, which required the programming of the whole calculator, she was also successful. The result is shown in Fig. 6.

The test session showed that the study design and procedure of the experiment worked well. Furthermore, the subject gave us some constructive feedback which will be considered in the main experiment.

References

1. Zimmer, H.D.: Memory for Action: A Distinct Form of Episodic Memory?. Oxford University Press, Oxford, England (2001)
2. Macedonia, M., Von Kriegstein, K.: Gestures enhance foreign language learning. Biolinguistics **6**, 393–416 (2012)
3. Fodor, J.A.: The Language of Thought. Harvester Press, Hassocks, England (1976)
4. Pylyshyn, Z.W.: Computation and Cognition: Toward a Foundation for Cognitive Science. MIT Press, Cambridge, MA (1984)
5. Barsalou, L.W.: Grounded cognition. Ann. Rev. Psychol. **59**, 617–645 (2008). https://doi.org/10.1146/annurev.psych.59.103006.093639
6. Fischer, M.H., Zwaan, R.A.: Embodied language: a review of the role of the motor system in language comprehension. Q. J. Exp. Psychol. **61**, 825–850 (2008). https://doi.org/10.1080/17470210701623605
7. Gallese, V., Lakoff, G.: The brain's concepts: the role of the sensory-motor system in conceptual knowledge. Cogn. Neuropsychol. **22**, 455–479. (2005). doi:714592738[pii] https://doi.org/10.1080/02643290442000310
8. Taylor, L.J., Lev-Ari, S., Zwaan, R.A.: Inferences about action engage action systems. Brain Lang. **107**, 62–67 (2008). https://doi.org/10.1016/j.bandl.2007.08.004

9. Macedonia, M., Klimesch, W.: Long-term effects of gestures on memory for foreign language words trained in the classroom. Mind Brain Educ. **8**(2), 74–88 (2014)
10. Pulvermüller, F.: Brain mechanisms linking language and action. Nat. Rev. Neurosci. **6**, 576–582 (2005)
11. Macedonia, M., Knösche, T.R.: Body in mind: how gestures empower foreign language learning. Mind Brain Educ. **5**(4), 196–211 (2011)
12. Macedonia, M., Müller, K., Friederici, A.D.: The impact of iconic gestures on foreign language word learning and its neural substrate. Hum. Brain Mapp. **32**(6), 982–998 (2011)

A Cloud-Based Lab Management and Analytics Software for Triangulated Human-Centered Research

Pierre-Majorique Léger, Francois Courtemanche, Marc Fredette and Sylvain Sénécal

Abstract This paper introduces a cloud-based lab management and analytics software platform that we have developed for triangulated human-centred research. The solution is designed to support three main requirements: (1) It enables accurate triangulation of enriched UX measures; (2) It produces triangulated enriched measures in a timely manner; and (3) it helps to generate meaningful recommendations. The application supports the key activities that are required to conduct an enriched UX research project: (1) Designing the UX test; (2) Planning and scheduling the test; (3) Executing the test; (4) Post-processing and triangulating of the collected data; (5) Analyzing and visualizing of the data and; (6) Lab Maintenance. At the time of publication, the application is currently at technology readiness level (TRL) 6 and we are currently conducting beta testing in academic and commercial lab to demonstrate the technology an operational environment.

Keywords UX · Neurophysiology · Triangulation
Lab management system (LMS) · Heatmaps

1 Introduction

Computerized interfaces are omnipresent in our daily lives and new technologies will continue to transform our lives in the future. What users experience when interacting with an interface is a key determinant of their intention to use and/or reuse a product or service. Research shows that user experience (UX) plays a crucial role in the success or failure of digital products and services [1]. UX is defined as the perceptions and responses that a person has either as a result of using or as a result of anticipating using an information technology (IT) product or service [2]. UX refers to far more than just the ergonomic aspect of an interface, for it encompasses the entire experience as it unfolds before, during, and after the interaction with the interface [3]. The experience

P.-M. Léger (✉) · F. Courtemanche · M. Fredette · S. Sénécal
Tech3Lab, HEC Montréal, Montréal, Canada
e-mail: pml@hec.ca; pierre-majorique.leger@hec.ca

F. D. Davis et al. (eds.), *Information Systems and Neuroscience*,
Lecture Notes in Information Systems and Organisation 29,
https://doi.org/10.1007/978-3-030-01087-4_11

is influenced by the usage context as well as by the user's cognitive and affective states.

Industry players and academic researchers are currently facing important scientific and industrial challenges that require urgent attention. Technology evolves rapidly, and so too must the methods used to evaluate UX. To capture the user's states in authentic usage contexts accurately, research is needed to develop enriched UX measures that enhance existing explicit measures with implicit ones. For example, self-reported measures such as questionnaires could be enhanced by taking account of automatic or subconscious reactions, such as eye movements and psychophysiological reactions, to increase the temporal resolution of measurements. This would capture the full extent of affective states like emotional valence, cognitive states like cognitive workload, and attentional states like divided attention throughout the experience of users. To use these tools and methods effectively, it is necessary to increase the speed and accuracy with which these enriched UX measures can be synchronized, processed, and visualized and to adapt these measures and processes so that they can be used in natural and real-time contexts that are ecologically valid.

The objective of this paper is to introduce a cloud-based lab management and analytics software platform that we have designed to specifically address these research needs. The platform is specifically designed to enable multimodal human computer interaction (HCI) as well as NeuroIS studies involving authentic research stimuli such as that related to the usability of mobile applications. In what follows, we begin by explaining why this tool is needed. After that, we detail the functionalities of this new application, and we conclude by discussing on its current technology readiness level (TRL).

2 The Requirement: The Need to Accelerate Insight Generation from Triangulated and Enriched UX Measures

2.1 Enabling Triangulation of Enriched UX Measures

Most of the literature on UX (either in information system (IS) or HCI research) report using explicit measures to assess the user experience [4, 5]. While informative, such measures have drawbacks that could be mitigated by the adjunction of implicit measures. For instance, users can hardly report on their cognitive and emotional states during the interaction without being inherently distracted by the interaction [6] The quasi-exclusive use of explicit measures and data in the field of UX research presents major limitations and potential biases, such as the difficulty of separating the emotional responses to adjacent stimuli, as well as retrospective and social desirability biases [7, 8]. This limitation represents a major industrial problem for diagnosing UX-related issues. Indeed, self-assessed questionnaires, intrinsically, cannot provide, on their own, unbiased longitudinal measures of the UX-related automatic cognitive and emotional mental states during the interaction without the subject being aware of them [4, 9]. For example, questionnaires cannot provide a temporally precise account of the frustration felt by a user when a problem arises at checkout, nor can they distinguish it from the overwhelmingly positive previous impressions left by an online shopping interface at the beginning of their shopping experience.

Research is needed to enrich measures by using implicit measures from which cognitive and emotional constructs can be inferred [4, 10]. Implicit approaches to the investigation of UX can complement current approaches because they address some of the limitations of the explicit measures. First, they enable real-time observation of the user's reactions to technology, as the person interacts with the interface. Second, they permit the capture of subconscious and automatic processes occurring without the individual being aware of them or realizing that they are indeed occurring, thus offering a more complete representation of what actually takes place within the brain, and at what moment. Third, since implicit antecedents need to be captured using methods (e.g., neurophysiological tools) that are different from those used to measure explicit ones (e.g., self-reports), the triangulation of structurally different methods reduces the common-method bias often suspected in UX research [7].

However, triangulation of implicit measures such as multimodal neurophysiological data is usually complex and thus prone to an array of statistical problems, such as high measurement errors and complex dependency patterns among the observations. To our knowledge, there are not commercially available technologies that provide accurate triangulation of multimodal data acquisition. Therefore, we pose our first requirement:

Requirement 1: The solution needs to enable accurate triangulation of enriched UX measures.

2.2 Processing Speed of Triangulated Enriched UX Measures

Now a mainstream approach, Agile Software Development is an umbrella term that refers to a set of methods in which requirements and solutions evolve rapidly over successive iterations. The main objective of this development method is to build useful software less prone to defects and with a shorter time-to-delivery. However, agile methods are not necessarily better at developing usable software that offers a rich experience to potential users [11]. User-centred design can ensure that UX is the focus in software development [12]. But in practice, applying agile development while maximizing UX can present many significant industrial challenges, such as constraints linked to the fast-paced iterations of the software development process [13, 14]. Researchers report on the difficulty that UX professionals have in informing developers in a timely manner, especially when using enriched measures, and show that UX methods are often called upon too late in the development process [15]. Thus, research on process innovation (e.g., innovation at the information system level to fully integrate the research process, end-to-end) is essential to accelerate the pace at which UX research is performed. An enriched UX research cycle that runs over a few weeks or months is simply not satisfactory when software development is scheduled in sprints of 30 days!

The need to accelerate processing speed is also very important in academic research. The imperative to shorten publication cycle time is increasing in the human-centred field such as HCI and psychology. Researchers need tools that will allow them to bring their research discovery sooner in the public space. Also, researchers also benefit from the increase automation in data processing. Relying on manual processing is prone to error and fast research assistant turnover contribute to slowing productive research time.

We therefore pose the second requirement.

Requirement 2: The solution needs to produce triangulated enriched measures in a timely manner.

2.3 Enabling Insight to Action

In a usability test, the most crucial take away for a UX designer is not results of the analysis that will come out of the project. The most crucial information is the insight (the aha moment) that one will gain out of this analysis that will to improve the design of an interface. Therefore, the solution must enable the generation of those insights and facilitate the identification of action that can be taken to make profit of those insights. In other words, the solution must inform the design in a meaningful and decisional way and should help, via data visualization methods, to reduce the time required to provide meaningful recommendations to UX professionals.

Here is an example of a novel triangulation and machine learning-based visualization technique that our research team has developed to represent the emotional state of users in a way to support the decision-making process of usability experts. Traditional gaze heatmaps are used in eye tracking as intuitive representations of aggregated gaze data [16]. Their main use is to help answer the question: "Where in the interface do people tend to look?" [17]. In our visualization method, the users' gaze now serves as a means of mapping physiological signals onto the user interface. The resulting heatmaps represent the physiological signals' distribution over the interface, and can help answer the following question: "Where in the interface do people tend to emotionally react?" In a recent publication, we detailed the four steps involved in the creation of physiological heatmaps: inference, normalization, accumulation, and colorization [18, 19]. Using the technique, developers can make a more comprehensive diagnosis of the ergonomic characteristics of an interface and propose an improvement thereof to the design team. This triangulated approach makes it possible to visually analyze users' various emotional states for specific areas of a given interface (e.g., cognitive load combined with emotional valence) [20]. The technique can be used to represent the emotional response to a naturally occurring stimulus on the interface, or to compare the response across multiple versions of the same interface (e.g., A/B testing) [21].

Requirement 3: The solution must help to generate meaningful recommendations.

3 The Solution

We have conducted more than 150 experiments and usability tests over the past 3 years with more than 3000 human subjects. Over time, we have designed processes and systems to meet the three requirements describe in the previous section. The solution we proposed, to meet the requirements described above, a cloud-based lab management and analytics software platform for triangulated human-centred research.

Our solution is a web application developed in a modern programming language designed by Facebook. The application is fully compatible with Noldus Observer XT. It currently natively handles scientific data acquisition equipment from equipment providers such as Brainvision, Tobii, SMI, Noldus Facereader, and Biopac; several other manufacturers are currently being made compatible. Our solution runs on in-memory analytics which allows for high speed analysis of high dimensional data such as neurophysiological, physiological, and eye tracking data.

There are six key activities that are required to conduct an enriched UX research project: (1) Designing the UX test, (2) Planning and scheduling the test, (3) Executing the test, (4) Post-processing and triangulation of the collected data, (5) Analyzing and visualization of the data. It was designed to support these activities in an integrated manner to increase the processing speed and accuracy of the research data.

The following Table 1 details the main functionalities enabled by our solution for each of these activities.

4 Discussion and Conclusion

At the time of publication, our application is currently in technology readiness level (TRL) 6, i.e., the technology has been tested in a high-fidelity laboratory environment to demonstrates its readiness. In this stage, we have been able to execute more than a dozen usability studies with several neurophysiological signals (including eyetracking, physiological measurements such as electrodermal activities and automatic facial analysis) and to produce actionable results in less than one week (from the moment the first user was tested to the moment where insights were provided to the UX team).

Our next step is to move to TRL 7, which involves demonstration of an actual system prototype in an operational environment. We have several of those tests planned in the next few months to ascertain the capabilities of our solution.

Table 1 Main functionalities of the solution

Activities	Description
Designing experiment	Define conditions, events, and stimuli; Define experimental metadata (e.g. file markers naming convention); Define recording equipment; Visualize experimental design
Scheduling experiment	Schedule experimental room; manage reservations; import data from external subject panel system (i.e., Sona System)
Executing experiment	Track participant; Write lab notes
Data postprocessing	Upload data from various data collection instruments; Upload data from Observer XT; Validate file according to experimental metadata; Parse data based on event markers; Built-in signal processing with R and python scripts (e.g., HR, HRV, GSR); Automatic data stream synchronization (using sync markers from syncbox)
Data analysis	Generate attentional, emotional, and cognitive heatmaps; Generate global experience map; Enable longitudinal and multiple project analysis; Export to statistical packages from one or multiple projects (e.g., cross-project analyses, compatibilities with third party visualization software such as Tableau software)

The technology roadmap of our solution is to continue to automate and reduce human intervention in the processing of triangulated and enriched UX measures. Building on new advances in data sciences and artificial intelligence, our goal is to achieve quasi-real-time analysis, enabling to produce results on the same day as a data collection.

References

1. Kuusinen, K.: Improving UX work in scrum development: A three-year follow-up study in a company. In: International Conference on Human-Centred Software Engineering. Springer, Berlin (2014)
2. Hartson, R., Pyla, P.S.: The UX Book: Process and Guidelines for Ensuring a Quality User Experience. Elsevier, New York (2012)
3. Bevan, N., et al.: New ISO standards for usability, usability reports and usability measures. In: International Conference on Human-Computer Interaction. Springer, Berlin (2016)
4. de Guinea, A.O., Titah, R., Léger, P.-M.: Explicit and implicit antecedents of users' behavioral beliefs in information systems: a neuropsychological investigation. J. Manage. Inf. Syst. **30**(4), 179–210 (2014)

5. Riedl, R., & Léger, P.M.: Fundamentals of NeuroIS Studies in Neuroscience, Psychology and Behavioral Economics. Springer, Berlin, Heidelberg (2016)
6. Ortiz de Guinea, A., Webster, J.: An investigation of information systems use patterns: technological events as triggers, the effect of time, and consequences for performance. Mis Q. **37**(4) (2013)
7. De Guinea, A.O., Titah, R., Léger, P.-M.: Measure for measure: a two study multi-trait multi-method investigation of construct validity in IS research. Comput. Hum. Behav. **29**(3), 833–844 (2013)
8. Malhotra, N.K., Kim, S.S., Patil, A.: Common method variance in IS research: a comparison of alternative approaches and a reanalysis of past research. Manage. Sci. **52**(12), 1865–1883 (2006)
9. Tams, S., et al.: NeuroIS-alternative or complement to existing methods? Illustrating the holistic effects of neuroscience and self-reported data in the context of technostress research. J. Assoc. Inf. Syst. **15**(10), 723 (2014)
10. Riedl, R., Davis, F.D., Hevner, A.R.: Towards a NeuroIS research methodology: intensifying the discussion on methods, tools, and measurement. J. Assoc. Inf. Syst. **15**(10), I (2014)
11. Brhel, M., et al.: Exploring principles of user-centred agile software development: a literature review. Inf. Softw. Technol. **61**, 163–181 (2015)
12. Abras, C., Maloney-Krichmar, D., Preece, J.: User-centred design. In: Bainbridge, W. (ed.) Encyclopedia of Human-Computer Interaction, vol 37, issue 4, pp. 445–456. Sage Publications, Thousand Oaks (2004)
13. Larusdottir, M., Gulliksen, J., Cajander, Å.: A license to kill—improving UCSD in agile development. J. Syst. Softw. **123**, 214–222 (2017)
14. Øvad, T., Larsen, L.B.: How to reduce the UX bottleneck–train your software developers. Behav. Inf. Technol. **35**(12), 1080–1090 (2016)
15. Kuusinen, K.: BoB: a framework for organizing within-iteration UX work in agile development. In: Integrating User-Centred Design in Agile Development, pp. 205–224. Springer, Berlin (2016)
16. Nielsen, J., Pernice, K.: Eyetracking Web Usability. New Riders (2010)
17. Wooding, D.S.: Fixation maps: quantifying eye-movement traces. In: Proceedings of the 2002 Symposium on Eye Tracking Research & Applications. ACM (2002)
18. Courtemanche, F., et al.: Physiological heatmaps: a tool for visualizing users' emotional reactions. Multimedia Tools Appl., 1–28 (2017)
19. Georges, V., et al.: Measuring visual complexity using neurophysiological data. In: Inf. Syst. Neurosci., 207–212 (Springer) (2015)
20. Courtemanche, F., et al.: Method of and System for Processing Signals Sensed From a User. Google Patents (2018)
21. Georges, V., et al.: The adoption of physiological measures as an evaluation tool in UX. In: International Conference on HCI in Business, Government, and Organizations. Springer, Berlin (2017)

brownieR: The R-Package for Neuro Information Systems Research

Sven Michalczyk, Dominik Jung, Mario Nadj, Michael T. Knierim and Raphael Rissler

Abstract Neuro-Information-Systems (NeuroIS) research has become an established approach in the information systems (IS) discipline for investigating and understanding user behavior. Our outlined package with the name *brownieR* is a freely-available open source R-package for analyzing NeuroIS data (i.e. the combination of physiological and behavioral data). The central purpose of this work is to instruct researchers how *brownieR* can be used in IS research by providing a practical guide on how to conduct the analysis of bio-physiological data combined with behavioral data (e.g., from the web, experimental tasks, or log files). Further, the article provides an analysis framework and covers the different stages involved in analyzing physiological data.

Keywords NeuroIS · Bio-physiology · Behavior analytics · Web analytics

1 Introduction

NeuroIS research has received much attention in recent years due to its potentials to investigate the neural and bio-physiological foundations of cognitive processes and corresponding behaviors, which offer important benefits in understanding the

S. Michalczyk · D. Jung (✉) · M. Nadj · M. T. Knierim · R. Rissler
Karlsruhe Institute of Technology (KIT), Karlsruhe, Germany
e-mail: d.jung@kit.edu

M. Nadj
e-mail: mario.nadj@kit.edu

M. T. Knierim
e-mail: michael.knierim@kit.edu

R. Rissler
e-mail: raphael.rissler@kit.edu

D. Jung
Karlsruhe Decision and Design Lab (KD2Lab), Karlsruhe, Germany

F. D. Davis et al. (eds.), *Information Systems and Neuroscience*,
Lecture Notes in Information Systems and Organisation 29,
https://doi.org/10.1007/978-3-030-01087-4_12

design, development, use, impact, and acceptance of IS [1]. Hereby, NeuroIS is an "interdisciplinary field of research that relies on knowledge from disciplines related to neurobiology and behavior, as well as knowledge from engineering disciplines" [2].

This contributes to a better understanding of the design, but also targets the positive bio-physiological outcomes of IS. The integration of bio-physiological data has many possible use cases in IS, for instance neuro-adaptive IS for emotion regulation [3], decision-support in financial decision-making [4], or emotion management in small, cooperative groups [5]. The NeuroIS methodology has also been proposed to help understand phenomenona like flow [6], or decision inertia [7].

As a foundation, NeuroIS research requires profound knowledge and tools to analyze bio-physiological data. However, only a very limited number of software packages are freely available for the analysis of bio-physiological data [8], especially at the interface of physiological and behavioral data. Having reviewed the R-archive network (CRAN) for the terms ("physiology", "heart rate", "electrodermal activity", "skin conductance", "electroencephalography", "electromyography", "pulse", "blood pressure", "BVP", "PPG", "ECG", "HR", "EDA", "SCR", "SCL", "EMG", "EEG"), only three related packages were found ("biosignalEMG", "eegkit", "RHRV"). A subsequent Google search for the same keys returned two more R-packages listed on Github ("EEK", "EEGUtils"). However, none of them combines IS user behavior data with the analysis of physiological data. Consequently, there remains a need for packages to integratively support NeuroIS research processes, and to provide a methodological toolbox in form of an R package. Such a package should provide a methodological guideline for analyses, but also support established research processes.

This article presents such an R-package. The package that we called *brownieR,* provides user-friendly functions for the selection, preprocessing, and analysis of physiological and behavioral data. It also provides functions to produce graphical representations and descriptive statistics of this data. In this article, our overall aim is to illustrate benefits and application of the package, based on a use case of NeuroIS research.

This article is organized along the stages of conducting a NeuroIS analysis, providing a practical guide for how each of the steps can be implemented in R. We illustrate our tutorial with a popular use case in IS, the analysis of behavioral web data for usability tests. Finally, we conclude by summarizing our work.

2 The Package *BrownieR*

To date, the R package is primarily supporting the data format from the NeuroIS experiment platform *Brownie* [9, 10]. However, an integration of further formats is planned. The *Brownie* platform is freely accessible at:

https://im.iism.kit.edu/1093_1100.php,

and the source files of the *brownieR* package are available at:

https://github.com/Fiddleman/NeuroIS.

In general, the package follows the standard data analysis process as proposed by Wickham and Grolemund [11], which is illustrated in Fig. 1. First of all, the data needs to be imported into R. Before we can start the modeling or visualizing phase, the data has to be transformed. Usually, before a model is built, we explore the data by plotting visualizations and by looking at descriptive statistics. After these steps, findings can be communicated. In this technical article, we follow this general process to structure the analysis of bio-physiological data. Furthermore, we use the user-called-functions for most of the naming of common functions in R like *plot()*, or *summary()*. Our analysis starts by using the *import()* function, which adds all log-files in a specified folder as objects to the global environment.

If an analysis on physiological data has to be conducted, a marker object for the relevant events has to be created. For that purpose we rely on the *marker()* function. To get an aggregated view on the data, the *summary()* function can be called on each of the objects. This view provides insights for plotting the data. In an iterative manner, we can dive deeper into the data by running the *summary()* and *plot*() function again. If we conducted the experiment on a website, we provide the *as_clickstream()* function to transform the data for modelling as Markov Chain, using the *clickstream* R-package by Scholz [12].

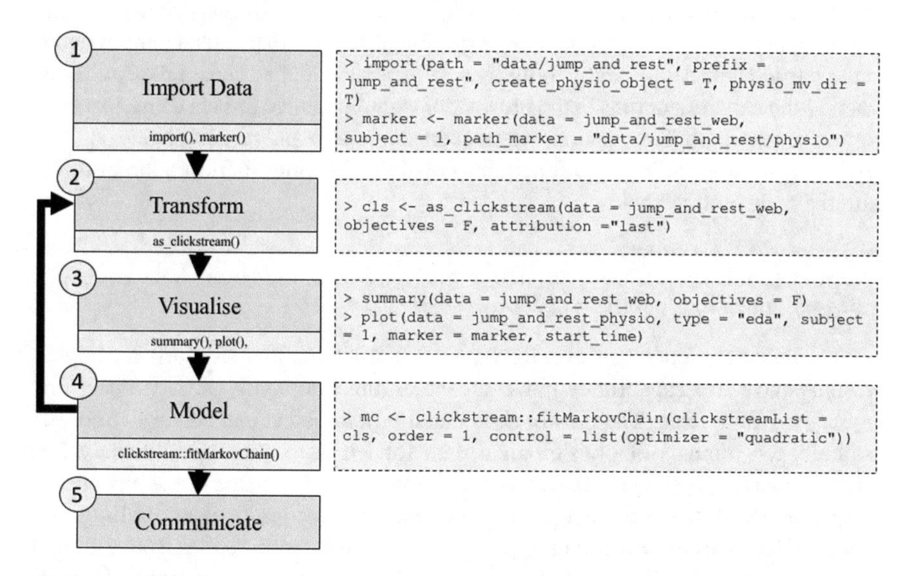

Fig. 1 The NeuroIS analysis process of this R-package based on Wickham [11]

3 Use Case: Analysis of Behavioral Web Data for Usability

In this chapter, we illustrate the package in action by relying on a popular use case in IS, the analysis of behavioral web data for usability tests. The data for our example is generated by the following setting: A user clicks randomly on different elements at the website http://iism.kit.edu. In order to introduce variance to the physiological data the user started jumping up and down for approximately 10 s from time to time. ECG, BVP and EDA data as well as web log-files are tracked by *Brownie* during the session. On this basis, several common research questions in NeuroIS research can be derived, which serve as a "route map" in the upcoming subchapters to explain our implementation:

1. Do central pages exist in terms of duration or click frequency?
2. At which pages did the user jump?
3. Which elements on a central page did the mouse last on?
4. What is the probability to go from one page to another, based on the user?

3.1 Data Import and Transformation

Depending on the kind of experiment, *Brownie* tracks the data in three different structures: *web* (web data), *physio* (bio-physiological data) and *exp* (experimental data). *Physio* data can be collected in every experiment in case that sensors are attached to the participant. Additionally, only *web* or *exp* data can be tracked depending on whether the experiment runs in the browser or is implemented in Java using *Brownie*. For our example implementation, we tracked solely *web* and *physio* data.

Firstly, we use the *import()* function to read the data into R from a target folder with the code outlined below:

```
library("brownieR")
import(path = "data/jump_and_rest", prefix = "jump_and_rest",
create_physio_object = T, physio_mv_dir = T)
```

The *import()* function has several arguments that are important for further analysis. We only have full capabilities if we set the arguments *create_physio_object* and *physio_mv_dir* to true. The drawback is that a *physio* object can be very large. For example, ten minutes of physiological data for ten users are approximately 500 Megabyte large (given the 1000 Hz sampling rate here). The name of the objects can be specified with the argument prefix. The *import()* function does not only import data into R. It also detects the data-type for each file and combines files from different users to the related object class. Additionally, the function removes empty columns to reduce size, whereas all time columns are converted to the common data format *POSIXct*.

3.2 Visualization and Data Understanding

After importing the data, we have a list of class type *physio* and another one of class type *web* in our global environment. To answer our first question about frequently clicked pages, we run the function *summary()* on the web object.

```
summary(data=jump_and_rest_web, objectives=F)
```

The argument *objectives* can be used to calculate web conversion rates. The appropriate use can be reviewed in the built-in R documentation by calling the function? *summary.web* (Table 1).

The column *Impression* shows, how often a URL (i.e. a single page), is shown to participants during the experiment. The column *Session* provides information on how many users invoke a URL. The column *Duration* offers the average length of stay at a URL over all impressions in seconds. In our example, the page 21.php has the longest durations of 10.6 s and with 28.php the most impressions. These two pages can be seen as central.

The *summary()* function can be run on a *physio* object as well. As we overload the functions with R's S3 methods, the function has a completely different output than the summary from base R. First of all, we need to create a marker object, which is necessary to see the URL changes (or screen changes in case of *exp* data) along the physio data. The command refers to the following:

```
marker <- marker(data = jump_and_rest_web,
subject = 1, path_marker = "data/jump_and_rest/physio")
```

We need to specify for which user the marker will be created with a subject. Every user has its unique marker file as all users execute screen or URL changes at different times. Additionally, we can specify a folder with *path_marker* where a marker file will be created. The creation of a marker file is only necessary in case you want to plot ECG data. Doing so, the marker file needs to be in the same folder as the *physio* data, usually a sub-folder named 'physio'.

Now we can run the summary on the *physio* data. Therefore, we need to specify for which type of physio data, ECG, EDA or BVP shall be summarized. Currently, *summary()* is only available for EDA data.

Table 1 Output of the *summary()* function (for our example data)

	URL	Impression	Session	Duration	Imp_per_Session
1	kd2lab.kit.edu/21.php	2	1	10.6	2
2	kd2lab.kit.edu/28.php	2	1	10.0	2
3	kd2lab.kit.edu/	1	1	5.7	1
4	kd2lab.kit.edu/index.php	1	1	9.3	1
5	kd2lab.kit.edu/26.php	1	1	7.0	1

```
start_time <-
min(jump_and_rest_web[jump_and_rest_web$SUBJECT_ID_SUBJEC
T == 1]$Time, na.rm = T)
summary(jump_and_rest_physio, type = "eda", subject = 1,
marker = marker, start_time = start_time)
```

Before running the summary, we need to determine the *start_time* of the experiment for the user. The *start_time* usually refers to the minimal time in *web* or *physio* data for the selected user (subject).

The output of the function is the following (Table 2):

With this view, we can answer the second question at which specific pages the user might have been jumping. It seems noticeable that the average EDA is clearly smaller at the page 21.php for the time of 14.75 and 35.76.

Furthermore, the package provides another visualization of this finding. For that purpose, we run the *plot()* function on the *physio* data for this user.

```
plot(data = jump_and_rest_physio, type = "eda",
subject = 1, marker = marker, start_time)
```

Alongside the *summary()* function, *type, subject, marker* and *start_time* need to be provided as arguments. We again plot the EDA data. Furthermore, plotting ECG data is possible by changing some arguments:

```
plot(jump_and_rest_physio, type = "ecg", subject = 1,
marker = "data/jump_and_rest/physio/marker1.csv",
start_time, physio_unisens_dir =
"data/jump_and_rest/physio/")
```

In the plot for EDA data, we get a visualization of what we have already investigated by running the *summary()* function (Fig. 2).

3.3 Modeling and Communication

In this chapter, we focus on the third question: Which elements on a central page did the mouse last on longest? We can answer this question by modeling the user

Table 2 Structure of the current physio data, after pre-processing

	Time	Event	Average EDA
1	0.76	https://www.kd2lab.kit.edu/	89.66
2	5.74	https://www.kd2lab.kit.edu/index.php	89.94
3	14.75	https://www.kd2lab.kit.edu/21.php	76.34
4	25.74	https://www.kd2lab.kit.edu/28.php	87.60
5	35.76	https://www.kd2lab.kit.edu/21.php	71.36
6	45.74	https://www.kd2lab.kit.edu/26.php	86.31
7	52.75	https://www.kd2lab.kit.edu/28.php	86.99

behavior as a Markov Chain and by running the *plot()* function on the resulting web object.

```
web_summary <- summary(data = jump_and_rest_web
, objectives = F)
urls <- web_summary\$URL
take_screenshot(urls = urls)
plot(jump_and_rest_web, url = urls[3], type =
"motion", subject = 1,
alpha = 0.1, size = 3, color = "purple")
```

Beforehand, some preparations are required: Firstly, we assign the output of the *summary.web()* function to an object to get a list of all URLs in the experiment. Afterwards, we make screenshots with the function *take_screenshots*. This function takes screenshots of all provided URLs and saves them in a sub-folder ‘screenshots’ in the working directory. Then, the *plot.web()* function has other arguments than the *plot.physio()* function. The arguments *alpha*, *size* and *color* change the appearance of the plot. As every time only one URL can be plotted, we need to specify the URL in the argument *url*. This is different with the argument *subject*. If we have an experiment with more than one user (which is usually the case), we can specify several or all users, by relying on a vector containing the user numbers. This is particularly useful to observe an area of interest, that all users moved their mouse over. In our example, the mouse was moved from the navigation bar to the lab address, which is an indicator that the user was interested in the location of the lab. Further, we can plot clicks on a page as well (Fig. 3).

To answer question four (transition probabilities from one page to another), we propose the usage of the package *clickstream* by Scholz [12].

```
Cls <- as_clickstream(data = jump_and_rest_web,
objectives = F, attribution ="last")
```

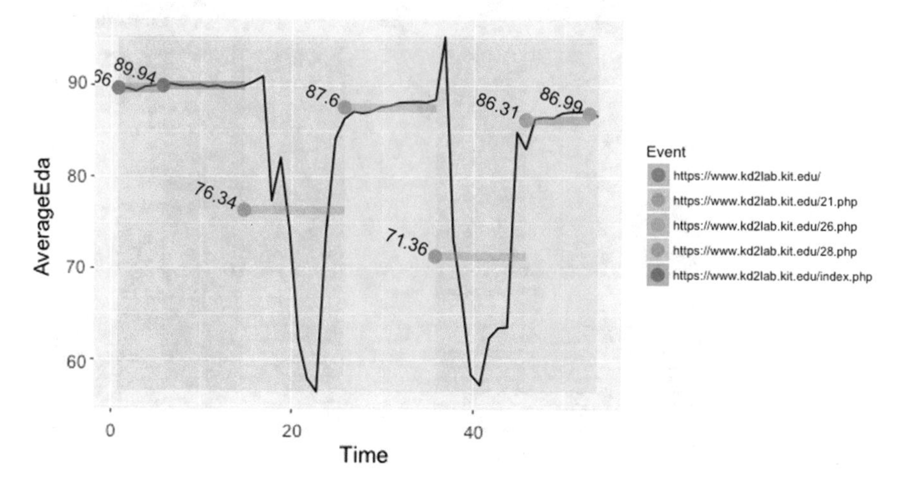

Fig. 2 EDA plot visualization of the *brownieR* package

```
mc <- fitMarkovChain(clickstreamList = cls, order = 1,
control = list(optimizer = "quadratic"))
summary(mc)
plot(mc, order = 1)
```

First of all, we need to transform the web data into a proper format for the *clickstream* package. For this purpose, the package provides a function called *as_clickstream()*. We have the option to specify goal pages of the website. A goal page of a website could refer to the thank-you-page after a successful purchase. As we do not offer a final end page in our example, we set the argument objectives to *false*. Additionally, we can define an attribution model with the argument attribution (first or last are possible values). By applying the function *fitMarkovChain()* to the output of the *as_clickstream()* function, we start to use the clickstream package and build a first order markov chain by using a quadratic optimizer. The model is saved to the class *mc*. On this basis, we can apply a summary or a plot function on this model to visualize it. For instance, the plot of the markov model allows visualizing the transition probability from one page to another.

4 Conclusion

With the advancements of NeuroIS research to examine and understand cognitive and physiological processes and associated user behaviors, this interdisciplinary research domain have become increasingly relevant at the nexus of IS artefact design in both

Fig. 3 Mouse movement visualization of the NeuroIS R package

academic and business communities. However, designing such artefacts and understanding the underlying phenomena, requires a profound knowledge and tool support to analyze the corresponding bio-physiological data, especially at the interface of behavioral data. By creating an R-package for the analysis of physiological and behavioral data, this piece of research provides a foundational step towards a toolbox supporting the need for an overarching NeuroIS research process. However some limitations are worth noting, in particularly the integration of more complex datatypes is still missing, and the toolbox does not support more complex data like fMRI data or eyetracking data at the moment. We are looking forward to integrate that in a future version of our toolbox.

In particular, we hope that this freely-available open source package can serve as a methodological guideline and analysis framework for scholars and practitioners in the broader landscape of NeuroIS research covering the various stages involved in analyzing bio-physiological data in combination with user behavior data.

References

1. Riedl, R., Léger, P.M.: Fundamentals of NeuroIS. Springer, Berlin (2016)
2. Riedl, R., Banker, R.D., Benbasat, I., Davis, F.D., Dennis, A.R., Dimoka, A., Gefen, D., Gupta, A., Ischebeck, A., Kenning, P.: On the foundations of NeuroIS: reflections on the Gmunden Retreat 2009. CAIS **27**, 15 (2010)
3. Astor, P.J., Adam, M.T., Jerčić, P., Schaaff, K., Weinhardt, C.: Integrating biosignals into information systems: a NeuroIS tool for improving emotion regulation. J. Manage. Inf. Syst. **30**(3), 247–278 (2013)
4. Jercic, P., Astor, P.J., Adam, M.T., Hilborn, O., Schaaff, K., Lindley, C., Sennersten, C., Eriksson, J.: A serious game using physiological interfaces for emotion regulation training in the context of financial decision-making. In: Proceedings of the 20th European Conference on Information Systems (ECIS) (2012)
5. Knierim, M.T., Jung, D., Dorner, V., Weinhardt, C.: Designing live biofeedback for groups to support emotion management in digital collaboration. In: International Conference on Design Science Research in Information Systems, pp. 479–484. Springer, Berlin (2017)
6. Knierim, M.T., Rissler, R., Dorner, V., Maedche, A., Weinhardt, C.: The psychophysiology of flow: a systematic review of peripheral nervous system features. In: Gmunden Retreat on NeuroIS (2017)
7. Jung, D., Dorner, V.: Decision inertia and arousal: using NeuroIS to analyze bio-physiological correlates of decision inertia in a dual choice paradigm. In: Gmunden Retreat on NeuroIS (2017)
8. Rosenberg, D.M., Horn, C.C.: Neurophysiological analytics for all! Free open-source software tools for documenting, analyzing, visualizing, and sharing using electronic notebooks. J. Neurophysiol. **116**(2), 252–262 (2016)
9. Hariharan, A., Adam, M.T.P., Lux, E., Pfeiffer, J., Dorner, V., Müller, M.B., Weinhardt, C.: Brownie: a platform for conducting NeuroIS experiments. J. Assoc. Inf. Syst. **18**(4), 264
10. Jung, D., Adam, M., Dorner, V., Hariharan, A.: A practical guide for human lab experiments in information systems research: a tutorial with Brownie. J. Syst. Inf. Technol. **19**(3/4), 228–256 (2017)
11. Wickham, H., Grolemund, G.: R for data science (2016)
12. Scholz, M.: Analyzes clickstreams based on Markov chains. R package version **1**(2) (2014)

Enhancing the Implicit Association Test: A Four-Step Model to Find Appropriate Stimuli

Gerhard Brenner, Monika Koller and Peter Walla

Abstract The Implicit Association Test (IAT) is a promising tool to assess implicit attitudes. Next to neuroscientific methods applied within the field of NeuroIS the IAT helps to overcome limits of traditional approaches, such as self-report studies. Introduced 20 years ago, it has been developed further within subsequent years. However, hardly any attention has been paid to optimize the stimuli sets. This is unfortunate, as if the time span participants need to decode the stimuli varies across the IAT, or if the subjects do not understand the stimuli equally, reaction times can be biased. As an IAT includes 120 measuring points per subject such biases might potentiate across all participants. The results might be biased and neither the researchers nor the participants would recognize such confounding effects. Thus, we focus on the time span between stimulus onset and response and develop a four-step model to create an optimized stimuli set including (1) brainstorming, (2) forming & performing (i.e. pretesting), (3) backward-brainstorming and (4) informing & interviewing.

Keywords Implicit Association test · Implicit measures · Indirect measures

G. Brenner (✉)
University of Applied Sciences, Johannes Gutenberg-Straße 3, 2700 Wiener Neustadt, Austria
e-mail: gerhard.brenner@fhwn.ac.at

M. Koller
WU Vienna, Welthandelsplatz 1, 1020 Vienna, Austria
e-mail: monika.koller@wu.ac.at

P. Walla
CanBeLab, Department of Psychology, Webster Vienna Private University, Praterstrasse 23, 1020 Vienna, Austria
e-mail: peter.walla@webster.ac.at

P. Walla
School of Psychology, Newcastle University, Newcastle, Callaghan, Australia

P. Walla
Faculty of Psychology, University of Vienna, Vienna, Austria

F. D. Davis et al. (eds.), *Information Systems and Neuroscience*, Lecture Notes in Information Systems and Organisation 29,
https://doi.org/10.1007/978-3-030-01087-4_13

1 Introduction

In both, academic as well as applied information systems research, measuring attitudes and emotions, e.g., associated with system use, has been important for decades. However, especially when it comes to the measurement of attitudes at an implicit level, traditional approaches have their limits. The field of NeuroIS has offered promising ways to help in this regard. For example, fMRI applications for instance to investigate psychological phenomena such as Dimoka (2010) did on trust [1] (also see a review by Riedl and Javor on this topic [2]) or other physiological approaches, such as Walla and Koller (2015) did on using startle reflex modulation [3] have been introduced to measure attitudes at an implicit level in the context of information systems. Besides methods from neursocience and/or psycho-physiological methods, the Implicit Association Test (IAT) has proven to be an alternative instrument to capture implicit attitudes in multiple disciplines. Today, the IAT is one of the most frequently applied methods to assess implicit mental processes such as implicit attitudes [4], not only in the field of social psychology, but also in consumer research [5–7]. Particularly, implicit attitude measurement methods are applied to overcome unwanted effects when participants answer questions of questionnaires. Such effects are often caused by social desirable answering, impression management, self-deviation, the incapability of introspection, and finally by low memory capacity [8]. Pretty often, when both implicit and explicit attitudes are measured, the results differ significantly from each other [9, 10]. Hence, in information systems research, applying the IAT could help to gain further insights whenever implicit attitudes are measured. For instance, when it comes to investigate current phenomena like the acceptance of new technologies, technostress or other negative consequences of overconsumption of digital technologies.

During the years following the introduction of the IAT as a method to measure implicit attitudes, the IAT has continuously been developed further [11–16]. Greenwald et al. [10] suggested an "improved scoring algorithm" as well as an enhanced IAT procedure. The new algorithm included the computation of a D-measure putting the mean reaction times in relation to standard deviations. The new IAT procedure comprised seven task sequences ("blocks"), compared to only five blocks in the original version of the IAT [17].

Compared to the number of blocks featured as well as the improved scoring algorithm, the IAT stimuli set has gained only little attention in previous research. This is unfortunate, as the stimuli set is crucial for the entire measurement approach incorporated in the IAT. As one example, Fiedler et al. [18] pointed out that single stimuli might be misunderstood by IAT subjects. Although, the creation of the stimuli set has already been named as critical for the entire procedure, comprehensive suggestions for improvement of the procedure are still missing.

2 The Implicit Association Test

In an IAT, the subjects are confronted with stimuli, either words or pictures representing four categories: Two of them are *target categories*—these are categories that are the targets to be measured (e.g. black vs. white people to assess prejudices of white over black people); and there are two *attribution categories,* representing emotional valence (positive vs. negative) [17].

The principle underlying the IAT is that people quickly respond (e.g., pressing a left or right key) to stimuli representing the *target categories* whenever these categories and the *positive* or *negative* emotional words (*attribution categories*) are implicitly understood congruently. On the other hand, the reaction times will extend when the underlying concepts do not fit. According to Greenwald et al. (2003) the IAT is performed by completing seven blocks (see Fig. 1). Reaction times are relevant for measurement in blocks 3, 4, 6, and 7. Reaction times following congruent stimuli (supposed to be shorter) are subtracted from reaction times following incongruent stimuli and subsequently divided by their pooled standard deviation [11].

Stimuli sets used in the IAT are usually created by brainstorming and pretests (e.g., Gattol et al. [19]). However, this procedure might be too simple and produce misunderstandings.

Since the IAT has been introduced by Greenwald et al. in 1998 [17], researchers also developed an IAT to generate results based on large numbers of participants. Greenwald, Banaji, and Nosek founded the "project implicit" on the Harvard University website (https://implicit.harvard.edu) to initiate web-based IATs with large numbers of participants [20]. In light of the previously mentioned issues of wrongly encoded stimuli [18], we argue that comprehensively preparing all single elements of an IAT is of utmost importance to avoid any biased or misleading results.

task	1	2	3	4	5	6	7
categories	**target categories** (20 trials)	**attribution categories** (20 trials)	**categories combined** (20 trials)	**categories combined** (40 trials)	**target categories (reversed)** (40 trials)	**categories combined (reversed)** (20 trials)	**categories combined (reversed)** (40 trials)
instructions	black people --> press "d"	positive --> press "d"	black people --> press "d" positive --> press "d"	black people --> press "d" positive --> press "d"	white people --> press "d"	white people --> press "d" positive --> press "d"	white people --> press "d" positive --> press "d"
	white people --> press "k"	negative --> press "k"	white people --> press "k" negative --> press "k"	white people --> press "k" negative --> press "k"	black people --> press "k"	black people --> press "k" negative --> press "k"	black people --> press "k" negative --> press "k"

Fig. 1 Seven IAT blocks: Combined task blocks 3, 4, 6, and 7 are relevant for measurement

3 Four-Step Model

To avoid wrongly decoded stimuli we developed a four-step model to find stimuli for the four IAT categories encompassing the steps: (1) brainstorming, (2) forming änd performing, (3) backward-brainstorming, and (4) informing and interviewing.

3.1 Criteria to Find Stimuli

The criteria the stimuli should meet are determined by their fit with the categories they represent: *First of all,* the stimuli should be *decoded* as fast as possible by the subjects and within the same time span, so that the difference of the reaction time span is determined by its evaluation only. What is ought to be measured by the IAT is the evaluation of a stimulus. However, the time span between stimulus onset and response (pressing a key) encompasses (a) identifying the stimulus, (b) understanding its meaning (including a phase of potentially wrong understanding), and (c) its evaluation. If a stimulus is decoded differently because of certain decoding problems (see as a hypothetical example depicted below), the evaluation of that IAT stimulus is altered (see Fig. 2).

Second, if possible, the stimuli representing the *target categories* should not be emotionally loaded. For instance, if the target category "GDR" (*German Democratic Republic*) is represented by the word "torture", it is the valence of the stimulus itself, which makes the response fitting the negative valence, but not its evaluation of the target category "GDR"—no matter whether the word "torture" can be representative for this category or not.

Third, the stimuli should be as specific as possible for the categories they represent. The word "motorcycle" for instance is not necessarily understood as "non-environmental" as suggested by the IAT of Gattol et al. [19]. Some participants might be bikers themselves, and thus might not associate "motorcycle" per se with "non-environmental" vehicles.

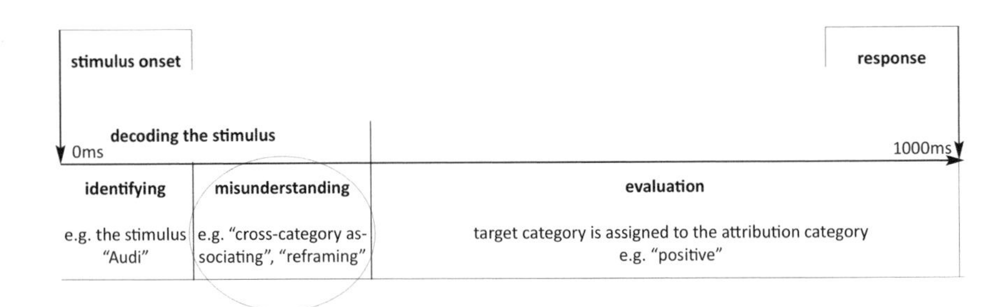

Fig. 2 Timeline (e.g., 1000 ms) between stimulus onset and response (e.g. press key "d")

Fourth, the researchers should seek to eliminate potential reframing effects. The problem of reframing occurs when—as in Fiedler et al's example—the category "Turkey" in one situation is framed as politically problematic, and in another situation it is framed as a popular holiday destination [18].

Fifth, the problem of "cross-category association" [18] occurs, when a stimulus fits both target categories or both attribution categories. For instance, the female prename "Katie", used by Greenwald et al. for *black* people, could be associated with *white* people just as well, at least if the IAT was performed today [17].

Sixth and finally, stimuli triggering "like-me" associations might be clearly positive, whereas "unlike-me" associations might be negative. Gattol et al. [19] used "Switzerland" as a stimulus representing the category "safety". However, the word "Switzerland" might be perceived more positively by participants coming from Switzerland, because of "like-me" associations, than for instance Germans or Austrians would, because of their "unlike-me" associations with "Switzerland".

To summarize, the aim of the present paper is to suggest a procedure for generating valid sets of stimuli through avoiding the six potential pitfalls outlined.

3.2 Four Steps

We suggest a 4-step model, in which those criteria and potential pitfalls are addressed: In *step 1,* a basic set of IAT stimuli is created by carrying out a *brainstorming* process. However, researchers are advised not to brainstorm on their own. They should include members of their planned sample population. In a first step, we recommend to run the brainstorming in written form by each team member on his or her own. In a second step, the team members should read out loud the words they have found, compare them and then add further associative words.

In *step 2,* the so called "forming and performing" step, problematic associative words are removed from the stimuli sets. As mentioned above, problems particularly occur with *emotionally valenced* stimuli representing one of the target categories. The same is true for *unspecific* stimuli (representing both, target and attribution categories), for stimuli potentially eliciting *reframing effects,* for stimuli potentially eliciting *cross-category associating effects,* and for stimuli which potentially elicit either a "like-me" or an "unlike-me" effect. When this procedure is completed, we recommend a first pretest. After each IAT, the participants are interviewed with the focus being on each of the stimuli [20]. If the stimulus set has not been reduced to a number of stimuli which is feasible for a pretest to be carried out (above 10 stimuli for each category, because too many stimuli might confuse the participants), after the forming procedure, pretesting can be postponed to step 3.

In *step 3,* a couple of new participants (again reflecting the planned sample population) is asked to undergo a "backward-brainstorming" process. The words which have been found in steps 1 and 2 are presented to the subjects and the participants are asked to spontaneously disclose what they associate them with. If they associate the words with the corresponding category, the words work properly—the quicker (first or later announced), the better. As a result, these associations are categorized by the criteria "first announcement" (category being announced as a first association) and "announced later on" (category being announced as a second or further association). We recommend to perform a couple of IAT pretests after having completed the backward-brainstorming procedure, to find out whether misunderstandings still might occur.

Step 4, finally, is carried out when the IAT is already designed for the experiment. It encompasses "informing" the participants before starting the IAT and "interviewing" them after having completed it. "Informing" means to tell the subjects which words represent which category and asking them to read those words out loud to make sure that they know the words and that they will associate them with the correct categories. This is kind of a learning procedure. In the interviews, after having completed the IAT, potential misunderstandings can be revealed if they have occurred despite the researchers best efforts to avoid them.

Particularly, the learning procedure ("informing") in step 4 allows the researcher to limit the presentation time of each stimulus to as low as 150 ms. This short presentation time and an inter-stimulus interval of 2000 ms guarantee affective reactions and should avoid unwanted cognitive processes within the subjects' minds. Cognitive processes might distract the affective nature of the IAT and should therefore be inhibited. Moreover, many IAT researchers give their participants the opportunity to correct false category assignments. If for instance a participant has pressed the left key instead of the right-one, he or she is given time to press the correct key. However, we suggest that the participants should not have the opportunity to correct erroneous reactions. Those errors are treated the same way as missing values are (mean of correct reaction times plus 600 ms), as suggested by Greenwald et al. [11]. By applying the four-step model to find stimuli the number of erroneous reactions can be reduced to a minimum. Error rates are one of the critical aspects of the IAT. Westgate et al. for example excluded data from participants who had produced an overall error rate of 30% or more, or an error rate of individual blocks of 40% or more [21]. We endorse this suggestion.

4 Conclusion

For researchers investigating attitudes, the IAT is a promising tool to overcome problems of social desirable answering, impression management, self-deviation, the incapability of introspection, and low memory capacity. However, if the IAT is not prepared diligently and performed carefully, its results might be biased. In this article, we focused on the time span between stimulus onset and response to the stimulus by pressing a key. Our proposed four-step model has the potential to optimize word stimuli sets to keep the decoding time span constant across all participants. If single decoding sequences are contaminated by distracting effects such as cross-category association effects, these biases may cause multiple contaminations of hundreds of milliseconds, which potentiate across a total of 120 measuring points in blocks 3, 4,

6, and 7. The crucial thing is that nobody will realize these contamination effects, because neither the researchers will be aware of what is going on within the subjects' minds nor the participants themselves will be aware of it.

The four-step model we developed is an exploratory suggestion on how the stimuli set used in an IAT can be created in a valid and comprehensive manner. Besides the merits of the suggested approach, there are still limitations subject to further research. For instance, the backward-brainstorming sequence might be calculated in a more sophisticated way by developing an algorithm to consider the weight of first-announcements and later-on-announcements.

References

1. Dimoka, A.: What does the brain tell us about trust and distrust? Evidence from a functional neuroimaging study. MIS Q. **34**(2), 373–396 (2010)
2. Riedl, R., Javor, A.: The biology of trust: Integrating evidence from genetics, endocrinology, and functional brain imaging. J. Neurosci. Psychol. Econ. **5**(2), 63–91 (2012)
3. Walla, P., Koller, M.: Emotion is not what you think it is: Startle reflex modulation (SRM) as a measure of affective processing in NeuroIS. In: Davis, F.D., Riedl, R., vom Brocke, J., Léger, P.-M., Randolph, A.B. (eds.) Information Systems and Neuroscience, Gmunden Retreat on Neuro IS 2015, pp. 181–186. Springer International Publishing Switzerland, Cham, Heidelberg, New York, Dortrecht, London (2015)
4. Gawronski, B., De Houwer, J.: Implicit measures in social and personality psychology. In: Reis, H.T., Judd, C.M. (eds.) Handbook of Research Methods in Social and Personality Psychology, pp. 283–310. Cambridge University Press, New York (2014)
5. Gregg, A.P., Klymowsky, J., Owens, D., Perryman, A.: Let their fingers do the talking? Using the Implicit Association Test in market research. Int. J. Market Res. **55**(4), 23–39 (2015)
6. Bosshard, S., Bourke, J.D., Kunaharan, S., Koller, M., Walla, P.: Established liked versus disliked brands: brain activity, implicit associations and explicit responses. Cogent Psychol. **3**, 1 (2016)
7. Walla, P., Koller, M., Brenner, G., Bosshard, S.: Evaluative conditioning of established brands: implicit measures reveal other effects than explicit measures. J. Neurosci. Psychol. Econ. **10**(1), 24–41 (2017)
8. Fazio, R.H., Olson, M.A.: Implicit measures in social cognition research: their meaning and use. Ann. Rev. Psychol. **54**, 297–327 (2003)
9. Gawronski, B., Strack, F.: On the propositional nature of cognitive consistency: dissonance changes explicit but not implicit attitudes. J. Exp. Soc. Psychol. **40**, 535–542 (2004)
10. Rydell, R.J., McConnell, A.R., Mackie, D.M., Strain, L.M.: Of two minds: forming and changing valence-inconsistent implicit and explicit attitudes. Psychol. Sci. **17**(11), 954–958 (2006)
11. Greenwald, A.G., Banaji, M.R., Nosek, B.A.: Understanding and using the Implicit Association Test: an improved scoring algorithm. J. Pers. Soc. Psychol. **85**, 197–216 (2003)
12. Greenwald, A.G., Nosek, B.A.: Health of the implicit association test at age 3. Zeitschrift für Experimentelle Psychologie **48**, 85–93 (2001)
13. Greenwald, A.G., Poehlman, T.A., Uhlmann, E.L., Banaji, M.R.: Understanding and using the Implicit Association Test: III. Meta-analysis of predictive validity. J. Pers. Soc. Psychol. **97**(1), 17–41 (2009)
14. Lane, K.A., Banaji, M.R., Nosek, B.A., Greenwald, A.G.: Understanding and using the Implicit Association Test: IV—what we know (so far) about the method. In: Wittenbrink, B., Schwarz, N. (eds.) From Implicit Measures of Attitudes, pp. 59–102. The Guilford Press, New York (2007)

15. Nosek, B.A., Greenwald, A.G., Banaji, M.R.: The Implicit Association Test: II. Method variables and construct validity. Pers. Soc. Psychol. Bull. **31**, 166–180 (2005)
16. Nosek, B.A., Greenwald, A.G., Banaji, M.R.: The Implicit Association Test at age 7: a methodological conceptual review. In: Bargh, J.A. (ed.) Automatic Processes in Social Thinking and Behavior, pp. 265–292. Psychology Press, New York (2007)
17. Greenwald, A.G., McGhee, D.E., Schwartz, J.L.K.: Measuring individual differences in implicit cognition: The Implicit Association Test. J. Pers. Soc. Psychol. **74**, 1464–1480 (1998)
18. Fiedler, K., Messner, C., Bluemke, M.: Unresolved problems with the "I", the "A", and the "T": a logical and psychometric critique of the Implicit Association Test (IAT). Eur. Rev. Soc. Psychol. **17**, 74–147 (2006)
19. Gattol, V., Sääksjärvi, M., Carbon, C.-C.: Extending the Implicit Association Test (IAT): assessing consumer attitudes based on multi-dimensional implicit associations. PloS ONE **6** (2011). https://doi.org/10.1371/journal.pone.0015849
20. Project Implicit: https://implicit.harvard.edu. Last accessed 21 Mar 2018
21. Westgate, E.C., Riskind, R.G., Nosek, B.A.: Implicit preferences for straight people over lesbian women and gay men weakened from 2006 to 2013. Collabra **1**(1), 1–10 (2015)

Facebrain: A P300 BCI to Facebook

Ben Warren and Adriane B. Randolph

Abstract Facebrain is a novel brain-computer interface utilizing the P300 signal as input for interacting with the Facebook social media platform. Electroencephalography along with the open-source BCI2000 software suite is used for both obtaining and processing the signals. Additionally, BCPy2000, an add-on allowing BCI2000 modules to be written in the scripting language Python, is utilized to allow for rapid interface generation, promoting extensibility, and a cross-platform solution. Users are able to select basic Facebook operations via a P300 matrix and then activate a P300 speller as needed for text input. Overall, the purpose of the system is to allow functional, hands-free, and voiceless access to Facebook's main features including, but not limited to, searching for and adding friends, making posts, using the chat system, and browsing profiles.

Keywords Brain computer interface · EEG · P300
BCI2000 · Python · Facebook
Communication

1 Introduction

Facebook (http://facebook.com), the social media platform, is a prime example of the role of technology in aiding communication, yet it is virtually inaccessible to those with severe motor impairments. Facebrain is an application designed to provide access to the major functionality of the popular social media platform using the principles behind P300 stimulus presentation. Such access would allow individuals who are locked-in the ability to proactively interact with friends and family using a medium now common to over a billion users [1]. For example, a survivor of a spinal

B. Warren · A. B. Randolph (✉)
Department of Information Systems, Kennesaw State University, Kennesaw, GA, USA
e-mail: arandol3@kennesaw.edu

B. Warren
e-mail: ben@warr.io

F. D. Davis et al. (eds.), *Information Systems and Neuroscience*,
Lecture Notes in Information Systems and Organisation 29,
https://doi.org/10.1007/978-3-030-01087-4_14

cord injury resulting in locked-in syndrome who was a previous user of Facebook may retain the ability to communicate with loved ones seamlessly via the same medium, though without the need for mechanical or vocal input into a computer. Those who have had long-term impairments would be given an avenue to stay informed and communicate with others in a way previously unavailable to them.

Brain-computer interfaces have provided a platform for individuals with long-term motor impairments to communicate and control their environments using neural input alone [2, 3]. Individuals with locked-in syndrome are paralyzed and unable to speak and yet otherwise cognitively in-tact and in need of a non-traditional communication system [4]. These interfaces have been shown to significantly increase quality of life and promote social inclusion [5, 6]. Further, researchers have shown the benefits of web-based and multimedia applications effectively accessed via P300 by healthy participants and participants with locked-in syndrome [7]. Here, we present a P300-based interface that integrates a tool for social inclusion in a manner appropriate for individuals with severe motor disabilities.

P300 is an event-related potential (ERP) present in an electroencephalograph (EEG) that corresponds with attention to infrequent presentation of a stimulus. Researchers have taken advantage of this signal to implement novel systems capable of discerning when a desired stimulus is indicated among an array of alternates. Spelling systems have been one of the most promising applications, allowing individuals to compose words and sentences by simply focusing on a particular character amongst a matrix of flashing letters [3].

We used the classic P300-spelling framework as a basis for our design where others have shown up to 90% accuracy with healthy participants on similar interface paradigms [7]. In addition, a visual interface, such as described in the following, has been shown to provide better results than an auditory paradigm for eliciting a P300 response with severely disabled participants [8]. This work presents an interface similar to most BCIs designed for synchronous use where a caregiver or research starts the system and the participant may then use it during the prescribed period of time [9].

2 System Design

2.1 BCI2000/BCPy2000

BCI2000 is an open-source application written in C++ used to design experiments for BCI research [10, 11]. BCI2000 supports a modular system consisting of three communicating units: a signal source, a signal processor, and an application module. Signals pass from the source module to the signal processor, where the signal is transformed to meaningful data which is then passed to the application module that acts on the data. Signal sources can be easily exchanged to allow several amplifier alternatives to be used with little to no modification of the other modules. Among the included prebuilt modules is a signal processing module for classifying P300 signals, and is used along with an included P300 speller application module.

The BCPy2000 framework is a community package enabling BCI2000 modules to be written using the Python scripting language. BCPy2000 modules can be mixed and matched with traditional BCI2000 modules, so that existing modules possessing the desired functionality would not need to be rewritten in Python. In this application, the BCPy2000 framework was used to design the application module, while the P3SignalProcessing module included with BCI2000 was used to classify P300 responses to stimuli presentation.

2.2 Application Design

Included with BCI2000 is a P3Speller application module along with a `P3SignalProcessing` module. The `P3Speller` application employs a parametric matrix, with a 6×6 alphanumeric matrix set as the default. It uses the oddball paradigm with Farwell and Donchin's classical row/column approach for stimulus presentation [12]. Due to the complexity of the design and the exploratory nature of the project, an entirely new BCPy2000 based application module was created rather than modifying the existing `P3Speller` application module. The `P3SignalProcessing` module is used for signal processing, and its output is utilized in the Facebrain application module.

To begin the program, the user navigates to the `BCI2000/batch` directory inside the Facebrain installation folder and executes the `Facebrain.bat` file. `Facebrain.bat` loads the appropriate signal source module (customizable by the user), the `P3SignalSource` module, and the `PythonApplication` module responsible for loading the Python source code present in `BCI2000/python/Facebrain.py`. A fourth unit named the `Operator` module coordinates the communication between those three modules.

Selecting Config brings up a Parameter Configuration panel displaying options for window sizing, matrix sizing, and stimulus/inter-stimulus durations, as well as various other customizable options for the varying components. Selecting Set Config from the `Operator` initializes the application with the parameters set during configuration, and brings up the blank application window. The Start button will become active, and selecting it will show the Home menu.

2.3 Menu System

Each menu and submenu of the application consists of an implementation of a P300-style speller. The interface consists of a matrix of selectable options to be chosen by the user via a flashing mechanism. The row/column paradigm has been implemented by default, but the architecture of the application allows for alternate flashing algo-

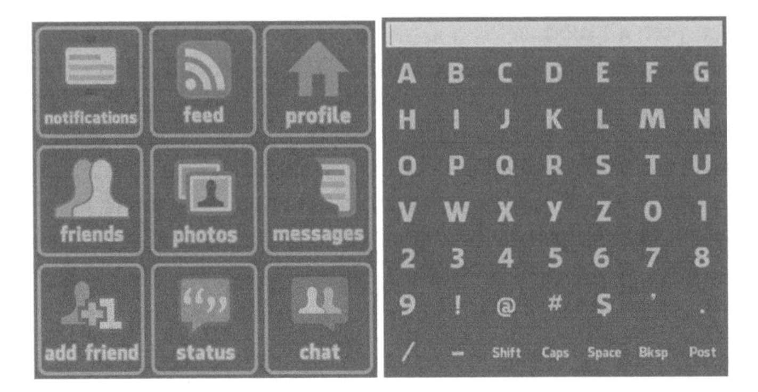

Fig. 1 Views of Facebrain's home menu (left) and an example 7 × 7 speller matrix used for text input (right)

rithms to be implemented and set via the configuration options if desired. The home menu consists of a 3 × 3 matrix of targets representing nine commonly used Facebook functions. A 7 × 7 matrix is also utilized for spelling words and sentences in the various submenus. Figure 1 illustrates the graphical interfaces for both the home menu and menu option for inputting text, an option similar to other researchers allowing additional freeform text by participants [7].

2.4 *Speller Functionality and Target Classification*

The Farwell and Donchin [13] row/column algorithm was chosen for stimulus presentation due to its ease of implementation and longstanding history. For every matrix of targets, the stimuli are grouped into sets of rows and columns, and each set is labeled with a distinct `StimulusCode`. The `StimulusCodes` are then compiled together into a single `Sequence` and randomly shuffled. A dialog is displayed to the user informing them to focus on a target and to count the number of flashes. A countdown is given, and after a short time the rows and columns are flashed in the order present in the sequence. After a sequence has completed, the sequence is randomly shuffled again such that the last stimulus of the completed sequence is not the first element of the new sequence. This continues until the total number of presented sequences equals the `NumberOfSequences` parameter.

At the start of each flash, the `P3SignalProcessing` module is informed of that flash's `StimulusCode` and begins recording an epoch of signal data. The length of the epoch is defined in the `EpochLength` parameter (default: 500 ms). The `NumberOfSequences` parameter defines how many epochs are to be averaged before reporting the results to the application module. The reported results contain a log-likelihood ratio that the `StimulusCode` being reported contains the desired P300 response. After all stimuli have been reported, the application module determines the two highest rated `StimulusCodes` and attempts to find an intersection between their corresponding sets of targets. If a single target intersects both sets, it is determined to be the most likely target and is indicated to the user. The system then performs the appropriate function associated with the target, sets appropriate variables, and continues with the program.

3 Summary

Facebrain as a P300-based brain-computer interface for accessing Facebook may provide a new level of social inclusion and enhancement to the quality of life for a person with severe motor impairments to the point of being locked-in. This interface incorporates the BCPy2000 framework for the BCI2000 open source package. This interface has been developed and tested in-lab with a signal generator module that mimics eight channels of real EEG but not yet more widely tested with live participants. Although still in its preliminary stages, a working prototype is available and shows promise that the P300 response may be utilized to provide much more than the ability to type. Future iterations may incorporate additional classifiers for asynchronous usage.

References

1. Vance, A.: Facebook: The Making of 1 Billion Users. Bloomberg Businessweek (2012)
2. Nijholt, A., et al.: Brain-computer interfacing for intelligent systems. IEEE Intell. Syst. **23**(3), 72–79 (2008)
3. Wolpaw, J.R., et al.: Brain-computer interfaces for communication and control. Clin. Neurophysiol. **113**(6), 767–791 (2002)
4. Randolph, A.B., Moore, M.M.: Jackson, assessing fit of nontraditional assistive technologies. ACM Trans. Accessible Comput. **2**(4), 1–31 (2010)
5. Holz, E.M., et al.: Long-term independent brain-computer interface home use improves quality of life of a patient in the locked-in state: a case study. Arch. Phys. Med. Rehabil. **96**(3), S16–S26 (2015)
6. Moore, M.M.: Real-world applications for brain-computer interface technology. IEEE Trans. Neural Syst. Rehabil. Eng. **11**(2), 162–165 (2003)
7. Halder, S., et al.: Brain-controlled applications using dynamic P300 speller matrices. Artif. Intell. Med. **63**(1), 7–17 (2015)
8. Kübler, A., et al.: A brain-computer interface controlled auditory event-related potential (p300) spelling system for locked-in patients. Ann. N. Y. Acad. Sci. **1157**, 90–100 (2009)
9. Pinegger, A., et al.: Control or non-control state: that is the question! An asynchronous visual P300-based BCI approach. J. Neural Eng. **12**(1), 014001 (2015)
10. Schalk, G., et al.: BCI2000: a general-purpose brain-computer interface (BCI) system. IEEE Trans. Biomed. Eng. **51**(6), 1034–1043 (2004)
11. Schalk, G.: Effective brain-computer interfacing using BCI2000. In: IEEE Engineering in Medicine and Biology Society (2009)
12. Aarts, H., Verplanken, B., Van Knippenberg, A.: Predicting behavior from actions in the past: repeated decision making or a matter of habit. J. Appl. Soc. Psychol. **28**(15), 1355–1374 (1998)

13. Farwell, L.A., Donchin, E.: Talking off the top of your head: toward a mental prothesis utilizing event-related brain potentials. Electroencephalogr. Clin. Neurophysiol. **70**(6), 510–523 (1988)

Sustained Attention in a Monitoring Task: Towards a Neuroadaptative Enterprise System Interface

Théophile Demazure, Alexander Karran, Élise Labonté-LeMoyne, Pierre-Majorique Léger, Sylvain Sénécal, Marc Fredette and Gilbert Babin

Abstract In today's data-driven information technology environment, the ability of humans to sustain attention over long periods of time has become an increasingly important skill. We report work in progress to create a novel passive brain computer interface (pBCI), designed to modulate a user's level of sustained attention in an ecologically valid information system (IS) context. To modulate sustained attention, we take measures of cognitive engagement and vigilance using electroencephalography (EEG) in real time, to form the basis of the BCI, and create a closed neurophysiological feedback loop which adapts elements of a dynamic user interface according to the user's level of sustained attention. The interface utilizes the ERPsim simulation engine to create an ecologically valid IS task supported by a real-life ERP framework.

Keywords Passive brain-computer interface · EEG · Vigilance · Adaptive system Human-computer interaction · ERP

T. Demazure (✉) · A. Karran · É. Labonté-LeMoyne · P.-M. Léger · S. Sénécal · M. Fredette G. Babin
HEC Montréal, Montréal, QC H3T 2A7, Canada
e-mail: theophile.demazure@hec.ca

A. Karran
e-mail: alexander-john.karran@hec.ca

É. Labonté-LeMoyne
e-mail: elise.labonte-lemoyne@hec.ca

P.-M. Léger
e-mail: pierre-majorique.leger@hec.ca

S. Sénécal
e-mail: sylvain.senecal@hec.ca

M. Fredette
e-mail: marc.fredette@hec.ca

G. Babin
e-mail: gilbert.babin@hec.ca

F. D. Davis et al. (eds.), *Information Systems and Neuroscience*,
Lecture Notes in Information Systems and Organisation 29,
https://doi.org/10.1007/978-3-030-01087-4_15

1 Introduction

Recent advances in information technologies, such as artificial intelligence and robotics are rapidly reshaping the way in which we interact with technology [1]. Tasks commonly performed through human labour are becoming increasingly automated, creating vast subsets of tasks that require a high level of human decision readiness, and a high degree of sustained attention (SA) to monitor complex information systems (IS), and the data they create. Users of modern IS, ranging from critical systems infrastructure, to business logistics, require the ability to quickly synthesize and interpret a wide variety of information, to make correct and timely decisions. However, the rapid adoption of automation for administration and analysis tasks has resulted in a potentially hazardous business mindset that considers the human element as a secondary function [2]. Studies have shown that while automation has increased productivity by reducing information-processing and cognitive load, it has decreased operator decision readiness and on-task safety, and that errors are often the result of a decrease in operator vigilance and SA [3, 4], which resulted in a call for research in the IS domain [5].

Previous Research has demonstrated that performance in long duration SA tasks is greatly reduced over prolonged and continuous periods of time [6]. This reduction in SA, termed the vigilance decrement [7] begins to manifest after 20–30 min of task engagement, whereupon reaction times and the probability of operator decision errors increase [8]. Thus, the vigilance decrement occurs when signals requiring detection are perceivable to operators, but do not compel changes in the operating environment. Our aim is to create a brain-computer interface that modulates a user's level of SA, by combining measures of task engagement, vigilance and an autoadaptive IS interface which creates attentional signals to encourage changes in the operating environment.

In the following sections of this manuscript we outline the design methodology and process framework for a BCI that monitors, classifies, and modulates a user's ability to maintain a steady state of SA in real time while monitoring a complex logistics task in enterprise system, and report on results from preliminary analysis.

2 Artefact Design and Requirement Analysis

Utilising design science methods [9], building upon application strategy 3 of NeuroIS [10] and synthesising previous work in the field of neuroscience concerned with task engagement [11–13], we created an iterative design and testing strategy that allowed for rapid development and testing, utilising both synthetic and real neurophysiological data. To capture the requirements of the BCI artefact, we first analysed the needs of an ecologically valid business task, then analysed the needs of an SA task. From this analysis we identified 3 primary requirements:

1. The artefact must represent a real information system (IS) monitoring task.

2. The IS task and its duration must induce and promote a vigilance decrement in the user.
3. The BCI component of the artefact should provide counter measures (CM), to modulate the level of SA, leading to a performance enhancement of its user without obstruction of the IS task.

With regard to the requirement 3, the use of electronic countermeasures within an interface or software artefact, to improve task performance or modulate cognitive workload is currently an area of active research within the human factors community (see NASA HRR [14]).

2.1 Implementation

To meet requirements 1 and 2, we created an ecologically valid IS task utilising an enterprise system [15, 16] offering the functionality, process, reports, and data to simulate a real-life organisation (i.e. SAP). To produce a vigilance decrement in the participant, we modified an enterprise system simulation called ERPsim (ERPsim-Lab, Montréal) so that time moves much slower than the speed suggested by the creators [15], extending the duration of the task to 90 min. The task itself involves maintaining stock levels in 3 locations, and participants are asked to make logistical decisions concerning stock allocation. During the simulation participants are required to perform 15 maintenance and 4 decision tasks each of 4 min duration. Stock depletion rates are non-uniform and dependent on different demand functions. A maximum stock capacity is provided to force decisions as soon as new stock is received, and all correct, incorrect, and missed decisions are logged for later analysis. Thus, the task was reduced to a monitoring task requiring a high level of sustained attention.

The architecture of the BCI artefact has two components, hardware (Sect. 2.2), and software (See Fig. 1), which consists of three elements: (1) NeuroRT software (Paris, France) implementing real-time data processing to extract SA according to chosen parameters. (2) SAP HANA (Walldorf, Germany), which deals with back-end operations such as, storing and serving the neurophysiological data, displaying the information dashboard, and running the simulation and storing the data it creates. Odata services allow the creation of the information dashboard which is automatically refreshed using asynchronous AJAX calls, the decision interface is provided using SAP JCO to directly call BAPI. SAP HANA is deployed to support the experiment via a HANA server owned by HEC Montreal. (3) the Feedback Controller provides the CM mechanism developed in Python, to classify a user's level of sustained attention provided by 1. and served to by 2. All information displayed within the interface follows the concepts of dashboard design [17, 18] and is provided through API queries. Furthermore, to modulate the attentional state of the user, interface CM are applied using a dynamic colour palette to indicate to the user their level of attention, such that the screen background is: white = high, amber = below optimum, red =

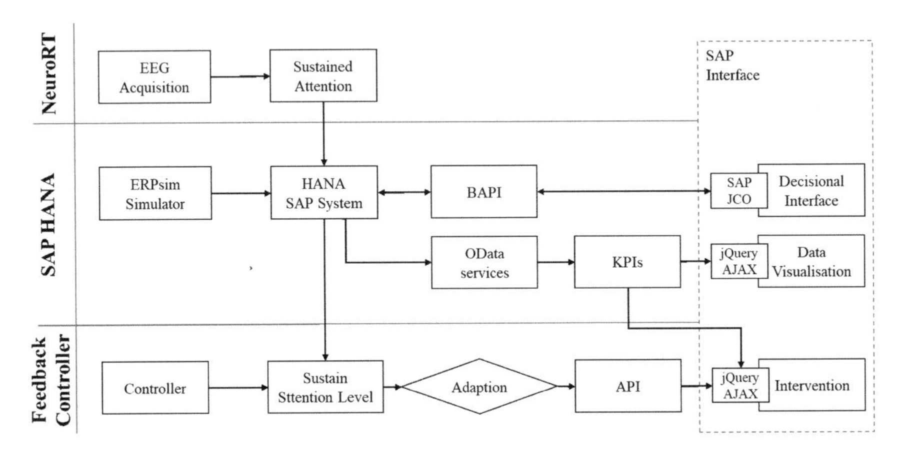

Fig. 1 BCI artefact software architecture schematic

poor. Thus, during active task phases, if the user remains in a heightened state of SA, there are no CM and the interface remains unchanged.

2.2 *Neurophysiological Methods*

We utilise Pope et al.'s [11] engagement index to provide real-time assessment of a user's attentional state and to drive the neurofeedback mechanism of the simulation task interface. This index has been previously used to observe a vigilance decrement [12, 13], making it an ideal candidate measure for this BCI. Following [12] we used a sensor hardware platform consisting of a 32 electrode EEG (Brainvision, Morrisville, NC), to measure variations in brainwave activity in the θ (4–7 Hz), α (8–12 Hz) and β (13–21 Hz) bands from F3, F4, O1, O2 on the international 10–20 system [19]. The SA index is calculated using β (power) divided by α (power) plus θ (power).

From a proposed participant pool of N = 30, 12 participants (6 female) aged 18–43 (Avg. = 24.89), have so far taken part in the study. Participants were of good health and possessed normal or corrected to normal vision, all signed consent in line with the University's ethics board and compensated 50 CAD. Participants were provided with a mouse and keyboard and sat approximately 80 cm from a 24″ computer screen.

The experimental task is split into 2 parts: calibration, lasting 22 min, and testing, lasting 90 min. During calibration the SA index is personalised to individual participants to create a spectrum scale of values ranging from high to low. Calibration is composed of a 1 min baseline (passive observation), then an engagement task of 10 min, then a 1 min baseline and a vigilance task of 10 min. We then compute thresholds for the individual that allow a variable spectrum of index values for user SA state, that fluctuates during part 2 of the experiment (see Fig. 2) in response to changes in mean SA threshold levels over time. These thresholds form the basis of

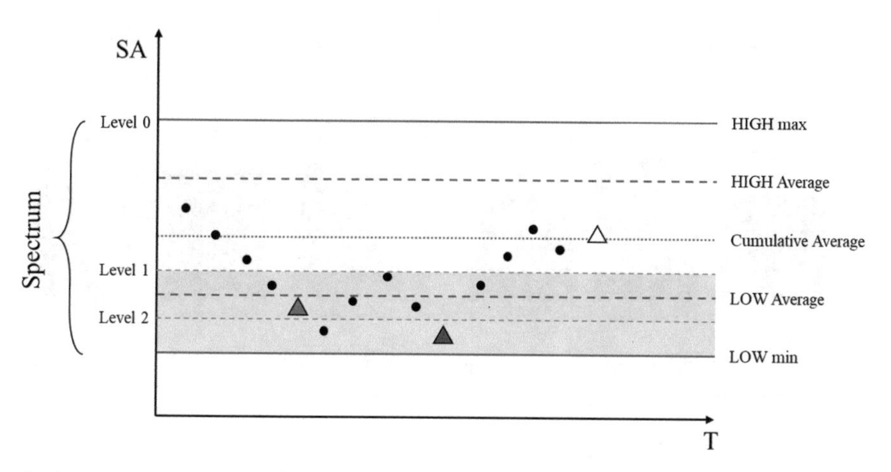

Fig. 2 User-specific attention spectrum

a set of rules that are the coded to create a "fuzzy logic" classifier that classifies SA into three levels where 0 = high SA, 1 = moderate SA and 2 = low SA. These classifications are then used by the feedback controller to produce CM (i.e. changes in the background colour) to modulate SA in the user.

3 Evaluation

Initial development of the BCI was completed iteratively using a combination of simulated EEG data, observation, and hands-on tests. The ERP simulation and task interface were built and tested, with and without adaptations, using test participants to provide feedback concerning the task, the simulation, and the overall experience.

The full experimental procedure (calibration and test) is run to evaluate the current BCI—IS artefact. Participants are randomly assigned to 1 of 3 conditions: no CM; continuous CM; event-based CM. During continuous CM, modulation occurs during the whole experiment. Event based CM consists of modulation only during event phases. User performance during the simulation is measured through actions per minute (APM), percentage of decision errors (PDE) and simulation score, this is then compared with SA level to assess the impact of SA modulations. Participants complete a questionnaire at the end of the procedure to provide a subjective assessment of perceived workload, level of boredom and affective response towards the task and interface. The questionnaire is composed of the raw-TLX [20], a shorter version of the NASA-TLX [21], the Boredom Experience Scale [22, 23], and a SAM Scale [24].

Figure 3 displays the mean SA and mean APM values for each group (n = 4) during task events for the currently available data, error bars represent variance within group. In this figure we observe almost no discernable mean difference in SA

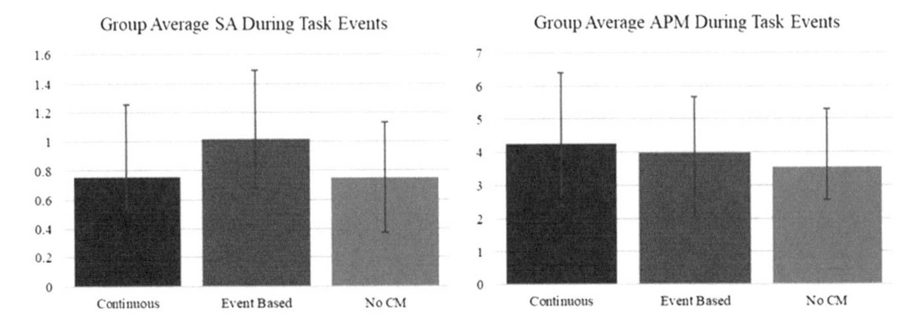

Fig. 3 Preliminary results n = 12, n = 4 per group showing mean SA and APM across conditions during task events

between the no CM (μ 0.75) and continuous CM (μ 0.76) groups, however given the within group variance, this may not reflect the overall strength of the effect once all data is collected. Looking at the APM difference between the same groups, we see that the trend indicates that those in the continuous CM group perform more task actions within active task events. However, if this effect correlates with less decision errors remains to be determined. Looking at the effect on SA and APM for the event-based CM group, shows a higher mean SA (μ 1.01) than both no CM, and continuous CM groups and a higher APM than the no CM group. This potentially indicates that event-based CM promotes higher SA, but not necessarily more actions within active task events, when compared to the continuous CM group, whether the effect of increased SA and APM equates to less decision errors remains to be determined after data collection is complete.

4 Work in Progress and Next Steps

The data so far indicate that the BCI artefact has a positive modulating effect on a user's level of SA, and positively influences the actions they take during active event periods during the enterprise system simulation. However, it is yet to be determined if this effect remains positive across both CM groups after data collection and analysis is complete. Furthermore, the relationship between modulated SA, increased APM and decision errors remains to be explored, as does the correlation between a qualitative assessment of performance, workload, and SA with quantitative observations.

In addition to utilising EEG to measure SA for this project, we also employed functional Near Infrared Spectroscopy (fNIRS) concurrently to measure changes in SA. fNIRS measures the haemodynamic response function of cortical areas to infer synaptic network activation. From this we seek to gain a deeper understanding of SA and apply machine learning to these data, to determine if fNIRS could also be used classify SA in real-time. Furthermore, we seek to utilise both fNIRS and more sophisticated derivatives of EEG data to create a hybrid BCI with the potential to provide a more granular, dynamic, interface environment and more robust SA assessment. We look forward to reporting our results once data collection and analysis is complete.

References

1. Autor, David H.: Why are there still so many jobs? The history and future of workplace automation. J. Econ. Perspect. **29**(3), 3–30 (2015)
2. Warm, J.S., Matthews, G., Finomore, V.S.: Workload and stress in sustained attention. In: Hancock, P.A., Szalma, J.L. (eds.) Performance Under Stress, pp. 115–141. Ashgate Publishing, Aldershot, UK (2008)
3. Molloy, R., Parasuraman, R.: Monitoring an automated system for a single failure: vigilance and task complexity effects. Hum. Factors **38**, 311–322 (1996)
4. De Boer, R., Dekker, S.: Models of automation surprise: results of a field survey in aviation. Safety **3**, 20 (2017)
5. Brocke, J.V., Riedl, R., Léger, P.M.: Application strategies for neuroscience in information systems design science research. J. Comput. Inf. Syst. **53**(3), 1–13 (2013)
6. Mackworth, N.H.: The breakdown of vigilance during prolonged visual search. Q. J. Exp. Psychol. **1**(1), 6–21 (1948)
7. Parasuraman, R., Warm, J.S., See, J.E.: Brain Systems of Vigilance (1998)
8. Grier, R.A., et al.: The vigilance decrement reflects limitations in effortful attention, not mindlessness. Hum. Factors **45**(3), 349–359 (2003)
9. Von Alan, R.H., et al.: Design science in information systems research. MIS Q. **28**(1), 75–105 (2004)
10. Brocke, J.V., Riedl, R., Léger, P.M.: Application strategies for neuroscience in information systems design science research. J. Comput. Inf. Syst. **53**(3), 1–13 (2013)
11. Pope, A.T.B., Edward H., Bartolonne, D.S.: Biocybernetic System Evaluates Indices of Operator Engagement in Automated Task (1995)
12. Mikulka, P.J., et al.: Effects of a biocybernetic system on vigilance performance. Hum. Factors J. Hum. Factors Ergon. Soc. **44**(4), 654–664 (2002)
13. Léger, P.M., Davis, F.D., Cronan, T.P., Perret, J.: Neurophysiological correlates of cognitive absorption in an enactive training context. Comput. Hum. Behav. **34**, 273–283 (2014)
14. NASA HRR: Development of a Software and User Interface to Support Scenario Modeling of Astronaut Schedules to Aid in the Selection of Fatigue Countermeasures Within the Behavioral Health and Performance Dashboard (BHP-DS)
15. Léger, P.-M.: Using a simulation game approach to teach enterprise resource planning concepts. J. Inf. Syst. Educ. **17**(4), 441 (2006)
16. Léger, P.-M., Robert, J., Babin, G., Pellerin, R., Wagner, B.: ERPsim. ERPsim Lab, HEC Montréal, Montréal, Qc (2007)
17. Few, S.: Information Dashboard Design (2006)
18. Few, S.: Show Me the Numbers: designing Tables and Graphs to Enlighten. Analytics Press (2012)
19. Jasper, H.H.: Report of the committee on methods of clinical examination in electroencephalography. Clin. Neurophysiol. **10**(2), 370–375 (1958)
20. Hart, S.G.: NASA-task load index (NASA-TLX); 20 years later. In: Proceedings of the Human Factors and Ergonomics Society Annual Meeting. Sage Publications, Sage, CA, Los Angeles, CA (2006)
21. Hart, S.G., Staveland, L.E.: Development of NASA-TLX (task load index): results of empirical and theoretical research. Adv. Psychol. **52**, 139–183 (1988). Elsevier
22. van Tilburg, W.A., Igou, E.R.: On boredom: lack of challenge and meaning as distinct boredom experiences. Motiv. Emot. **36**(2), 181–194 (2012)

23. Vodanovich, S.J., Watt, J.D.: Self-report measures of boredom: an updated review of the literature. J. Psychol. **150**(2), 196–228 (2016)
24. Bradley, M.M., Lang, P.J.: Measuring emotion: the self-assessment manikin and the semantic differential. J. Behav. Ther. Exp. Psychiatry **25**(1), 49–59 (1994)

Towards Designing Robo-advisors for Unexperienced Investors with Experience Sampling of Time-Series Data

Florian Glaser, Zwetelina Iliewa, Dominik Jung and Martin Weber

Abstract We propose an experimental study to examine how to optimally design a robo-advisor for the purposes of financial risk taking. Specifically, we focus on robo-advisors which are able to (i) "speak" the language of the investors by communicating information on the statistical properties of risky assets in an intuitive way, (ii) "listen" to the investor by monitoring her emotional reactions and (iii) do both. The objectives of our study are twofold. First, we aim to understand how robo-advisors affect financial risk taking and the revisiting of investment decisions. Second, we aim to identify who is most affected by robo-advice.

Keywords Robo-advisory · Financial risk · Taking · Emotion regulation Biofeedback · Physiological arousal

---

F. Glaser · D. Jung (✉)
Karlsruhe Institute of Technology (KIT), Karlsruhe, Germany
e-mail: d.jung@kit.edu

F. Glaser
e-mail: florian.glaser@kit.edu

Z. Iliewa
Max Planck Institute for Research on Collective Goods, Bonn, Germany
e-mail: iliewa@coll.mpg.de

D. Jung
Karlsruhe Decision and Design Lab (KD2Lab), Karlsruhe, Germany

M. Weber
University of Mannheim, Center for European Economic Research (ZEW), Mannheim, Germany
e-mail: weber@bank.bwl.uni-mannheim.de

F. D. Davis et al. (eds.), *Information Systems and Neuroscience*,
Lecture Notes in Information Systems and Organisation 29,
https://doi.org/10.1007/978-3-030-01087-4_16

1 Introduction

Without robo-advisors, taking financial risks is a challenging task for retail investors. Consequently, a new form of financial advisory, so called robo-advisors have become more and more popular in retail and private banking [1], due to their ability to provide unbiased financial advice for retail investors at low cost [2].

Robo-advisors are digital platforms using information technology to guide customers through an automated (investment) advisory process based on interactive and intelligent user assistance components [1, 3]. In particular, they can be helpful tools for investors with low financial knowledge, as well as investors who are susceptible to making financial mistakes [1].

Hence, robo-advisors might have the potential to be useful investment aids to guide people to a share of risky investments which fits their needs and their preferences [3]. Unexperienced investors are not only the largest part of the population; they are also subject to numerous known fallacies in financial decision-making. Across countries, households shy away from risky financial assets [4] and those who take financial risks, often make costly financial mistakes (see, e.g., [5]). In Germany, only 14% of the population own shares [6], a number which is considerably lower than in the UK or US. Consequently, there is considerable research interest in financial advisory because these investors need good decision support.

However, the design of such systems is not an easy matter, and we argue that the sources of financial mistakes vary for different individuals (e.g. low statistical skills) and in different situations (e.g. stress, cognitive load etc.). Consequently, computer-based decision support systems like robo-advisors should take into account (i) individual user characteristics and (ii) situational factors such as the user's internal state. Previous research in information systems has illustrated the importance of considering individual user characteristics when designing decision support systems. The importance of leveraging user models to individualize decisional guidance and explanations has been recognized many years ago [7, 8]. Much work has been done in this field and the stream of research on designing "knowledge interfaces" is still evolving [9], providing new functions, for example, gamification features [10], advanced presentation (virtual and augmented reality) as well as intelligent adaptation leveraging user models on the basis of multi-modal user monitoring. Multi-modal user monitoring includes measurement of users' physiological states (e.g., arousal or cognitive load) which help to assess the user's internal states.

In conclusion, recent information systems research has recognized user characteristics as an important factor for adoption and use of different IS types, for example, in collaboration systems [11], group-decision-making [12], or multi-channel financial services [13, 14]. Users' internal states play a crucial role in information processing and decision behaviour [15–17]. Thus, to successfully support computer-supported investment decisions, we consider a deeper understanding of individual characteristics and internal states as an important research gap.

The present paper presents an experimental study to examine how to optimally design a robo-advisor for the purposes of financial risk taking, thus targeting this research gap. On the basis of a short literature review, we present our experimental design. The work concludes with a summary and an outlook.

2 Designing Financial Decision Support Systems

2.1 Financial Decision Support Systems

Some studies have attempted to design financial information systems that, for instance, restrict reinvestment or provide decisional guidance [18]. Another stream of research is taking a closer look at how well individuals actually understand the decisions they are asked to make (e.g. [19]). These studies show that individuals generally find it difficult to process risk and probabilities (see e.g. [20]) which often leads to a skewed view of the risk or volatility associated with an investment alternative. By altering the way information on the underlying statistical distribution is presented from a mere description towards experiencing the distribution through simulation (i.e. experiential presentation), individuals obtain a better understanding of the investment alternatives and are more inclined to make a risky investment [21, 22].

2.2 Robo-advisory

Initial research in the area of robo-advisory has, for instance, focused on the design of the risk election and modelling of robo-advisors [23]. Tertilt and Scholz reviewed the profiling steps of existing robo-advisory solutions, investigating the risk profiling. Jung et al. derived different design principles based on literature review and expert interviews, and evaluated them in a design science approach [2]. Other work has focused on the legal limitations of robo-advisory [24], or comparing the cost and quality of robo-advisors and human advisory [25].

Nevertheless, only a few studies have addressed the design of robo-advisor user interfaces so far [3]. We argue that there is an increasing need to target this shortcoming and focus on the design of robo-advisory platforms. For instance, service and user interface design, customer behavior, and risk measurement and modelling have been identified as areas with a pressing need for investigation [3].

3 Proposed Experimental Study

In this initial project we plan to design and implement a prototypical robo-advisor, which provides users with basic and advanced experientially presented risk information. In the laboratory experiment, we will compare how well different types of individuals understand risk, what their investment decisions are, and how satisfied and confident they are depending on whether they use this new system or a system that provides the same information in a simpler way. Prior to the experimental study, we will conduct a small pilot test with members of both groups to elicit additional presentation design requirements and suggestions.

The experiment will be a repeated between-subject design with two groups of participants with low and high levels of expertise in statistics, respectively. Participants in both groups are randomly assigned to one of the two treatment conditions, basic and advanced experiential risk information presentation. In particular, we use a 2×2 between-subject design to test for the impact of two potentially important characteristics of the design of robo-advisors, separately and in combination, against the benchmark of no advice. Subjects are randomly distributed in the four treatments. The first characteristic of the design of a robo-advisor is connected to its ability to "speak" the language of the investors. It consists in communicating the information about the probability distribution of the risky asset by allowing the investors to sample from it instead of showing them a formal description. The second characteristic is related to the ability of the robo-advisor to "listen" to the investor. It consists of continuously monitoring the emotional reaction of the investors over all stages of the decision making process and training the investors how to regulate their emotions. The experiment will be conducted with 200 subjects and divided into four treatments of 50 subjects respectively. A pretest will be conducted prior to the experiment for the purposes of calibrating the duration of filler tasks and testing for sufficient cross-sectional variability of our measures of personal characteristics. The experimental procedure is illustrated in Fig. 1.

To gain insight on why robo-advice might potentially affect financial risk taking, we measure several covariates of the decision-making process. On the asset level, we obtain subjects' understanding of the probability distribution and their risk perception. For every decision, we elicit subjects' confidence and decision time. For every stage of the decision-making process (information acquisition, decision, realization of outcome), we measure subjects' heart rates (physiological arousal) with the aim of identifying changes in their internal states and relating them to individual charac-

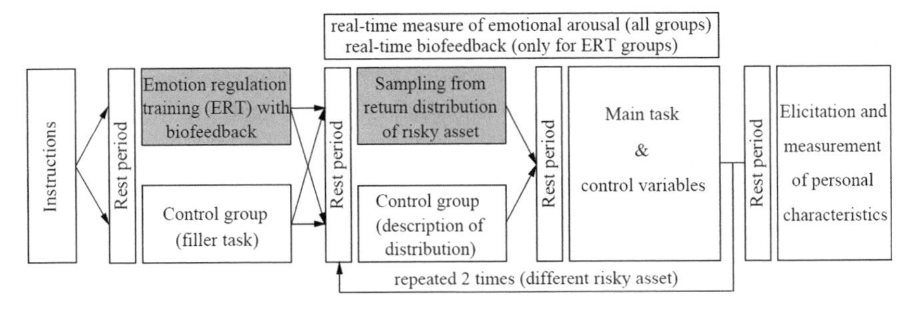

Fig. 1 Proposed design of the initial experimental study

teristics and treatments. For that purpose, we will rely on the experimental platform Brownie, following the Brownie guideline for experimental research in NeuroIS [26]. Participants fill in a questionnaire about their statistics expertise (possibly using an objective measure as the Berlin numeracy task), investment experience, demographic information, risk aversion etc. They are then presented with the treatment (risk information presentation) and subsequently given the task to judge several investment alternatives with respect to their risk/volatility and to indicate which investment alternative they would choose. The final questionnaire surveys decision confidence, satisfaction, etc.

We expect that our experiments help explain how different individuals need to be presented with risk information such that they understand it better and, as a result, are able to make better investment decisions. Our research also helps to develop recommendations for designing better financial information systems for retail investors.

4 Expected Results and Conclusion

Decision support systems like robo-advisors have the potential to be useful investment aids and more research should be devoted to examine how to design them. This experimental study is intended to build the basis for future research at the intersection of Finance and Information Systems. Intelligent systems that adapt to individual users to provide advanced support for investment decisions are considered as an important research domain with a strong impact on society.

In particular, we are interested in examining the temporal stability of learning and the effect of stress. The quality of and capacity for information processing decreases with higher cognitive load and at higher stress levels. For instance, high investments likely lead to higher stress levels for individuals, and that could argue for the need to present information differently (e.g. design a stress-sensitive adaptive system). In future research, we plan to add one treatment variable (stress) and the corresponding treatments. With regard to the temporal stability of learning, we plan to conduct repeated measures studies, for example by inviting participants to multiple experimental sessions in which they are repeatedly given the task to judge several investment alternatives.

References

1. Sironi, P.: FinTech Innovation: from Robo-Advisors to Goal Based Investing and Gamification. Wiley (2016)
2. Jung, D., Dorner, V., Weinhardt, C., et al.: Designing a robo-advisor for risk-averse, low-budget investors. Electron. Markets. Special Issue on FinTech and the transformation of the Financial Industry (2017)
3. Jung, D., Dorner, V., Glaser, F., et al.: Robo-advisory: digitalization and automation of financial advisory. Bus. Inf. Syst. Eng. **60**(1), 81–86 (2018)

4. Badarinza, C., Campbell, J.Y., Ramadorai, T.: International comparative household finance. Ann. Rev. Econ. **8**, 111–144 (2016)
5. Calvet, L.E., Campbell, J.Y., Sodini, P.: Down or out: assessing the welfare costs of household investment mistakes. J. Polit. Econ. **115**(5), 707–747 (2007)
6. Deutsches Aktieninstitut: Aktionärszahlen des Deutschen Aktieninstituts. https://www.dai.de/files/dai_usercontent/dokumente/studien/2016-02-09%20DAI%20Aktionaerszahlen%202015%20Web.pdf (2015)
7. Gregor, S., Benbasat, I.: Explanations from intelligent systems: theoretical foundations and implications for practice. MIS Q., 497–530 (1999)
8. Silver, M.S.: Decisional guidance for computer-based decision support. MIS Q., 105–122 (1991)
9. Gregor, S., Maedche, A., Morana, S., et al. (eds.): Designing knowledge interface systems: past, present, and future. DESRIST (2016)
10. Feil, S., Kretzer, M., Werder, K., et al. (eds.): Using gamification to tackle the cold-start problem in recommender systems. ACM (2016)
11. Schacht, S., Morana, S., Urbach, N., et al.: Are you a Maverick? Towards a Segmentation of Collaboration Technology Users (2015)
12. Knierim, M.T., Jung, D., Dorner, V., et al. (eds.): Designing Live Biofeedback for Groups to Support Emotion Management in Digital Collaboration. Springer (2017)
13. Hummel, D., Schacht, S., Maedche, A.: Determinants of Multi-Channel Behavior: exploring Avenues for Future Research in the Services Industry (2016)
14. Hummel, D., Schacht, S., Maedche, A.: Designing Adaptive Nudges for Multi-Channel Choices of Digital Services: a Laboratory Experiment Design (2017)
15. Astor, P.J., Adam, M.T.P., Jerčić, P., et al.: Integrating biosignals into information systems: a NeuroIS tool for improving emotion regulation. J. Manage. Inf. Syst. **30**(3), 247–278 (2013)
16. Dimoka, A., Davis, F.D., Gupta, A., et al.: On the use of neurophysiological tools in IS research: developing a research agenda for NeuroIS. MIS Q., 679–702 (2012)
17. Riedl, R., Davis, F.D., Hevner, A.R.: Towards a NeuroIS research methodology: intensifying the discussion on methods, tools, and measurement. J. Assoc. Inf. Syst. **15**(10), I (2014)
18. Looney, C.A., Hardin, A.M.: Decision support for retirement portfolio management: overcoming myopic loss aversion via technology design. Manage. Sci. **55**(10), 1688–1703 (2009)
19. Ehm, C., Kaufmann, C., Weber, M.: Volatility inadaptability: investors care about risk, but cannot cope with volatility. Rev. Finance **18**(4), 1387–1423 (2014)
20. Jung, D., Dorner, V.: Decision inertia and arousal: using NeuroIS to analyze bio-physiological correlates of decision inertia in a dual-choice paradigm. Inf. Syst. Neurosci., 159–166 (2017)
21. Bradbury, M.A.S., Hens, T., Zeisberger, S.: Improving investment decisions with simulated experience. Rev. Finance **19**(3), 1019–1052 (2014)
22. Kaufmann, C., Weber, M., Haisley, E.: The role of experience sampling and graphical displays on one's investment risk appetite. Manage. Sci. **59**(2), 323–340 (2013)
23. Tertilt, M., Scholz, P.: To Advice, or not to Advice—How Robo-Advisors Evaluate the Risk Preferences of Private Investors (2017)
24. Fein, M.L.: Robo-Advisors: a Closer Look. Available at SSRN 2658701 (2015)
25. Fisch, J.E., Turner, J.A., Center, P.P.: Robo Advisers vs. Humans: which Make the Better Financial Advisers? (2017)
26. Jung, D., Adam, M., Dorner, V., et al.: A practical guide for human lab experiments in information systems research: a tutorial with Brownie. J. Syst. Inf. Technol. **19**(3/4), 228–256 (2017)

Neural Correlates of Human Decision Making in Recommendation Systems: A Research Proposal

Naveen Zehra Quazilbash, Zaheeruddin Asif and Syed Asil Ali Naqvi

Abstract Significant research has been conducted on human decision making behavior in recommendation systems during the past decade, yet it remains a challenge to design effective and efficient recommendation systems so that they not only produce useful suggestions and ease the decision making task but also turn it into a pleasurable experience. Algorithms have been designed based on research that highlight individual theoretical constructs yet there is an absence of a comprehensive model of human decision-making. This research offers an insight into the core of this issue by examining the neural correlates of human decision-making using Electroencephalography (EEG). The insights generated maybe used to construct a comprehensive model of human decision making in recommendation systems and generate new design principles for the same.

Keywords Decision making · Recommendation systems · EEG
Source localization

1 Introduction

Picking an option out of a set of recommendations is a crucial skill for users aiming to solve real-life decision-making problems. Which book to read, which movie to watch, who to be friends with, which gadget to buy and what items to shop for grocery are a few examples where people may choose to go for a particular suggestion. While taking complex decisions, people usually rely on advice provided by others [1]. Whereas, in online settings, Recommendation Systems (RS) do the job [2, 3]. The software that facilitates human decision making by providing useful suggestions is called a recommendation system.

N. Z. Quazilbash (✉) · Z. Asif · S. A. A. Naqvi
Faculty of Computer Science, Institute of Business Administration, Karachi, Pakistan
e-mail: nquazilbash@iba.edu.pk

Z. Asif
e-mail: zasif@iba.edu.pk

F. D. Davis et al. (eds.), *Information Systems and Neuroscience*,
Lecture Notes in Information Systems and Organisation 29,
https://doi.org/10.1007/978-3-030-01087-4_17

In the context of recommendation systems, the process of arriving at a decision is a two-way collaborative process. It is collaborative in a way that the recommender suggests something and the chooser decides or may give some feedback on it, based on which the recommender may improve the suggestion, thereby helping the chooser reach a decision. Hence, it is necessary to maintain this collaboration between the user and recommendation system [4–7]. A better understanding of how people decide or make choices may help in achieving this goal.

Since the process of choosing or decision making is tiring and effort-intensive [8], a good choice would refer to the one that involves little effort and time [7]. Therefore, it would be useful to determine the factors influencing the choice behavior, because this information can be used for a reduction in the effort applied in decision making and an improvement in generating accurate suggestions [9].

Previous research has shown that there are four major categories of brain processes, namely, social, emotion, cognitive and decision making processes; each involving specific brain areas. Neurological studies have shown that a combination of brain areas get activated when a certain process is being performed. The specific brain areas map onto specific constructs such as ambiguity, risk, emotion and cognitive calculations [10]. However, there is no one-to-one mapping between these processes and the brain areas that get activated.

Since, there does not exist a one-to-one mapping between the brain areas and processes i.e. only one factor contributing towards any particular process, say decision making, it is pertinent to explore the network of factors that contribute towards human decision making process in RSs. This network, referred to as "model" in the proposed study, will help in opening up the black box of human brain processes, which in turn may provide useful insights for improving the design of recommendation systems. This conforms to the call by Taylor to design effective ways of refining information systems by modelling internal neurological actions [11].

Developing a better understanding of human decision making in recommendation systems by using data collected through surveys via questionnaires and interviews is a difficult task. Most of the studies exploring constructs like emotion and trust possess self-reporting bias, owing to their behavioral nature, a case in point being study conducted by Hu and Pu [12]. Self-reported data using questionnaires is limited to conscious perceptions and thoughts, whereas in real life information processing and unconscious perceptions impact human behavior [13]. Due to this influence it is hard to develop an accurate understanding of the IT related human behavior using only the self reported data.

In order to address the challenge of self-reporting bias, researchers are now using neurophysiological tools to measure brain signals directly instead of asking the subjects [10, 14, 15]. Functional brain imaging has proven to be a promising area in explicating the unanswered questions in fields of psychology, marketing and economics and recently has found its way into Information Systems as well [10, 15, 16]. In the proposed study, Electroencephalography (EEG) is used to capture electrical brain signals, which are then used for localization of their source. This brain source will be mapped onto an IS construct which reflects the participation of this construct in the decision making process.

Combining the knowledge of HDM from social sciences and tools from cognitive neuroscience this study aims to develop a model of constructs that pertain to human decision making in recommendation systems thereby answering the following research questions:

- What are the individual neural correlates of human decision making in content based book recommendation systems?
- Which dimensions of HDM manifest themselves simultaneously when a decision making activity is performed in content-based book RSs?

In order to address afore mentioned research questions, we intend to conduct a study comprised of complementary empirical and behavioral parts. In this way, we plan to explore the neural correlates of human decision making in recommendation systems. It would help researchers understand how recommendation agents influence individual cognition during multiple choice decision making.

The remaining paper is structured as follows. The following section presents and discusses the proposed hypotheses. Research methodology is discussed in Sect. 3. Section 4 presents a discussion on subjective justification of methodology along with expected contributions in various directions.

2 Research Model

Since there does not exist a one-to-one mapping of brain areas contributing toward a particular brain process, instead multiple brain areas get activated when some activity is performed by a human subject, therefore, we expect that decision making activity in content-based book recommendation systems would span simultaneously over different brain areas.

H1 *decision making activity in content-based book recommendation systems span simultaneously over different brain areas.*

H2 *decision making activity elicits more cerebral activity in some brain areas than the others.*

Since any activity excites multiple brain areas simultaneously, so we propose that multiple IS constructs play their role in it. These constructs are referred to as dimensions of human decision making in RSs. Prior research has shown that these constructs tend to include risk, uncertainty, and ambiguity, theory of mind, calculation, distrust, risk and emotion.

The basis of social adaptive learning, a categorical way of following advice is fundamentally rewarding [5] i.e. the brain process and/or IS construct behind advice following is that of reward. The experimental fMRI study of human subjects shows that greater positive BOLD responses were generated in reward sensitive brain areas when recommendations were given to support decision making in comparison to decision making that was not supported by recommendations. Two outcomes can be derived from these results that could benefit the design of our study. Firstly, it can be deduced from this study that choice decision making in recommendation systems is a type of social learning activity. Secondly, it links **reward** construct with the choice decision making in advice taking.

H3a Reward is a dimension of decision making in recommendation systems

People rely on advice from others for reaching good choices therefore role of recommendation systems providing suggestions to the user can be termed as a social learning activity [5]. This interaction between the user and the recommendation system reflects the approach of deducing how the others are thinking and further predicting their behavior. This theoretical concept is referred to as "*theory of mind*" [17]. The choice activity in recommendation systems is not only social but also calculative in nature therefore Bhatt and Camerer [18] proposed an integration of these two components of "**theory of mind**" by identifying a distinctive neural correlate for each of them.

H3b Theory of Mind is a dimension of decision making in recommendation systems

One of the most widely explored constructs in NeuroIS research is Trust. There had been several studies that identify the role of trust in disparate information systems and these studies contribute some fine findings related to the influence of trust measure in different scenarios. (For example René Riedl et al. [19] study on trust in anthropomorphic decision aids and (Dimoka [20]) exploration of trust and distrust constructs). Familiarity with RS builds users' trust; users' beliefs about the degree to which the RSs understand them and are personalized for them are key factors in RS adoption [21]. Whereas on the other hand a lesser emphasis has been laid on exploring the dynamics of distrust. While inferring how other people will think, brain areas pertaining to distrust may also get activated [17, 22].

H3c Distrust is a dimension of decision making in recommendation systems.

3 Proposed Methodology

This proposal makes use of Electroencephalography (EEG) (a neurophysiological tool) Event Related Potential (ERP), to capture brain signals of human subjects. It is to be done by presenting a stimulus multiple times and then averaging the output signal. This averages out the noise signal and leaves only the signal that was generated in response to the stimulus. Afterwards, sLoreta [23, 24] is used for localization of sources of the signals. These sources can be mapped onto constructs that contribute toward decision making activity. After the empirical part the subjects are supposed to undergo a complementary behavioral part of the study to enhance ecological and external validity of the study. In the behavioral part the participants will be required to fill out a questionnaire related to the activity.

3.1 Participants

We plan to recruit thirty participants for the study by posting around the campus. Right-handed healthy individuals with no reported history of any sort of brain injury and perfect and/or corrected-to-perfect eyesight will be recruited for the experiment. These subjects will be aged between 20 and 45 years old book seekers. The study will be reviewed by Board of Advanced Studies and Research (BASR) before commencement. This sample demographic is based on the analysis of the book-crossing[1] dataset.

3.2 Experimental Procedure

Every participant will be tested individually and will be briefed about the study beforehand. The instructions will be shared with the participant in written as well. Subjects will be asked about their favorite book through a subject information form before starting the experiment. The subject will be shown a fixation cross for 10 s. The content-based recommendations from an online customized book store will then be showed to the subject. The EEG signals will be recorded starting from the time the fixation cross is shown. The recommendations will be shown as visual stimulus for 15 s. Altogether each subject will go through 5 runs of activity with a wash-out time of 5 min in between runs.

3.3 EEG Acquisition and Analysis

The electroencephalography (EEG) signals will be acquired via MDX Neuro-Pro 32 equipment. 19 channels will be used and the data will be acquired at 200 Hz/s of sampling rate. The electrodes will be positioned as per international 10–20 system. The acquired data will be preprocessed by average referencing.

Finally, source localization is performed by comparing fixation cross and recommendation files in Loreta [23, 24]. This will reveal the brain areas (Brodmann Areas[2]) activated during the decision making activity. In particular, we plan to employ sLoreta [23, 24] for this purpose as it identifies the source with the strongest signal activation.

---

[1]The Book Crossing dataset was collected by Cai-Nicolas Ziegler in 2004. It contains 278,858 users' anonymized demographic data about books.

[2]German scientist Korbinian Brodmann named different regions of the brain based on the cytoarchitectural structure of neurons. These areas are referred to as Brodmann Areas.

4 Discussion and Expected Contributions

Human decision making in recommendation systems is a debated issue in literature. However, little attention has been diverted toward uncovering the neural underpinnings of such activity. In this proposal we suggest that development of a model of neural correlates of human decision making in recommendation systems will contribute towards better design of RSs. Despite the development of research in this area, our understanding would remain constrained if we do not explore the role of the constructs involved in this activity.

Our objective is to develop this model by complementing NeuroIS data with behavioral data to not only to avoid self-reporting bias but also ensure ecological validity via triangulation of measures. Our study is likely to deliver insights leading to better comprehension of the psychological underpinnings of recommendation influenced decision making. Another expected contribution of this study is to provide insights into implementation of embedded systems since the world is moving towards mind controlled gadgets and robots and decision is one of the normal human conduct that will be shown by the robots in future. Finally, we expect that our work will contribute towards progressing NeuroIS research by using neurophysiological tools to report human behavior in IS thereby help in opening the black box of human brain and thus in designing better systems with improved perceived usefulness.

Acknowledgements We are grateful to the BCI Lab at Department of Physics, University of Karachi for helpful suggestions in drafting this paper and providing facilities.

References

1. Bonaccio, S., Dalal, R.S.: Advice taking and decision-making: an integrative literature review, and implications for the organizational sciences. Organ. Behav. Hum. Decis. Process. **101**(2), 127–151 (2006)
2. Xiao, B., Benbasat, I.: E-commerce product recommendation agents: use, characteristics, and impact. MIS Q. **31**(1), 137–209 (2007)
3. Xiao, B., Benbasat, I.: Research on the use, characteristics, and impact of e-commerce product recommendation agents: a review and update for 2007–2012. In: Handbook of Strategic e-Business Management, pp. 403–431. Springer Berlin Heidelberg (2014)
4. Chen, L., de Gemmis, M., Felfernig, A., Lops, P., Ricci, F., Semeraro, G.: Human decision making and recommender systems. ACM Trans. Interact. Intell. Syst. (TiiS) **3**(3), 17 (2013)
5. Biele, G., Rieskamp, J., Krugel, L.K., Heekeren, H.R.: The neural basis of following advice. PLoS Biol. **9**(6), e1001089 (2011)
6. Jannach, D., Hegelich, K.: A case study on the effectiveness of recommendations in the mobile internet. In Proceedings of the third ACM conference on Recommender Systems, pp. 205–208. ACM (2009)
7. Jameson, A., Willemsen, M.C., Felfernig, A., de Gemmis, M., Lops, P., Semeraro, G., Chen, L.: Human decision making and recommender systems. In Recommender Systems Handbook, pp. 611–648. Springer, Boston, MA (2015)
8. Cerf, M., Garcia-Garcia, M., Kotler, P.: Consumer Neuroscience. MIT Press (2017)

9. Jannach, D., Zanker, M., Ge, M., Gröning, M.: Recommender systems in computer science and information systems—a landscape of research. In: International Conference on Electronic Commerce and Web Technologies, pp. 76–87. Springer, Berlin, Heidelberg (2012)
10. Dimoka, A., Pavlou, P.A., Davis, F.D.: Research commentary—NeuroIS: the potential of cognitive neuroscience for information systems research. Inf. Syst. Res. **22**(4), 687–702 (2011)
11. Taylor, J.G.: Future directions for neural networks and intelligent systems from the brain imaging research. In: Future Directions for Intelligent Systems and Information Sciences, pp. 191–212. Physica, Heidelberg (2000)
12. Hu, R., Pu, P.: A comparative user study on rating vs. personality quiz based preference elicitation methods. In: Proceedings of the 14th International Conference on Intelligent User Interfaces, pp. 367–372. ACM (2009)
13. Lieberman, M.D.: Social cognitive neuroscience: a review of core processes. Annu. Rev. Psychol. **10**(58), 259–289 (2007)
14. Dimoka, A., Davis, F.D., Gupta, A., Pavlou, P.A., Banker, R.D., Dennis, A.R., Ischebeck, A., Müller-Putz, G., Benbasat, I., Gefen, D., Kenning, P.H.: On the use of neurophysiological tools in IS research: developing a research agenda for NeuroIS. MIS Q. **1**, 679–702 (2012)
15. Brocke, J.V., Riedl, R., Léger, P.M.: Application strategies for neuroscience in information systems design science research. J. Comput. Inf. Syst. **53**(3), 1–3 (2013)
16. Riedl, R., Léger, P.M.: Fundamentals of NeuroIS. In: Studies in Neuroscience, Psychology and Behavioral Economics. Springer, Berlin, Heidelberg (2016)
17. Pavlou, P., Davis, F., Dimoka, A.: Neuro IS: the potential of cognitive neuroscience for information systems research. In: ICIS 2007 Proceedings, p. 122. (2007)
18. Bhatt, M., Camerer, C.F.: Self-referential thinking and equilibrium as states of mind in games: fMRI evidence. Games Econ. Behavior. **52**(2), 424–459 (2005)
19. Riedl, R., Mohr, P.N., Kenning, P.H., Davis, F.D., Heekeren, H.R.: Trusting humans and avatars: a brain imaging study based on evolution theory. J. Manage. Inf. Syst. **30**(4), 83–114 (2014). Apr 1
20. Dimoka, A.: What does the brain tell us about trust and distrust? Evidence from a functional neuroimaging study. MIS Q. **1**, 373–396 (2010)
21. Komiak, S.Y., Benbasat, I.: The effects of personalization and familiarity on trust and adoption of recommendation agents. MIS Q. **1**, 941–960 (2006)
22. Winston, J.S., Strange, B.A., O'Doherty, J., Dolan, R.J.: Automatic and intentional brain responses during evaluation of trustworthiness of faces. Nat. Neurosci. **5**(3), 277 (2002)
23. Pascual-Marqui, R.D.: Standardized low-resolution brain electromagnetic tomography (sLORETA): technical details. Methods Find. Exp. Clin. Pharmacol. **24**(Suppl D), 5–12 (2002)
24. Pascual-Marqui, R.D.: Discrete, 3D distributed, linear imaging methods of electric neuronal activity. Part 1: exact, zero error localization. arXiv preprint arXiv:0710.3341 (2007)

Cognitive Fit and Visual Pattern Recognition in Financial Information System: An Experimental Study

Jérôme Martin, Martin Boyer, Pierre-Majorique Léger and Laurence Dumont

Abstract This experimental study uses traders to understand the effect of cognitive fit on the performance of decision makers for pattern recognition tasks using financial information systems. Building upon signal detection theory, we find that performance is affected by both attention level and working memory capacity while the level of knowledge in finance and experience in finance have no significant impact. Our results also suggest that overconfidence has a detrimental effect on performance.

Keywords Decision making · Behavioral finance · Cognitive fit
Pattern recognition

1 Introduction

Despite the rise of machine learning and artificial intelligence, there are still numerous IT jobs that rely on visual pattern recognition in monitoring tasks. This is the case of traders using financial information system (FIS) who must be able to recognize recurrent patterns in visual task monitoring. By recognizing those patterns, technical traders are making investment decisions based on their anticipation of market reaction. The objective of this paper is to study the effect of cognitive fit [1] on the performance of decision makers for pattern detection in a FIS. Specifically, we conducted a laboratory experiment to examine the effect of attention and working memory on the performance of traders in a recall financial task.

J. Martin (✉) · M. Boyer · P.-M. Léger · L. Dumont
HEC Montreal, Montreal, Canada
e-mail: jerome.martin@hec.ca

M. Boyer
e-mail: martin.boyer@hec.ca

P.-M. Léger
e-mail: pierre-majorique.leger@hec.ca

L. Dumont
e-mail: laurence.dumont@hec.ca

F. D. Davis et al. (eds.), *Information Systems and Neuroscience*,
Lecture Notes in Information Systems and Organisation 29,
https://doi.org/10.1007/978-3-030-01087-4_18

2 Literature Review

Attention refers to the selectivity mechanisms that allow us to process the information we classify as important [2]. Selective visual attention allows us to ignore irrelevant information and direct our processing capacities to the stimuli that are aligned with our goals. Only relevant information based on pattern recognition is processed. In connection with decision making, visual attention narrows attention to fixate on stimuli required for decision making. Fixating a stimulus improves perceptual representation by strengthening its visual attributes and location, thus reinforcing the influence of fixated stimuli as compared to the non-fixated stimuli [3]. Attention also plays an important role for pattern recognition [4].

Working memory refers to the cognitive system responsible for manipulating information temporarily available for processing in short term memory [5]. It has been shown that performance on working memory predicts performance on a wide range of cognitive tasks such as reasoning, reading, and more [6–8]. However, this system can only store a limited amount of information [9, 10]. Different types of errors can occur when trying to recall information from working memory. Such errors are accounted for in the signal detection theory, which sorts pattern recognition trials into one of four categories (Hit: pattern present and respond present, Miss: pattern present and respond absent, Correct rejection: pattern absent and respond absent, False alarm: pattern absent and respond present) [11].

Further, a strong relationship between attention and working memory has been demonstrated by the eye-mind assumption [12]. This theory holds that working memory's limited capacity leads individuals to rely on fixations to reduce working memory's cognitive load during challenging tasks [13–15]. Eye-tracking studies testing the eye-mind assumption have demonstrated a significant positive relationship between working memory load and both the number of fixations [16] and fixation durations [17].

3 Hypotheses Development

Before developing our hypotheses, we will make two assumptions. Since attention and working memory are useful mechanisms in problem representation as well as in pattern recognition, we expect that they will be positively correlated to the performance of decision makers.

H1 *The performance of decision makers is positively correlated to their working memory capacity.*

H2 *The performance of decision makers is positively correlated to their attention level.*

We also expect that decision makers will perform differently depending on their signal classification trials by the signal detection theory [6–8]. More precisely, we expect that decision makers with higher propensity to think they face a seen pattern will perform better in general but will have a lower performance for false alarms trials. Indeed, these participants will rely on what was previously stored in their working memory to solve the task while no trace of the actual pattern will be present in it. Thus, decision makers with biased working memory will think they saw previous patterns more often than unbiased decision makers. This situation could represent overconfidence resulting in poor decision-making [18, 19].

H3 The performance of decision makers is impacted negatively when facing a new task which they think they have previously faced.

Finally, it is expected that decision makers will show different attention levels based on their trial classification by the signal detection theory [3, 4]. More specifically, we expect that they will show a lower attention level for hit trials, as they will accurately recognize a pattern they have previously been exposed to. This implies that a lower attention level is required to make an appropriate decision.

H4 The performance of decision makers is less impacted by their attention level for a task they know they previously faced.

4 Methodology

To answer the research question, we conducted a within subject experiment. Thirty individuals (22 males) aged between 18 and 42 years (average 24.63, std. dev 5.90) were recruited from our institution's research panel. Ethics approval was obtained by our institution's Research Ethics Board. Since our study focuses on novice investors, participants needed to have completed a maximum of three finance courses to be eligible. In exchange for their participation, individuals received a 20$ gift card that could be exchanged at the university's bookstore. Their performance was incentivized by a chance to win an additional 200$ gift card to the bookstore.

The experiment consisted of four separate blocks that were conducted in a financial information system (FIS) context. First, participants answered a questionnaire gathering demographic information, such as gender, age, self-reported knowledge in finance, working experience in finance, and their experience investing in capital markets.

Two five-minute trading simulation tasks were then presented to the participants so that they could become familiar with the trading platform and the behavior of the fictive market index we used in this experiment. During these simulations, participants were given the objective to generate the highest profit possible.

In the third block, participants were exposed to 16 scenarios which were part of what we call the "Investment Survey". Each scenario featured a chart showing the evolution of the price of the same fictive market index as in the second task. Half of the scenarios were taken from the second task and the other half were new scenarios. The charts displayed a price path length equivalent to one fifth of simulation. For each scenario, participants had to make an investment decision and to determine the number of contracts they wanted to trade. They also had to determine if they recall seeing the price path shown by the chart during the second task. The scenarios were presented in two random groups containing eight different scenarios. For each scenario, we calculated the participant performance based on a predetermined ending price. The ending price for scenarios taken from the second task were simply the price value 15 s after the end of each chart shown in the scenarios, while the ending price for the scenarios participants did not see were designed like scenarios that could be expected in the trading simulations and were all the same for each participant.

Finally, for the fourth task, participants underwent a n-back 1 test to assess their working memory [20]. During this test, small white squares appear on the screen at one of the 15 possible locations and participants had to determine if the square they saw was at the same place as a previous square. This test gave a n-back 1 score which represents working memory capacity and a n-back 1 bias which measures the propensity to answer positively (that the square is in the same place as the previous one) [21].

We used eye tracking to record the participant's eye movement during the experiment. The eye movements were tracked using an infrared pupil reflection system (SMI RED250) with a sampling rate of 60 Hz. We constructed one area of interest (AOI) which captured the whole chart for every scenario of the Investment Survey allowing us to generate eye fixation related data for the chart area.

5 Results

To test our hypotheses, we performed two independent linear mixed model regressions shown in Table 1. For each of these regressions, we included five control variables (gender, age, level of knowledge, job in finance and investment experience of participants) and they were not significant for all regressions. The first regression tested the impact of the n-back 1 score on the total profit in the Investment Survey with the control variables. We found that the n-back 1 score positively predicts the performance. The second regression tested the impact of the number of fixations in the chart on the total profit in the Investment Survey with control variables. We found that the number of fixations in the chart also affects positively the total profit.

We next performed linear mixed model regressions to see the interaction between n-back measures and the type of scenario based on whether participant saw it previously or not, and whether he reported it as seen or unseen. We also used the same control variables (see Table 2). We found a positive relationship between n-back 1 bias and total profit, but this effect is less important for false alarms trials (unseen scenarios that were reported as seen). We also did the same regressions on total profit with the number of fixations in the chart as well as with the fixations duration in the chart. Both regressions showed the same results; a positive impact on total profit with a less important effect for hit trials (seen scenarios that were reported as seen).

Table 1 Summary of linear mixed model regressions of working memory and attention on total profit (with control variables)

Effect on total profit	B	SE B	DF	t	*p*
(H1) N-back 1 score	22.938	12.921	450	1.43	0.038*
(H2) Number of fixations in the chart	1.041	0.614	449	1.34	0.045*

*$p < 0.05$ (one-tailed)

Table 2 Summary of linear mixed model regressions of the interaction between working memory or attention and the type of scenario on total profit (with control variables)

Effect	B	SE B	DF	t	*p*
(H1) N-back 1 bias	64.501	25.548	448	2.52	0.012*
(H1) N-back 1 bias with scenario not seen but reported as seen	−82.598	26.478	448	−3.12	0.002*
(H2) Number of fixations in the chart	1.625	0.647	447	2.51	0.012*
(H2) Number of fixations in the chart with scenario seen and reported as seen	−1.702	0.628	447	−2.71	0.007*
(H2) Fixations duration in the chart	0.004	0.002	448	2.17	0.031*
(H2) Fixations duration in the chart with scenario seen and reported as seen	−0.006	0.002	448	−2.81	0.005*

*$p < 0.05$ (two-tailed)

6 Discussion and Conclusion

These first results show that both attention and working memory are positively correlated to performance of decision makers, which supports H1 and H2. They also support previous findings for both working memory-performance relationship [6, 8], and attention-performance relationship [22]. Having a higher working memory capacity and looking longer at the chart result in a superior performance. Additionally, the smaller effect of the n-back 1 bias on total profit for false alarm trials (not seen but reported as seen) supports H3 and could indicate that participants were affected by overconfidence. It has already been demonstrated that investors affected by overconfidence make poor investment decisions which results in lower investment returns [18, 19]. We also found that the longer and the more frequently participants look at the charts, the higher the profit they generate, but this effect is less important for hit trials supporting H4. This finding could indicate that participants should rely more on their working memory when making investment decisions in presence of scenarios they know they already saw rather than putting too much weight on their external memory. Sadly, this situation is not representative of real investment decisions when capital markets are efficient. The situation does correspond, however, to markets where technical analysis is extensively used.

In conclusion, the objective of this study was to determine how attention and working memory affect the performance of decision makers and more specifically beginner investors. We found that both attention and working memory are positively correlated to performance which confirms the importance of using the right type of external representation as stated by the cognitive fit theory [1]. It would be interesting to incorporate a function reporting the resemblance degree of visual patterns when creating new FIS. We also found that beginner investors are subject to overconfidence. These results are limited to novice investors and therefore should not be considered as representative of all decision makers. Future research should focus on expert decision makers to see if we will find the same results. It would also be interesting test the impact of decision makers risk profile on problem-solving performance. Finally, it would be interesting to evaluate the impact of overconfidence on decision makers' problem-solving performance.

References

1. Vessey, I.: Cognitive fit: a theory-based analysis of the graphs versus tables literature. Decis. Sci. **22**(2), 219–240 (1991)
2. Carrasco, M.: Visual attention: the past 25 years. Vision. Res. **51**(13), 1484–1525 (2011)
3. Orquin, J.L., Loose, S.M.: Attention and choice: a review on eye movements in decision making. Acta Physiol. (Oxf) **144**(1), 190–206 (2013)
4. Posner, M.I., Petersen, S.E.: The attention system of the human brain. Annu. Rev. Neurosci. **13**(1), 25–42 (1990)
5. Miyake, A., Shah, P. (eds.): Models of working memory: mechanisms of active maintenance and executive control. Cambridge University Press, Cambridge (1999)

6. Daneman, M., Carpenter, P.A.: Individual differences in working memory and reading. J. Verbal Learn. Verbal Behav. **19**(4), 450–466 (1980)
7. Ackerman, P.L., Beier, M.E., Boyle, M.O.: Working memory and intelligence: the same or different constructs? Psychol. Bull. **131**(1), 30–60 (2005)
8. Kane, M.J., Hambrick, D.Z., Conway, A.R.: Working memory capacity and fluid intelligence are strongly related constructs: comment on Ackerman, Beier, and Boyle (2005). Psychol. Bull. **131**(1), 66–71 (2005)
9. Cowan, N.: The magical number 4 in short-term memory: a reconsideration of mental storage capacity. Behav. Brain Sci. **24**(1), 87–114 (2001)
10. Miller, G.A.: The magical number seven, plus or minus two: some limits on our capacity for processing information. Psychol. Rev. **63**(2), 81 (1956)
11. Tanner Jr., W.P., Swets, J.A.: A decision-making theory of visual detection. Psychol. Rev. **61**(6), 401 (1954)
12. Just, M.A., Carpenter, P.A.: Using eye fixations to study reading comprehension. New Methods Reading Comprehension Res., 151–182 (1984)
13. Droll, J.A., Hayhoe, M.M.: Trade-offs between gaze and working memory use. J. Exp. Psychol. Hum. Percept. Perform. **33**(6), 1352 (2007)
14. Hayhoe, M.M., Bensinger, D.G., Ballard, D.H.: Task constraints in visual working memory. Vision. Res. **38**(1), 125–137 (1998)
15. Karn, K.S., Hayhoe, M.M.: Memory representations guide targeting eye movements in a natural task. Vis. Cogn. **7**(6), 673–703 (2000)
16. Just, M.A., Carpenter, P.A.: Eye fixations and cognitive processes. Cogn. Psychol. **8**(4), 441–480 (1976)
17. Gould, J.D.: Eye movements during visual search and memory search. J. Exp. Psychol. **98**(1), 184 (1973)
18. Odean, T.: Do investors trade too much? Am. Econ. Rev. **89**(5), 1279–1298 (1999)
19. Barber, B.M., Odean, T.: Trading is hazardous to your wealth: the common stock investment performance of individual investors. J. Finan. **55**(2), 773–806 (2000)
20. Kirchner, W.K.: Age differences in short-term retention of rapidly changing information. J. Exp. Psychol. **55**(4), 352 (1958)
21. Kramer, J.H., Mungas, D., Possin, K.L., Rankin, K.P., Boxer, A.L., Rosen, H.J., Widmeyer, M.: NIH EXAMINER: conceptualization and development of an executive function battery. J. Int. Neuropsychol. Soc. **20**(1), 11–19 (2014)
22. Prinzmetal, W., McCool, C., Park, S.: Attention: reaction time and accuracy reveal different mechanisms. J. Exp. Psychol. Gen. **134**(1), 73–92 (2005)

How Attention Networks Can Inform Research in Information Systems

Colin Conrad and Aaron Newman

Abstract Attention is a construct that has been pursued throughout the information systems literature. It is also a topic that has been extensively studied in the cognitive neuroscience literature. To our knowledge there has not been any comprehensive work to bridge these two bodies of work. This idea paper introduces the Attention Networks model, which is one of the prominent models of attention in cognitive neuroscience. We also introduce the Attention Network Test, one of the prominent measures of attention networks. We explore two ways that the model can inform information systems research and conclude that there are many other potential ways that the study of attention networks can advance research in information systems.

Keywords Attention · Attention networks test · NeuroIS research methods

1 Introduction

Information Systems researchers have identified a number of contexts where the study of attention is relevant. For example, attentional capacity has been identified as relevant to optimal virtual workplaces [1] or the sorts of ideas that are generated when brainstorming [2]. In the context of NeuroIS research, attentional processes have recently been investigated for the role they play in e-commerce decision making [3] its relationship with affective states in the context of user assistance systems [4], and has been identified as an area of interest among the NeuroIS community [5]. It is likely that attention will continue to be a relevant topic of interest to information systems in the future. However, despite the interest in attention, to the best of our knowledge there has not been comprehensive work describing the role of attention in IS research. Though there is a significant literature in cognitive neuroscience, key findings from

C. Conrad (✉) · A. Newman
Dalhousie University, Halifax, Canada
e-mail: colin.conrad@dal.ca

A. Newman
e-mail: aaron.newman@dal.ca

F. D. Davis et al. (eds.), *Information Systems and Neuroscience*,
Lecture Notes in Information Systems and Organisation 29,
https://doi.org/10.1007/978-3-030-01087-4_19

this field have not been influential on information systems research to date. In this paper we discuss the Attention Networks model developed by Michael Posner and colleagues [6], one of the dominant attention models in cognitive psychology and cognitive neuroscience. We then propose some ways that this model can inform and extend the understanding of attention in information systems research.

2 The Study of Attention in Cognitive Sciences

Attention is among the most enduring subjects of inquiry in psychology and neuroscience. William James, one of the pioneers of psychology, investigated the phenomenology attention and identified it as a process to focus on "one out of what seem several simultaneously possible objects or trains of thought." [7]. Rather than a single mechanism as identified by James however, modern cognitive science identifies attention as a number of cognitive processes that work together to yield the attention phenomenon, and could even reflect different mechanisms for different domains (e.g. auditory, visual) [8]. Though there are different models of attention, we will focus on the well-established Posner attention networks model in the context of visual attention [9, 10].

2.1 *Attention Networks*

Attention networks describe the networks of neurons that govern the functions of attention. The original Posner attention model was imagined based on cognitive functions observed by psychologists in the 1970s and 1980s. These accounts distinguish three fundamental functions that are essential to the experience of attention: alerting, orienting and executive control. *Alerting* describes the function of maintaining a high degree of sensitivity to stimuli and is often distinguished from general arousal. *Orienting* describes the process of aligning with the source of sensory signals. *Executive control* describes the resolution of conflict among stimuli, including selecting some stimuli for attentional focus while inhibiting responses to other stimuli. Though each of these functions were envisioned based on research in cognitive psychology, they form the foundation for many ongoing research programs in neuroimaging and are foundational to much of the applied work on attention in clinical applications.

In the original attention networks model, the alerting network was originally identified by observing sustained vigilance in behavioral studies and was later correlated with brainstem activity and networks in the right hemisphere [8]. Knowledge of the alerting functions have significantly expanded since the publication of the original model but have largely corroborated alerting as a distinct network [10]. For instance, the effects of neuromodulator norepinephrine have been studied in monkeys and were observed influencing orienting functions, but not alerting, which supports this distinction [11]. However, in most real-world scenarios, the alerting function is observed in conjunction with orienting, leading some to question the independence of the networks [12]. Nevertheless, alerting is still commonly studied as a distinct phenomenon.

Orienting was originally distinguished by Posner in his works on attentional shifts [8]. In its original conception as a network, orienting functions were observed in association with the pulvinar and superior colliculus. However, more recent work suggests that orienting is more complex and involves multiple brain areas including the dorsal system [10, 13, 14]. Orienting continues to be a subject of considerable interest among cognitive neuroscientists not least because it governs the fundamental mechanism of feature selection, the process of recognizing visual patterns or relevant visual stimuli. Orienting is often further divided into *overt* and *covert* orienting, which rely on different observations. Where overt orienting is typically associated with eye movements or other overt behaviour in the direction of the attended stimulus/location, covert orienting does not necessarily evoke eye movements or other motor activity towards the attended stimulus/location but nonetheless engages similar neural networks [15–17]. Recent work on orienting networks have continued to explore this overt/covert distinction and its implications for attention networks research [18, 19].

Executive control is a function that was originally conceptualized to describe target detection and explain the limited capacity of attention. Models have found this function to be associated with connections between the medial frontal and anterior cingulate cortex. Recent understandings of executive control have expanded on this original conception. The original conception of executive control considered it to associate with focal attention. Recent theories suggest two separate executive control networks, as evidenced by neuroimaging studies which reveal distinct frontoparietal and cingulo-opercular networks [10, 20]. Other conceptions of executive control identify it with the same network as working memory or as a component of working memory [21, 22] or recognize it as many distinct networks for different domains (i.e. visual, auditory) [23]. Though the extension of executive control networks continues to be a live topic of inquiry, the original conception of executive control continues to play a significant role in attention research today. Though the Posner three systems model has been arguably the most historically influential model, there is significant ongoing work in attention networks to move beyond this model, especially in the space of the executive brain. Contemporary models have introduced other networks that have been observed since and have incorporated them into an extended attention networks model [24].

2.2 *Measuring Attention Networks Using the Attention Networks Test*

Attention networks performance are often measured through neurocognitive tests that are designed to separably tap each of the three independent networks. The Attention

Network Test (ANT) is the most prominent example of such a test [25, 26]. The ANT measures the three attention networks through a combination of flanker tests, which are tasks designed to test response inhibition, and reaction time from cuing tasks, which were designed to measure attentional shifts [27]. In the ANT, the participants' task is to respond as quickly and accurately as possible indicating the direction of an arrow (left or right). The attention networks are differentially engaged by, on different trials, preceding the target with either spatially informative (orienting) or uninformative (alerting) cues, and by arrows flanking the target that are either congruent (same direction) or incongruent (opposite direction) stimuli flanking the target when it appears (executive control). Differences in reaction time can be used to measure the efficiency of alerting and orienting, while executive control is examined by measuring successful responses to the cues.

Though the ANT is the dominant test to advance the study of attention and population research, it has limitations. In response, researchers have investigated expanded measures to explain functioning of attention networks. Studies of the ANT have found weaker associations between alerting and orienting network scores and other attention measures such as those used in the Dalhousie Computerized Attention Battery (DalCAB) [24]. DalCAB is an example of a new attention battery, which uses eight reaction time tests to improve on the ANT by introducing additional measures such as vigilance [24]. As research in this field continues to advance, NeuroIS can benefit by observing the advances in neurocognitive tests and adapt them to IS contexts.

2.3 Measuring Attention Networks Using EEG and MEG

Much of attention networks theory has been validated using neuroimaging, notably electroencephalography (EEG), magnetoencephalography (MEG) and functional magnetic resonance imaging (fMRI). Many studies observe correlations between attention task performance, such as the ANT, and the neurophysiological indicators of attention [12, 28]. While fMRI research and attention networks is an active area of inquiry and has been demonstrated in the context of attention networks and the ANT [12], there are also common EEG (and consequently MEG) correlates that are observed, particularly with event related potentials, which are short changes in electrical potential on the scalp triggered by neural activity. We introduce two EEG correlates because EEG has been identified as an accessible technology to IS researchers and is applicable to many IS contexts [29], while noting that considerable work has been done on identifying Attention Networks using fMRI and other neuroimaging tools.

The P1-N1-P2 complex is a mandatory response triggered by early attention control mechanisms in the occipital regions of the brain and is sensitive to both visual and auditory stimuli [30]. When a stimulus is detected by the auditory or visual system, this pattern of electrical potentials can be observed at 100–220 ms. Attended stimuli can be observed having higher electrical amplitudes from this response. Early negative electrical potential responses have been found to be associated with alerting and have be observed during the Attention Networks Test [10]. The P1-N1-P2 complex is thus a useful neurophysiological response that can be used to observe alerting and orienting networks in the context of human-computer interactions.

A second EEG component that is often studied in attention research is the P3 component. The P3 response occurs immediately following the P1-N1-P2 response, typically between 250 and 500 ms, but only in response to task-related, attended stimuli. The P3 is known to be driven by the activation of executive attention and contextual updating in working memory [28, 31]. In the context of the ANT, the P3 is evoked during the cuing and can be observed having lower amplitudes depending on attention capacities [28]. Study of the P3 response can thus also be a useful neurophysiological indicator to observe executive functions or dysfunction in IS contexts.

3 Improving IS Measures with Attention Networks

Though there are many potential applications of this research [29], perhaps the most promising contribution of attention networks to the information systems field is in the improvement of IS measures. As mentioned, though alerting, orienting and executive control networks can be identified as separate phenomena, they are often examined in conjunction. A number of research topics in information systems such as awareness displays [1], visual search in web/e-commerce [32–34], electronic brainstorming [35], and online wait times [36, 37] have examined topics where the alerting and orienting networks may play a role in the phenomena observed. The methods used in these works included construct questionnaire measures [32, 37], comprehension measures [33] or task success measures [34–37]. These represent constructs that could be examined using neuroimaging to determine the impact of attention networks on the tasks, particularly by observing EEG event related potentials such as the P1-N1-P2 response or the P3. By doing so, we can improve the attention-related IS constructs perhaps most noticeably by adding *specificity* and *temporality* to the measures.

Though we are not aware of any extant work in the information systems literature that leverages the neuroscience of alerting or orienting, some work considers the role of arousal, which has long been noted for having common psychophysiological correlates [8]. Electrodermal activity and EEG oscillatory activity has been employed to observe changes in users' cognitive states and to observe flow, which may have some similarities [38]. Considering orienting networks, notable recent work has been conducted by Léger et al., which [39] established the P3 ERP and eye fixation-related potentials as significant measures in information systems research. These methods reflect the state of the art in overt orienting research [40] and open a new area of inquiry for the field with applications to is research with a visual component. Covert orienting, by contrast, remains a potential topic of interest for IS research involving this type of attention without a visual component. Such questions might benefit by leveraging covert orienting measures such as auditory event related potentials [41].

We anticipate that this line of reasoning presents a larger research project on the topic of attention which has the potential to advance IS and human-computer interaction research. In this paper, we discussed the neuroscience of attention networks, an important concept in the neuroscience of attention that, to our knowledge, has not been addressed in the information systems or NeuroIS literature. We also propose a potential application where attention networks can advance research in information systems. However, we conclude that there are also many other potential applications of the attention literature that remain to be seen. For instance, attention networks could inform the creation of new IT artifacts or could inform the creation of brain-computer interfaces [29]. We anticipate that deeper understandings of attention will not only help advance the field but offers the potential to raise entirely new domains of inquiry into the interaction between humans and information technology.

Acknowledgements This research is supported by the Killam and NSERC Doctoral scholarships to Colin Conrad and an NSERC Discovery Grant to Aaron Newman. We also thank the participants of the 2017 NeuroIS training course for their feedback on these ideas.

References

1. Dabbish, L., Kraut, R.: Awareness displays and social motivation for coordinating communication. Inf. Syst. Res. **19**(2), 221–238 (2008)
2. Potter, R.E., Balthazard, P.: The role of individual memory and attention processes during electronic brainstorming. MIS Q. **28**(4), 621–643 (2004)
3. Shamy, N., Hassanein, K.: The impact of age and cognitive style on e-commerce decisions: the role of cognitive bias susceptibility. In: Davis F., Riedl, R., vom Brocke, J., Léger PM., Randolph, A. (eds.) Information Systems and Neuroscience. Lecture Notes in Information Systems and Organisation, vol. 25, pp. 73–83 (2018)
4. Friemel, C., Morana, S., Pfeiffer, J., Maedche, A.: On the role of users' cognitive-affective states for user assistance invocation. In: Davis F., Riedl, R., vom Brocke, J., Léger PM., Randolph, A. (eds.) Information Systems and Neuroscience. Lecture Notes in Information Systems and Organisation, vol. 25, pp. 37–46 (2018)
5. Riedl, R., Davis, F.D., Banker, R., Kenning, P.: Neuroscience in information systems research. In: Lecture Notes in Information Systems and Organisation, vol. 21. (2018)
6. Posner, M.I., Petersen, S.E.: The attention system of the human brain. Annu. Rev. Neurosci. **13**, 25–42 (1990)
7. James, W.: The Principles of Psychology. Henry Holt and Company, New York (1890)
8. Montemayor, C., Haladjian, H.H.: Consciousness, Attention and Conscious Attention. MIT Press, Cambridge (2015)
9. Posner, M.I., Rothbart, M.K.: Research on attention networks as a model for the integration of psychological science. Annu. Rev. Psychol. **58**, 1–23 (2007)
10. Petersen, S.E., Posner, M.I.: The attention system of the human brain: 20 years after. Annu. Rev. Neurosci. **35**, 73–89 (2012)
11. Beane, M., Marrocco, R.T.: Norepinephrine and acetylcholine mediation of the components of reflexive attention: implications for attention deficit disorders. Prog. Neurobiol. **74**(3), 167–181 (2004)
12. Fan, J., Gu, X., Guise, K.G., Liu, X., Fossella, J., Wang, H., Posner, M.I.: Testing the behavioral interaction and integration of attentional networks. Brain Cogn. **70**(2), 209–220 (2009)
13. Corbetta, M., Shulman, G.L.: Control of goal-directed and stimulus-driven attention in the brain. Nat. Rev. Neurosci. **3**(3), 201–205 (2002)

14. Thompson, K.G., Biscoe, K.L., Sato, T.R.: Neuronal basis of covert spatial attention in the frontal eye field. J. Neurosci. **25**(41), 9479–9487 (2005)
15. Klein, R., Kingstone, A., Pontefract, A. Orienting of visual attention. In: Eye Movements and Visual Cognition, pp. 46–65. Springer Series in Neuropsychology. Springer, New York, NY (1992)
16. Corbetta, M., Akbudak, E., Conturo, T.E., Snyder, A.Z., Ollinger, J.M., Drury, H.A., Linenweber, M.R., Peterse, L.S., Raichle, M.E., van Essen, D.C., Shulman, G.L.: A common network of functional areas for attention and eye movements. Neuron **21**(4), 761–773 (1998)
17. Smith, D.T., Schenk, T.: The premotor theory of attention: time to move on? Neuropsychologia **50**(6), 1104–1114 (2012)
18. Harrison, W.J., Mattingley, J.B., Remington, R.W.: Eye movement targets are released from visual crowding. J. Neurosci. **33**(7), 2927–2933 (2013)
19. Armstrong, T., Olatunji, B.O.: Eye tracking of attention in the affective disorders: a meta-analytic review and synthesis. Clin. Psychol. Rev. **32**(8), 704–723 (2012)
20. Dosenbach, N.U., Fair, D.A., Cohen, A.L., Schlaggar, B.L., Petersen, S.E.: A dual-networks architecture of top-down control. Trends Cogn. Sci. **12**(3), 99–105 (2008)
21. Engle, R.W.: Working memory capacity as executive attention. Curr. Dir. Psychol. Sci. **11**(1), 19–23 (2002)
22. McCabe, D.P., Roediger III, H.L., McDaniel, M.A., Balota, D.A., Hambrick, D.Z.: The relationship between working memory capacity and executive functioning: evidence for a common executive attention construct. Neuropsychology **24**(2), 222–243 (2010)
23. Luck, S.J., Vogel, E.K.: Visual working memory capacity: from psychophysics and neurobiology to individual differences. Trends Cogn. Sci. **17**(8), 391–400 (2013)
24. Jones, S.A., Butler, B., Kintzel, F., Salmon, J.P., Klein, R.M., Eskes, G.A.: Measuring the components of attention using the Dalhousie Computerized Attention Battery (DalCAB). Psychol. Assess. **27**(4), 1286–1300 (2015)
25. Fan, J., McCandliss, B.D., Sommer, T., Raz, A., Posner, M.I.: Testing the efficiency and independence of attentional networks. J. Cogn. Neurosci. **14**(3), 340–347 (2002)
26. Fan, J., McCandliss, B.D., Fossella, J., Flombaum, J.I., Posner, M.I.: The activation of attentional networks. Neuroimage **26**(2), 471–479 (2005)
27. Posner, M.I.: Orienting of attention. Q. J. Exp. Psychol. **32**(1), 3–25 (1980)
28. Petley, L., Bardouille, T., Chiasson, D., Froese, P., Patterson, S., Newman, A., Omisade, A., Beyea, S.: Attentional dysfunction and recovery in concussion: effects on the P300 m and contingent magnetic variation. Brain Inj. **32**, 1–10 (2018)
29. Riedl, R., Léger, P.M.: Fundamentals of NeuroIS. Springer, Berlin, Heidelberg (2016)
30. Hillyard, S.A., Vogel, E.K., Luck, S.J.: Sensory gain control (amplification) as a mechanism of selective attention: electrophysiological and neuroimaging evidence. Philos. Trans. Royal Soc. London B: Biol. Sci. **353**(1373), 1257–1270 (1998)
31. Polich, J.: Updating P300: an integrative theory of P3a and P3b. Clin. Neurophysiol. **118**(10), 2128–2148 (2007)
32. Hong, W., Thong, J.Y., Tam, K.Y.: Does animation attract online users' attention? The effects of flash on information search performance and perceptions. Inf. Syst. Res. **15**(1), 60–86 (2004)
33. Tam, K.Y., Ho, S.Y.: Understanding the impact of web personalization on user information processing and decision outcomes. MIS Q. **30**(4), 865–890 (2006)
34. Tan, B.C., Yi, C., Chan, H.C.: Beyond annoyance: the latent benefits of distracting website features. ICIS 2008 Proc., 188 (2008)
35. Potter, R.E., Balthazard, P.: The role of individual memory and attention processes during electronic brainstorming. MIS Q. **28**(4), 621–643 (2004)
36. Hong, W., Hess, T.J., Hardin, A.: When filling the wait makes it feel longer: a paradigm shift perspective for managing online delay. MIS Q. **37**(2), 621–643 (2013)
37. Lee, Y., Chen, A.N., Ilie, V.: Can online wait be managed? The effect of filler interfaces and presentation modes on perceived waiting time online. MIS Q. **36**(2), 365–394 (2012)
38. Léger, P.M., Davis, F.D., Cronan, T.P., Perret, J.: Neurophysiological correlates of cognitive absorption in an enactive training context. Comput. Hum. Behav. **34**, 273–283 (2014)

39. Léger, P.M., Sénecal, S., Courtemanche, F., de Guinea, A.O., Titah, R., Fredette, M., Labonte-LeMoyne, É.: Precision is in the eye of the beholder: application of eye fixation-related potentials to information systems research. J. Assoc. Inf. Syst. **15**(10), 651–678 (2014)
40. Geva, R., Zivan, M., Warsha, A., Olchik, D.: Alerting, orienting or executive attention networks: differential patters of pupil dilations. Frontiers Behav. Neurosci. **7**, 145 (2013)
41. Hillyard, S.A., Hink, R.F., Schwent, V.L., Picton, T.W.: Electrical signs of selective attention in the human brain. Science **182**(4108), 177–180 (1973)

A Domains Oriented Framework of Recent Machine Learning Applications in Mobile Mental Health

Max-Marcel Theilig, Kim Janine Blankenhagel and Rüdiger Zarnekow

Abstract This research illustrates how the interdisciplinary integration of mobile health (mHealth) and Machine Learning (ML) can contribute to implementing mobile care for mental health. 94 articles were identified in a literature review to derive functional domains and composing information items improving the comprehension of ML benefits with mHealth integration. Identified items of each domain were pooled into clusters and information flow was quantified according to prevailing occurrence of included articles. We derive a comprehensive domains oriented framework (DF) and visualize an information flow graph. The DF indicates that the utilization of ML is well established (e.g. stress detection, activity recognition). Because deployment and data acquisition currently relies heavily on mobile phones, only 65% of current applications make fully integrated use of data sources to assert patient's mental state. Big data integration and a lack of commercially available devices to measure physiological or psychological parameters represent current bottlenecks to leverage synergies.

Keywords Machine learning · Application · Mobile health · Mental health
Framework

1 Introduction

Rates of diagnoses for mental and behavioral disorders have increased substantially worldwide resulting in an expanding demand for treatments and interventions [1].

M.-M. Theilig (✉) · K. J. Blankenhagel · R. Zarnekow
Technical University Berlin, Chair of Information and Communication Management, Berlin, Germany
e-mail: m.theilig@tu-berlin.de

K. J. Blankenhagel
e-mail: k.blankenhagel@tu-berlin.de

R. Zarnekow
e-mail: ruediger.zarnekow@tu-berlin.de

F. D. Davis et al. (eds.), *Information Systems and Neuroscience*,
Lecture Notes in Information Systems and Organisation 29,
https://doi.org/10.1007/978-3-030-01087-4_20

Traditionally, on-site monitoring devices, such as electroencephalogram (EEG) monitor, and medical appointments have been used to sense and store medical data relevant for mental health. Additionally, assessment by a medical professional or psychiatrist is intensive human labor.

The major strength of Machine Learning (ML) is to generalize beyond the examples given from an observation [2]. That seems to be a good fit for medical applications due to the fact that no matter how many medical resources there are, it is very unlikely that one will see those exact medical symptoms, with the same severity, under the same circumstances again at time of medical examination. New technologies such as wearable computing and the prominent development of sensing devices have facilitated the process of collecting attributes related to the individuals. At the same time, the amount of healthcare data gets bigger and much more difficult to handle and to process, because the number of combinations of attribute values and activities can be very large. The raw data provided by sensors are often useless and therefore systems make use of ML tools, which enable to recognize activities and to build patterns to describe, analyze and predict data [3].

In the scope of this paper, we will have the high-level view of mental disorders as abstract descriptions, which makes ML the perfect diagnosis tool for the aspects of mental disorders and is shown by an increase in computational approaches [4]. In addition, with the pervasive use of connected technologies (above all the internet), wearable medical sensors (WMS) and wireless body networks (WBN) emerged in the last decade [5]. mHealth accordingly no longer only relates to telemedicine but manifests itself in WBN, WMS and mobile phones. Still, using data to predict mood or depression, is an active research field that at the moment largely relies on self-reported surveys (e.g. PHQ-9 responses) in order to obtain and assess the underlying mental state [6].

Hence, introducing ML algorithms provides immense potential to leverage synergies of those three fields and by that to increase a patient's autonomy and safety through mobile mental health applications while improving quality of diagnosis, treatment and enhanced clinical trials as well as reducing medical errors and costs. For this purpose, the following contribution derives a domains oriented framework (DF) in order to get a better understanding of information items and obstacles for interdisciplinary integration of mHealth and ML within the increasingly important field of mental health.

2 Methodology

This paper follows the idea of the Theoretical Domains Framework [7], derives a framework for implementation research and provides a theoretical lens through which one can view the applications and their recent influence in mental health [8]. First off, keyword combinations (see Table 1) suitable to find recent application cases and derive a representative and recent (initially Jan 1st, 2012 or newer) overview were developed. We decided to search for recent cases that focused on the description of a

Table 1 Keyword combinations

Mental health keyword			and	mHealth application keyword	and	ML keyword
General	or	Specific				
Mental health, disorder, psychology	Depression, anxiety, attention-deficit, hyperactivity, oppositional defiant, conduct, disruptive behavior, behavior			Smart phone, cell phone, android, iPhone, app, sensor, wearable, mobile and [device, phone, technology, sensor]	Machine learning, decision support, pattern recognition, supervised, classification, unsupervised	

conducted study or an application case while not being an overview of a set of applications, because we wanted to visualize a quantified information flow afterwards (cf. [9–11]). The underlying set to the DF after screening was supposed to consist of literature that met the following inclusion criteria: (1) The article implies that the case provides diagnosis, treatment or support for (2) mental health as its main goal. (3) The article implies that the application case uses ML for handling or providing information. (4) The data was acquired by a system developed for mobile native use and compatible with smartphones, wearables (e.g. smartwatches) or sensors somehow gathering or requesting subject's individual input data. A subject may not only be a patient but also includes users seeking for support in stress management or for the prevention of mental health issues. Application cases using ML for comorbidities (e.g., depressed people with heart diseases) were excluded because different care needs or outputs dilute the model to be built. An analogous situation was identified for Internet of Things devices, where i.e. webcams are not classified as mobile because they belong to a specific stationary environment even if that environment is deployed at home.

3 Results

The identified applications and studies represent a functional layered hierarchical DF (Fig. 1), as this illustrates a number of processes, issues, and algorithms currently being treated from a research perspective. Due to the strong ability of ML to generalize [2], direct transformation of electric signals into mental health features are possible as well as taking input parameters directly to be processed by ML without knowing what mental feature they represent. The generalization aspect is visualized

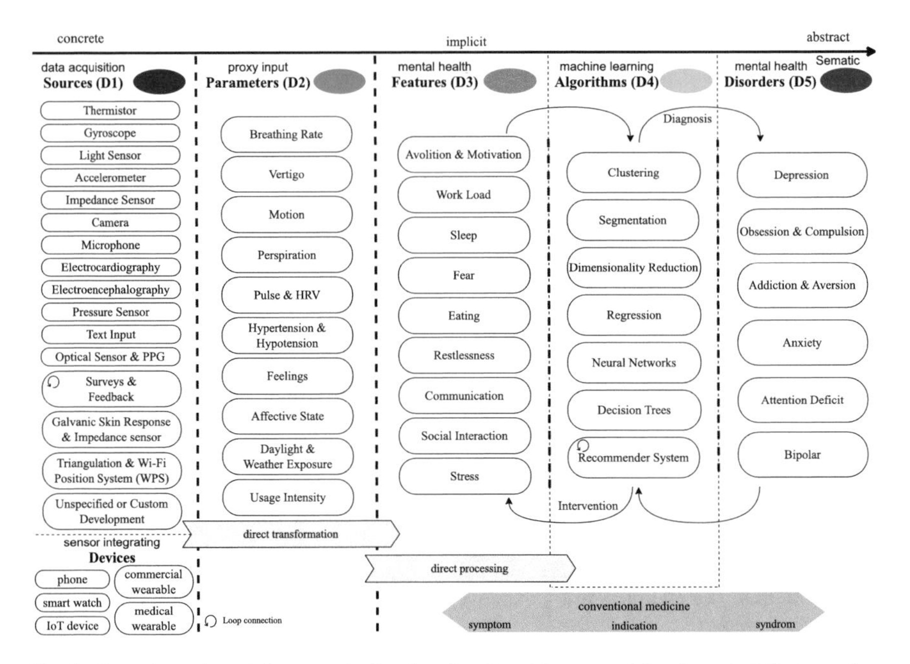

Fig. 1 Domains oriented framework showing functional layers and its characteristic items in a hierarchy from concrete to abstract mental disorder

by the decreasing number of items in each domain when going from left to right. Going through the characteristics, we will describe the clusters of items that are necessary to look at the information flow in a quantitative way in Fig. 2. The domains are characterized as follows:

Sources D1. Data sources found in application cases vary the most and represent different format, mean of mobility, and granularity. While per search strategy all devices are wearables, they can be separated into commercially available *Wearables* and *Medical* grade devices. *Medical* grade devices in our definition are used under medical supervision and not accessible for patients outside a hospital environment (i.e. [12]). Driven by the trend of the quantified-self ubiquitous sensors are commercially available. The last cluster for the general *Input* consists of diary studies and surveys like the Perceived Stress Questionnaire [13] or the Depression Anxiety Stress Scale [14, 15].

Parameter D2. The biggest group describes the *Activity* of the patient most commonly also referring to vital parameters or its current motion (e.g. sitting or walking). The aspect of heart rate is very central here and might get directly processed by ML [16]. With the help of galvanic skin response, skin temperature or electrocardiography (EEG) feelings and states of affection are described [17] representing the cluster of *Emotion* of D2. The *Environment* itself on the other hand influences the mental health as a stressor [18, 19]. The *Interaction* cluster represents to what intensity a

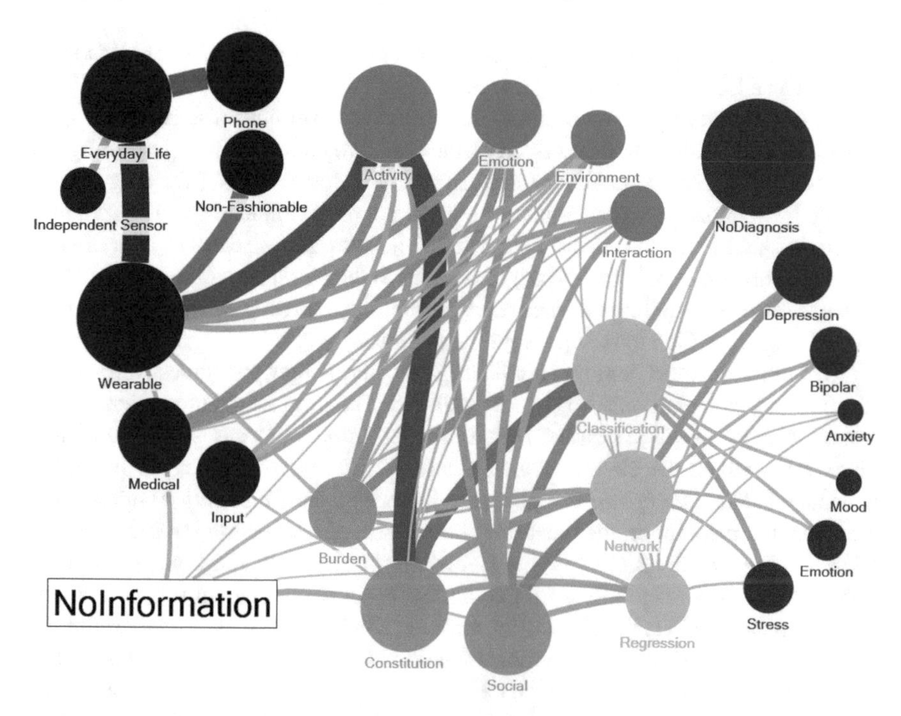

Fig. 2 Application cases flow graph. Clusters were pooled from the DF items. Linearity applies to vertices as well as edges (i.e. double the sizes follows double the observations)

patient interacts with an object or situation. Main items are addiction and usage [20, 21].

Features D3. Mental health features compose disorders and might be seen as parts of the diagnosis for an actual medical mental disorder (symptom). i.e., motion from D2 is mainly followed by a general description of the physical *Constitution* of patient. The *Burden* of a patient aggregates features like restlessness, fear or bad sleep [22]. *Social* aspects are the characteristic feature for mental diseases, which we decided on since depression is associated with social isolation [23, 24].

Algorithms D4. As expected included articles display a variety of ML algorithms [2] (SVM, NB, GMM, J48 etc.). We decided to cluster them based on their approach and intention into *Regression* (i.e. predicting [25]), *Classification* [26] and *Network* approaches like Artificial Neural Networks or Decision Trees, that are widely used for descriptive reasons when input is of high-dimensionality [27, 28]. We found that the input for the ML algorithms is widely spread across D1 to D3, which is explained by the generalization ability of ML approaches.

Disorders D5. The disorders domain is dominated by the prevalent diseases of mental health [1]. Most frequently mentioned were *Depression*, *Anxiety* and *Bipolar* disorder (e.g. [29–31]). However, we found that in most cases there is no output of an actual mental health disorder (53 cases) rather than diagnosis of *Stress*, *Mood*, or *Emotion* as a kind of sickness [16, 26, 32].

Information flow. For a substantial discussion and overview of research activities as well as gaps regarding the derived framework, we created a flow graph (Fig. 2). For descriptive reasons, not every item from every domain of the DF (Fig. 1) was included, but items were pooled in clusters. Size depicts simple observation count of an article classified in the according cluster by its item. As one can see, a lot of cases were categorized, except for the edges leading to the *NoInformation*-Vertex. Those edges are also representing the *direct transformation* or direct processing aspect from the DF (Fig. 1), which make up for 12 cases only.

To identify the most common application one may follow the biggest vertices and strongest edges from domain D1 to domain D5. The most frequent application is a commercial *Wearable* (51 cases) used to model *Activity* (41 cases; 25 from *Wearable*), from which the *Constitution* (34 cases; 25 from *Activity*) of the individual is deduced. The *Classification* (23 cases; 17 cases from *Constitution*) approach is implemented resulting in the most found diagnosis of *Depression* (16 cases; 6 from *Classification*) or Stress (11 cases; 6 from *Classification*). In addition, selecting the right currently commercially available hardware for mental health seems problematic: In 18 cases of 51, the chosen *Wearable* is a bulky or at least *Non-Fashionable* device, whereas fashionable wearables would have been beneficial (see [24]). This is leading to the conclusion that the desired data of handy or fashionable sensors is either not measured, not detailed enough or not directly (only via cloud) accessible. The most important example on this issue are ECG and EEG measurements, which are very important for diagnosis of mental disorders. Sensors are available only in a *Non-Fashionable* mean and commercially available devices are not really at a consumer level of usability. Mentionable *Non-Fashionable* devices from the included literature are the Shimmer 3, ActiGraph GT3X+, Zephyr BioPatch™, Empatica E3 wristband, Q-Sensor Affectiva and EMOTIVE EPOC+ Headset. The remaining 36 devices suitable for *Everyday Life* consist of 27 cases where solely *Phones* were used and 9 *Independent Sensors*.

Success rate of domain transition. For further verification and reasons of logical conclusion we bring together the content of Fig. 1 plus Fig. 2 to follow a potential application success rate in Fig. 3. Therefore, we calculated the share of applications that were successfully grouped into that category for every domain:

$$P_{\mathrm{Dn}-1\rightarrow Dn} = \frac{\text{successfully categorized in } Dn-1}{\text{successfully categorized in } Dn} \quad \forall n \in \{2, 3, 4, 5\} \tag{1}$$

For D1 an extra condition applies, that the data source is not part of a mobile phone, since this paper originally was intended to include true wearable sensors. Direct transformation and direct processing have been calculated analogous. The interpretation of Fig. 3 is, that the more abstract the starting domain for an application is, the higher is the probability that the next domain can be clearly assigned. For example, if one wants to develop an application and starts with the raw skin conductance data there is an 84% chance ($P_{D1\rightarrow D2}$) he is able to relate this data to an affective

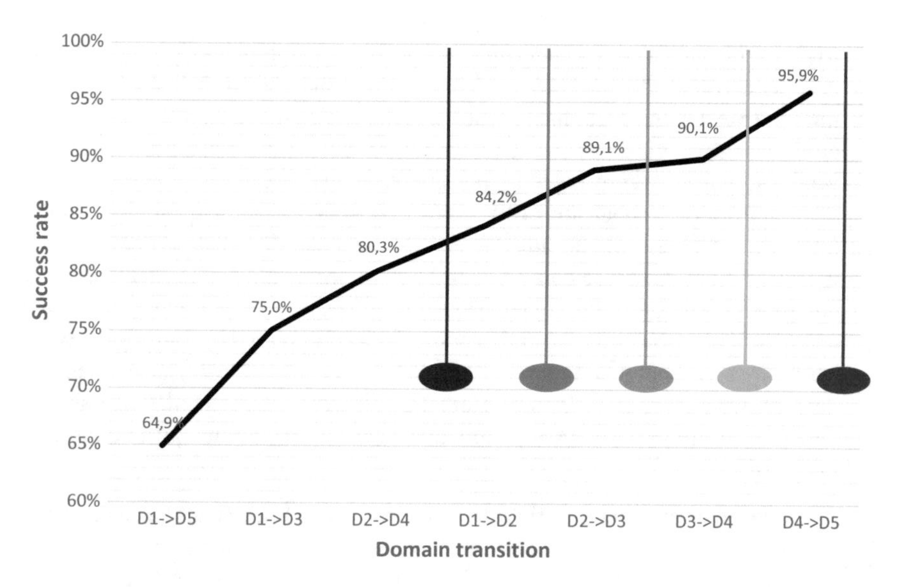

Fig. 3 Success rate of domain transition

state successfully. If this developer would start with information about a parameter like the affective state, the chance to relate this information to stress would increase to 89% ($P_{D2\rightarrow D3}$). Implementing the *direct transformation* from skin conductance to stress has a success rate of only 75% ($P_{D1\rightarrow D3}$), as our data indicates. This finding is congruent to the assumption that more concrete values (i.e. from D1) are harder to relate to mental health disorders, since there are more steps in between, that one has to derive (i.e. stress from D3). If someone is able to directly access feature from D3 relating those to mental health disorders is significantly easier ($P_{D3\rightarrow D5} = 86\%$). To get an overall probability to successfully explain a mental health related state of a patient one may simply multiply the probabilities of all domains:

$$P_{D1\rightarrow D5} = P_{D1\rightarrow D2} \times P_{D2\rightarrow D3} \times P_{D3\rightarrow D4} \times P_{D4\rightarrow D5} \tag{2}$$

According to the DF we built and the categorization process, this means: If one wants to develop a new application for mental health using a wearable sensor this has a 65% ($P_{D1\rightarrow D5}$) chance to result in a meaningful assertion about patient's mental state.

4 Discussion

While analyzing the included literature, we found reoccurring arguments that help explaining obstacles of the DF and its representation in the information flow graph:

1. Big data or multimodal sensor integration (like weather data [33] or pollution [34]) as an input for ML falls short in mobile mental health (see *Environment*, *Interaction*). On the other hand, when making use of big data, most cases focus on describing syndromes (i.e. *Classification*) rather than recommending interventions, which is represented by the smallest cluster of *Regression* in Fig. 2.
2. We see from the proportion of the disorders group in the flow graph of Fig. 2 that in most cases ML algorithms do not diagnose classical disorders rather than conditions. Of those being *Stress* the most occurring. Additionally it seems complex to develop a comprehensive application that uses physical sensors and derives an assertion of the mental health, which is supported by $P_{D1 \rightarrow D5} = 65\%$.
3. There are not enough commercially available (labeled *Everyday Life*) wearable devices, whereas mobile *Phones* are not a sufficient replacement. They are not able to measure physiological parameters, because they do not have continuous body contact.

The issues identified in this article are in consensus with recent research and challenges faced by mobile wearable systems [35]. ML employment was not yet able to overcome all technical and integrative challenges. Only 65% of current applications make fully integrated use of data sources and we found that the more semantically abstract information is, the easier it is to derive an assertion of a patient's mental health.

Limitations. This framework derivation has some limitations. We systematically scanned for ML approaches, whereas there is the possibility that an unneglectable set of application cases uses ML without mentioning it. The more modern eHealth makes use of ML, the more it becomes a background technology by default. In addition, the used keyword combinations may not cover the huge interdisciplinary area that was to investigate. Secondly, this review focuses only on published literature cases in ML in context and therefore does not consider the wide field of applications such as app stores. There are still means to be found to academically assess prototypes and software.

Implications. The derived DF for mobile application scenarios based ML shows the potential to remove barriers to care, adjust to users individually, and alleviate suffering for a large number of people with mental disorders at a modest cost and with minimal daily life intrusion. In summary, research efforts towards improving measurements quality (especially EEG and ECG), mobile and environmental data integration, process partitioning, deployment to smart watches or commercial wearables and extending the ML into the field of intervention recommendation are necessary. To relate information to a mental health assertion successfully, it is recommended to start with the least concrete described (raw) data set or information. Information science should address the topic of workload distribution to maximize utilization.

References

1. Polanczyk, G.V., Salum, G.A., Sugaya, L.S., Caye, A., Rohde, L.A.: Annual research review. A meta-analysis of the worldwide prevalence of mental disorders in children and adolescents. J. Child Psychol. Psychiatry **56**, 345–365 (2015)
2. Domingos, P.: A few useful things to know about machine learning. Commun. ACM **55**, 78 (2012)
3. Lara, O.D., Labrador, M.A.: A survey on human activity recognition using wearable sensors. IEEE Commun. Surv. Tutorials **15**, 1192–1209 (2013)
4. Torous, J., Baker, J.T.: Why psychiatry needs data science and data science needs psychiatry. Connecting with technology. JAMA Psychiatry **73**, 3–4 (2016)
5. Iqbal, M.H., Aydin, A., Brunckhorst, O., Dasgupta, P., Ahmed, K.: A review of wearable technology in medicine. J. R. Soc. Med. **109**, 372–380 (2016)
6. Saeb, S., Zhang, M., Karr, C.J., Schueller, S.M., Corden, M.E., Kording, K.P., Mohr, D.C.: Mobile phone sensor correlates of depressive symptom severity in daily-life behavior. An exploratory study. J. Med. Internet Res. **17**, e175 (2015)
7. Atkins, L., Francis, J., Islam, R., O'Connor, D., Patey, A., Ivers, N., Foy, R., Duncan, E.M., Colquhoun, H., Grimshaw, J.M., et al.: A guide to using the theoretical domains framework of behaviour change to investigate implementation problems. Implementation Sci. **12**, 1–18 (2017)
8. Michie, S., Johnston, M., Abraham, C., Lawton, R., Parker, D., Walker, A.: Making psychological theory useful for implementing evidence based practice. A consensus approach. BMJ Qual. Saf., **14**, 26–33 (2005)
9. Vom Brocke, J., Riedl, R., Léger, P.-M.: Application strategies for neuroscience in information systems design science research. J. Comput. Inf. Syst. **53**, 1–13 (2015)
10. Vom Brocke, J., Simons, A., Niehaves, B., Riemer, K., Plattfaut, R., Cleven, A.: Reconstructing the giant. On the importance of rigour in documenting the literature search process. In: ECIS 2009 Proc., **9**, 2206–2217 (2009)
11. Wohlin, C.: Guidelines for snowballing in systematic literature studies and a replication in software engineering. In: Shepperd, M., Hall, T. (eds.) EASE 2014, pp. 1–10. ACM (2014)
12. Setz, C., Arnrich, B., Schumm, J., La Marca, R., Tröster, G., Ehlert, U.: Discriminating stress from cognitive load using a wearable EDA device. IEEE Trans. Inf Technol. Biomed. **14**, 410–417 (2010)
13. Cohen, S., Kamarck, T., Mermelstein, R.: A global measure of perceived stress. J. Health Soc. Behav. **24**, 385–396 (1983)
14. Parkitny, L., McAule, J.: The depression anxiety stress scale (DASS). J. Physiotherapy **56**, 204 (2010)
15. Farhan, A.A., Lu, J., Bi, J., Russell, A., Wang, B., Bamis, A.: multi-view bi-clustering to identify smartphone sensing features indicative of depression. In: 2016 IEEE First International Conference on Connected Health: Applications, Systems and Engineering Technologies, pp. 264–273. IEEE, Piscataway, NJ (2016)
16. Sioni, R., Chittaro, L.: Stress detection using physiological sensors. Computer **48**, 26–33 (2015)
17. Gravina, R., Fortino, G.: Automatic methods for the detection of accelerative cardiac defense response. IEEE Transac. Affect. Comput. **7**, 286–298 (2016)
18. Howarth, E., Hoffman, M.S.: A multidimensional approach to the relationship between mood and weather. Br. J. Psychol. **75**(Pt 1), 15–23 (1984)
19. Sanders, J.L., Brizzolara, M.S.: Relationships between weather and mood. J. Gen. Psychol. **107**, 155–156 (1982)
20. LiKamWa, R., Liu, Y., Lane, N.D., Zhong, L.: MoodScope. In: Chu, H.-H. (ed.) MobiSys '13 Proceeding of the 11th International Conference on Mobile Systems, Applications, and Services, p. 389. ACM (2013)
21. Ahsan, G.M.T., Addo, I.D., Ahamed, S.I., Petereit, D., Kanekar, S., Burhansstipanov, L., Krebs, L.U.: Toward an mHealth intervention for smoking cessation. In: Proceedings of the Annual International Computer Software and Applications Conference. COMPSAC (2013)

22. Sano, A., Phillips, A.J., Yu, A.Z., Mchill, A., Taylor, S., Jaques, N., Czeisler, C., Klerman, E., Picard, R.: Recognizing academic performance, sleep quality, stress level, and mental health using personality traits, wearable sensors and mobile phones. In: International Conference on Wearable and Implantable Body Sensor Networks, pp. 1–6 (2015)
23. Sanders, C.E., Field, T.M., Diego, M., Kaplan, M.: The relationship of Internet use to depression and social isolation among adolescents. Adolescence **35**, 237–242 (2000)
24. Cacioppo, J.T., Hawkley, L.C., Thisted, R.A.: Perceived social isolation makes me sad. 5-year cross-lagged analyses of loneliness and depressive symptomatology in the Chicago health, aging, and social relations study. Psychol. Aging **25**, 453–463 (2010)
25. Valenza, G., Nardelli, M., Lanata', A., Gentili, C., Bertschy, G., Kosel, M., Scilingo, E.P.: Predicting mood changes in bipolar disorder through heartbeat nonlinear dynamics. IEEE J. Biomed. Health Inform. (2016)
26. Zhu, Z., Satizabal, H.F., Blanke, U., Perez-Uribe, A., Troster, G.: Naturalistic recognition of activities and mood using wearable electronics. IEEE Trans. Affect. Comput. **7**, 272–285 (2016)
27. Maxhuni, A., Hernandez-Leal, P., Sucar, E.L., Osmani, V., Morales, E.F., Mayora, O.: Stress modelling and prediction in presence of scarce data. J. Biomed. Inform. **63**, 344–356 (2016)
28. Faurholt-Jepsen, M., Busk, J., Frost, M., Vinberg, M., Christensen, E.M., Winther, O., Bardram, J.E., Kessing, L.V.: Voice analysis as an objective state marker in bipolar disorder. Transl. Psychiatry **6**, e856 (2016)
29. Frost, M., Doryab, A., Bardram, J.: Disease insights through analysis. Using machine learning to provide feedback in the MONARCA system. In: Czerwinski, M., Staff, I. (eds.) 7th International Conference on Pervasive Computing Technologies for Healthcare (PervasiveHealth 2013). ICST (2013)
30. Katsis, C.D., Katertsidis, N.S., Fotiadis, D.I.: An integrated system based on physiological signals for the assessment of affective states in patients with anxiety disorders. Biomed. Signal Process. Control **6**, 261–268 (2011)
31. Faedda, G.L., Ohashi, K., Hernandez, M., McGreenery, C.E., Grant, M.C., Baroni, A., Polcari, A., Teicher, M.H.: Actigraph measures discriminate pediatric bipolar disorder from attention-deficit/hyperactivity disorder and typically developing controls. J. Child Psychol. Psychiatry **57**, 706–716 (2016)
32. Grünerbl, A., Muaremi, A., Osmani, V., Bahle, G., Ohler, S., Tröster, G., Mayora, O., Haring, C., Lukowicz, P.: Smartphone-based recognition of states and state changes in bipolar disorder patients. IEEE J. Biomed. Health Inform. **19**, 140–148 (2015)
33. Bogomolov, A., Lepri, B., Ferron, M., Pianesi, F., Pentland, A.: Daily Stress Recognition from Mobile Phone Data, Weather Conditions and Individual Traits. In: Hua, K.A. (ed.) MM '14: Proceedings of the 22nd ACM international conference on Multimedia, pp. 477–486. ACM (2014)
34. Liu, H.-Y., Dunea, D., Oprea, M., Savu, T., Iordache, S.: Improving the protection of children against air pollution threats in Romania—the RokidAIR project approach and future perspectives. Nukleonika -Original Edition- **68**, 841–846 (2017)
35. Baig, M.M., GholamHosseini, H., Moqeem, A.A., Mirza, F., Lindé, M.: A systematic review of wearable patient monitoring systems—current challenges and opportunities for clinical adoption. J. Med. Syst. **41**, 115 (2017)

Microsaccades as a Predictor of a User's Level of Concentration

Ricardo Buettner, Hermann Baumgartl and Daniel Sauter

Abstract In comparison to voluntary eye movements (saccades), micro-saccades are very small, jerk-like and involuntary. While microsaccades and cognition has become one of the most rapidly growing areas of study in visual neuroscience [Trends Neurosci. 32: 463–475], microsaccades are still neglected in NeuroIS. Using experimental data by Walcher et al. [Conscious Cogn. 53:165–175; Data Brief 15:18–24] we demonstrate the potential of microsaccades to evaluate the level of concentration a user perceives during task fulfillment. As a result we found a substantial negative relationship between the magnitudes of the microsaccades and the level of concentration ($p < 0.01$).

Keywords NeuroIS · Eye-tracking · Microsaccades · Concentration

1 Introduction

Microsaccades are the very small and fast movements of a human's eyes including tremor, drift and correcting saccades when fixing on a visual target [1–3]. Microsaccades can be defined as "small, fast, jerk-like eye movements that occur during voluntary fixation" [3]. In medicine and psychology microsaccades play an important role in helping us understand visual attention, mental concentration and information processing [4–17]. From a physiological point of view microsaccades correct and stabilize visual attention [1] and optimize the sampling for visual scenes [4]. From an information processing perspective it has been found that microsaccades are linked to mental concentration related concepts such as attention [13], concentration [14] and memory load [15], workload [16], task difficulty [10, 11, 15, 17], and mental fatigue [12].

Despite the important role of microsaccades in mental concentration processes, NeuroIS related eye-tracking research only uses (regular) eye saccades [18], gaze fix-

R. Buettner (✉) · H. Baumgartl · D. Sauter
Aalen University, Aalen, Germany
e-mail: ricardo.buettner@hs-aalen.de

F. D. Davis et al. (eds.), *Information Systems and Neuroscience*,
Lecture Notes in Information Systems and Organisation 29,
https://doi.org/10.1007/978-3-030-01087-4_21

ations [19], and several pupil derivatives [20–25], while microsaccades are neglected [26].

That is why—using existing data for secondary analysis [27, 28]—this paper evaluates the potential of microsaccades to determine the level of concentration a user perceives during task fulfillment.

2 Methodology

2.1 Data

The data used for this analysis were hosted by Walcher et al. [27, 28] at the Open Science Framework (OSF) (https://osf.io/fh66g/).

2.2 Participants

Forty-eight healthy participants (19–27 years old; 32 females, 15 males, 1 NA; normal or corrected-to-normal vision), mostly university students, participated in the experiment conducted by Walcher et al. [27, 28].

2.3 Instruments and Devices

Binocular eye data was recorded using 500 Hz EyeLink 1000 Plus Tower Mount eye-tracker (SR Research) [27, 28]. Stimuli were presented on a 19-in LG flatroon L1920P monitor run a 60 Hz and a 1240 × 1024 pixels resolution [27, 28]. EyeLink Experiment Builder software (SR Research) was used for stimulus presentation and response recording [27, 28].

The level of concentration a user perceived during task fulfillment was measured using the six-point Likert scale by Walcher et al. [27, 28].

2.4 Stimuli and Procedure

Data comprises eye-tracking information from eight idea generation tasks (alternate uses task by Guilford [29]) and eight letter-by-letter reading tasks (by Walcher et al. [27]) under two background brightness conditions (RGB color codes: 204, 204, 204 and 102, 102, 102) [28]. All stimuli were presented as counterbalanced (Fig. 1).

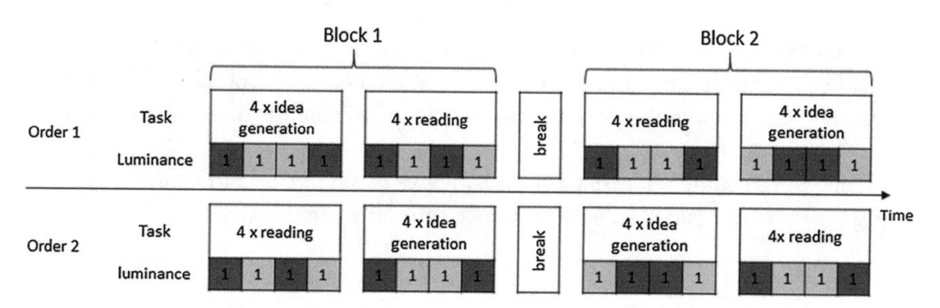

Fig. 1 Test procedure; from [28, p. 168]

3 Results

In both types of task (idea generation, and reading) we found a substantial negative relationship between the magnitudes of the microsaccades and the level of concentration (Table 1).

4 Discussion, Limitations and Future Research

Our results are in line with the results by Steinman et al. [14] who found lower microsaccadic magnitudes when the participants have to concentrate themselves (holding their eyes still).

The results demonstrate the potential of microsaccades for evaluating the level of a user's concentration. Using the microsaccades bio-data NeuroIS scholars can better understand a user's concentration as an Information Systems construct (cf. NeuroIS guideline 4 [30, 31] in general).

4.1 Limitations and Future Work

Since the concept of concentration is not clearly defined and it is not distinguished from to other concepts such as attention, flow [32] or boredom [33], a validity-

Table 1 Relationship between the magnitudes of the microsaccades and the level of concentration (spearman test)

Type of task	Level of correlation	Level of significance
Idea generation task by Guilford [29]	−0.32	$p < 0.01$
Letter-by-letter reading tasks by Walcher et al. [27]	−0.27	$p < 0.01$

related measurement weakness exists in our analysis. Future work could use scales measuring concentration related concepts. Another measurement related weakness is concerned with the sole use of a one-item scale to indicate the level of concentration [27, 28].

Future work should address these weaknesses using multi-item instruments measuring the level of concentration and concentration-related concepts.

References

1. Rolfs, M.: Microsaccades: small steps on a long way. Vis. Res. **49**, 2415–2441 (2009)
2. Riggs, L.A., Ratliff, F., Cornsweet, J.C., Cornsweet, T.N.: The disappearance of steadily fixated visual test objects. J. Opt. Soc. Am. **43**(6), 495–501 (1953)
3. Martinez-Conde, S., Macknik, S.L., Hubel, D.H.: The role of fixational eye movements in visual perception. Nat. Rev. Neurosci. **5**, 229–240 (2004)
4. Martinez-Conde, S., Macknik, S.L., Troncoso, X.G., Hubel, D.H.: Microsaccades: a neurophysiological analysis. Trends Neurosci. **32**(9), 463–475 (2009)
5. Martinez-Conde, S., Otero-Millan, J., Macknik, S.L.: The impact of microsaccades on vision: towards a unified theory of saccadic function. Nat. Rev. Neurosci. **14**, 83–96 (2013)
6. Otero-Millan, J., Troncoso, X.G., Macknik, S.L., Serrano-Pedraza, I., Martinez-Conde, S.: Saccades and microsaccades during visual fixation, exploration, and search: foundations for a common saccadic generator. J. Vis. **8**(14, 21), 1–18 (2008)
7. Donner, K., Hemilä, S.: Modelling the effect of microsaccades on retinal responses to stationary contrast patterns. Vis. Res. **47**, 1166–1177 (2007)
8. Ko, H.-K., Poletti, M., Rucci, M.: Microsaccades precisely relocate gaze in a high visual acuity task. Nat. Neurosci. **12**, 1549–1554 (2010)
9. Ahissar, E., Arieli, A., Fried, M., Bonneh, Y.: On the possible roles of microsaccades and drifts in visual perception. Vis. Res. **118**, 25–30 (2016)
10. Benedetto, S., Pedrotti, M., Bridgeman, B.: Microsaccades exploratory saccades in a naturalistic environment. J. Eye Mov. Res. **4**(2, 2), 1–10 (2011)
11. Chen, Y., Martinez-Conde, S., Macknik, S.L., Bereshpolova, Y., Swadlow, H.A., Alonso, J.-M.: Task difficulty modulates the activity of specific neuronal populations in primary visual cortex. Nat. Neurosci. **11**(8), 974–982 (2008)
12. Di Stasi, L.L., McCamy, M.B., Catena, A., Macknik, S.L., Cañas, J.J., Martinez-Conde, S.: Microsaccade and drift dynamics reflect mental fatigue. Eur. J. Neurosci. 1–10 (2013)
13. Pastukhov, A., Braun, J.: Rare but precious: microsaccades are highly informative about attentional allocation. Vis. Res. **50**, 1173–1184 (2010)
14. Steinman, R.M., Cunitz, R.J., Timberlake, G.T., Herman, M.: Voluntary control of microsaccades during maintained monocular fixation. Science **155**(3769), 1577–1579 (1967)
15. Siegenthaler, E., Costela, F.M., McCamy, M.B., Di Stasi, L.L., Otero-Millan, J., Sonderegger, A., Groner, R., Macknik, S., Martinez-Conde, S.: Task difficulty in mental arithmetic affects microsaccadic rates and magnitudes. Eur. J. Neurosci. **39**, 287–294 (2014)
16. Gao, X., Yan, H., Sun, H.-J.: Modulation of microsaccade rate by task difficulty revealed through between- and within-trial comparisons. J. Vis. **15**(3), 1–15 (2015)
17. Xue, L., Huang, D., Wang, T., Hu, Q., Chai, X., Li, L., Chen, Y.: Dynamic modulation of the perceptual load on microsaccades during a selective spatial attention task. Sci. R. **7**(16496), 1–13 (2017)
18. Buettner, R.: Cognitive workload of humans using artificial intelligence systems: towards objective measurement applying eye-tracking technology. In: KI 2013 Proc., Ser. LNAI, 8077, pp. 37–48 (2013)

19. Buettner, R.: Asking both the user's brain and its owner using subjective and objective psychophysiological NeuroIS instruments. In: ICIS 2017 Proceedings: 38th International Conference on Information Systems, Seoul, South Korea, 10–13 December 2017
20. Buettner, R., Daxenberger, B., Eckhardt, A., Maier, C.: Cognitive workload induced by information systems: introducing an objective way of measuring based on pupillary diameter responses. In: Pre-ICIS HCI/MIS 2013 Proc., paper 20 (2013)
21. Buettner, R.: The relationship between visual website complexity and a user's mental workload: a NeuroIS perspective. In: Information Systems and Neuro Science, vol. 10 of LNISO, pp. 107–113. Gmunden, Austria (2016)
22. Buettner, R.: Analyzing mental workload states on the basis of the pupillary hippus. In: NeuroIS'14 Proc., p. 52 (2014)
23. Buettner, R., Timm, I.J., Scheuermann, I.F., Koot, C., Rössle, M., Stationarity of a user's pupil size signal as a precondition of pupillary-based mental workload evaluation. In Information Systems and Neuro Science, vol. 25 of LNISO, pp. 195–200. Gmunden, Austria (2018)
24. Buettner, R., Sauer, S., Maier, C., Eckhardt, A.: Towards ex ante Prediction of User Performance: a novel NeuroIS methodology based on real-time measurement of mental effort. In: HICSS-48 Proc., pp. 533–542 (2015)
25. Buettner, R., Sauer, S., Maier, C., Eckhardt, A.: Real-time prediction of user performance based on pupillary assessment via eye-tracking. AIS Trans. Hum. Comput. Interact. **10**(1), 26–56 (2018)
26. Riedl, R., Fischer, T., Léger, P.M.: A decade of NeuroIS research: status quo, challenges, and future directions. In: ICIS 2017 Proceedings: 38th International Conference on Information Systems, Seoul, South Korea, 10–13 December 2017
27. Walcher, S., Körner, C., Benedek, M.: Data on eye behavior during idea generation and letter-by-letter reading. Data in Brief **15**, 18–24 (2017)
28. Walcher, S., Körner, C., Benedek, M.: Looking for ideas: eye behavior during goal-directed internally focused cognition. Conscious. Cogn. **53**, 165–175 (2017)
29. Guilford, J.: The Nature of Human Intelligence. McGraw-Hill, New York (1967)
30. vom Brocke, J., Liang, T.-P.: Guidelines for neuroscience studies in information systems research. JMIS **30**(4), 211–234 (2014)
31. Riedl, R., Banker, R.D., Benbasat, I., Davis, F.D., Dennis, A.R., Dimoka, A., Gefen, D., Gupta, A., Ischebeck, A., Kenning, P., Müller-Putz, G., Pavlou, P.A., Straub, D.W., Vom Brocke, J., Weber, B.: On the foundations of NeuroIS: reflections on the gmunden retreat 2009. CAIS **27**, 243–264 (2010)
32. Ghani, J.A., Deshpande, S.P.: Task characteristics and the experience of optimal flow in human-computer interaction. J. Psychol. **128**(4), 381–391 (1994)
33. Vodanovich, S.J., Watt, J.D.: Self-report measures of boredom: an updated review of the literature. J. Psychol. **150**(2), 196–228 (2014)

Tracking and Comparing Eye Movements Patterns While Watching Interactive and Non-interactive Videos

Ananda Rohit Daita, Bin Mai and Kamesh Namuduri

Abstract In this paper, we demonstrate how our eye moments differ when we are watching non-interactive videos (sports clips) versus interactive videos (video games clips). We obtained the eye tracking data from Collaborative Research in Computational Neuroscience's (CRCNS) data sharing set; we analyzed the subsets of eye movement data which were tracked while the test subjects were watching sports clips (videos whose contents are non-interactive) and video game clips (videos whose contents are interactive). We then compare the variations of both x- and y-coordinate eye movements between watching real videos and watching animated videos to identify the difference in eye movement patterns between the cases. Moreover, we also conducted tests on to see if there exists any difference in the distribution of the eye status of fixation or saccade between the cases. Our results provide insights into the cognitive processes when people are watching videos. We also discuss the implications of our results to the various applications in IS field.

Keywords Video watching · Fixation · Saccade · Eye tracking · Eye movements

1 Introduction

In recent years, two research streams have obtained increasing interests from researchers who want to gain insights into people's cognitive processes. One is the eye movement data acquisition and analysis, and the other is video game playing/video viewing analysis. Eye tracking has become a conventional technique for

A. R. Daita · K. Namuduri
University of North Texas, Denton, USA
e-mail: AnandaRohitDaita@my.unt.edu

K. Namuduri
e-mail: kamesh.namuduri@unt.edu

B. Mai (✉)
Texas A&M University, College Station, USA
e-mail: binmai@tamu.edu

F. D. Davis et al. (eds.), *Information Systems and Neuroscience*,
Lecture Notes in Information Systems and Organisation 29,
https://doi.org/10.1007/978-3-030-01087-4_22

studying cognition processing for visualization (e.g., [1–3]). This technique is applied in some research studies which involves reading sentences [4], comparing reading speed between low-frequency words and high-frequency words and missing letters [5], some real-time situations like how a batsman hit the ball [6], and social desirability bias [7]. The insights obtained from the eye tracking data analytics facilitate our understanding of human cognitive architecture, and eye tracking information can serve "as a window onto cognitive processes in dynamic visualization environments" [1]. On the video game playing/viewing stream, one crucial insight is from [8], which identifies the sense and practice of control during video game playing versus regular video viewing as an essential factor that impacts people's cognitive processes during the respective experiences. However, there are not sufficient literature that systematically integrates these two promising research streams. And there are much exciting research questions remained to be addressed. There were some studies which compared animation/dynamic pictures with static pictures (e.g., [9, 10]). The purpose of these surveys is straightforward; the authors want to investigate if there is any difference in learning between animation/dynamic picture presentation and static picture presentation. In neither of these experiments did they use any eye tracking data.

In this paper, we attempt to integrate the research in eye movement and video game playing/video viewing and address the question whether the eye movements differ during video game playing and regular video viewing. We compare the patterns of human's eye movement while watching video clips of sports events and video clips of video games. We hypothesize about the possible relationships between the type of the videos people are watching and their eye movement patterns and collect data to test the relationships. Our results provide insights into how people's eye movements differ when watching sports videos versus video game clips. Consequently, our results have significant implications to various IS applications, such as the design of IT artifacts, the design of learning technologies, and the design of online advertisement.

The rest of the paper proceeds as follows: in Sect. 2, we review relevant literature and formulate our hypotheses. In Sect. 3 we describe our data, followed by the data analyses and results in Sect. 4. And finally, we discuss the implications of our work and conclude in Sect. 5.

2 Literature and Hypothesis

2.1 Literature Survey About Eye Movement

There has been increasing interests among researchers in neuroscience and psychology to understand how human's eye movements can tell us about how our brain works [11, 12]. Besides experiments from a strict laboratory environment, researchers have also conducted experiments with natural stimuli. In [4] we can see how mind wandering effects one's eye movements. They found that brain wandering effects the speed of reading and also the ability to understand a sentence. Also, for persons who were into the task took less time to go through less frequent words somewhat compared to mind wandering persons. The same was also stated in [13] where the researchers had found less time for frequently used words when compared to rarely used words. They also observed fewer fixations before mind wandering. In [14] they found a contradicting result saying that mind wandering persons have taken less time for going through to difficult words than of the persons who are in focus. They argue that the persons who are not concentrating do not care to understand or give some extra time for understanding where the persons who are focused on task will try to spend more time on understanding. In [15] researchers found a difference in understanding a concept with both text and pictures among different age groups. More specifically, they have discovered that fourth-grade students faced more significant difficulty in interpreting the pictorial representation (in this case flower and its description). In addition, they found that they spent much less time on pictorial representation when compared to text, whereas the adults have spent a considerably equal amount of time looking at both text and pictures.

There is also some research to analyze eye tracking and differentiate between good batsmen in cricket compared to regular batsmen [6]. Concentrating on a right place with efficient timing is very important for any ball sports such as cricket or tennis. "Keep your eye on the ball" is the first advice given by the coaches when you are playing cricket or tennis. In cricket, batsmen can somehow anticipate the ball by some predictions based on the finger moment of bowler or pitch behavior. However, they found that a short latency in the first saccade in the batsmen eye movement differentiates good batsmen from standard batsmen. The critical factor is how fast you understand the line and length of the ball.

2.2 *Literature Survey About Video Game Playing/Regular Video Viewing*

Video gaming as a mechanism to understand the human cognitive processes has garnered significant interests from the researchers [16, 17]. In addition, eye movement data collection and analyses during video games have been implemented to obtain further insights on the player's cognitive processes [18, 19]. One common theme identified in these video games literature is the sense and practice of control for the players. In [8] the author pointed out that the interactive nature of video games usually results in the player's sense and practice of control over the contents and development of the game, and this is one major factor that differentiates peoples' cognitive processes while watching movies (i.e., non-interactive) versus playing video games (i.e., interactive). [8] pointed out that compared to regular videos, video games are transformed "into an interactive form that enables the player actively to participate in shaping the games". This interactivity afforded by video games is the most significant factor that distinguishes video games clips and regular videos such as sports clips.

Many other aspects of the two types of video clips remain quite similar. For example, while researchers have noticed that for many video games, user's eye movement focus more on the center areas of the screen (e.g., [20, 21]), researchers also notice that for regular video clips (e.g., [22]), including the sports clips (e.g., [23]), the viewers' eye movement also have a high degree of fixation with the center of the screen, presumably due to video making conventions placing the most relevant information in the center of the screen. Thus the sense and practice of control derived from the interactivity nature of video games should be the main factor that could potentially impact the eye movement patterns.

It is reasonable to believe that in the scenario of watching a video game clip, a person would also display the sense and practice of control, even though the person is not interactively playing the video games. We believe that this element of control due to the interactivity nature of the video game would play a role in the viewer's eye movement. In the existing literature, we are not aware of any research that specifically studies this possible variation between the eye movements when watching video clips of various levels of interactivity. We fill this void in this paper.

2.3 Hypothesis

We study two possible aspects the levels of video interactivity can impact the eye movement. The first aspect is the variation of the eye movement, represented by the x- and y-coordinates of the eye movement data. When watching video game clips, people will have a higher level of sense of control due to the interactivity nature, and as a result, their visual attention would be more focused, compared to when they watch non-interactive sports clips. Thus, their eye movement data should display various levels of variations between the two types of video watching. Therefore, we obtain the following Hypothesis:

H1: Variation of eye movement co-ordinates does not significantly differ between watching non-interactive videos (sports clips) and watching interactive videos (video games clips).

For this null hypothesis, we formulate our alternative hypothesis as: variation of eye movement co-ordinates is higher in the case of watching non-interactive videos (sports clips) than watching interactive videos (video games clips).

The second aspect is the distribution of eye status. Eye tracking data usually include the statuses of the eye. Those statuses can be categorized into 6 groups, which will be explained in the next section. We believe that when watching non-interactive videos (sports clips) versus watching interactive videos (video games clips), there should be a difference in the distribution of the eye status. Therefore, we obtain the following Hypothesis:

H2: The distribution of statuses of the eye is not significantly different between watching non-interactive videos (sports clips) and watching interactive videos (video games clips).

For this null hypothesis, our alternative hypothesis is that the distribution of eye statuses is affected by whether people are watching non-interactive or interactive videos.

3 Data Description

The data used herein is obtained from Collaborative Research in Computational Neuroscience's (CRCNS) data sharing set [24]. The data consists of not only eye tracking movements but also provide us with pictures and videos of how the eye coordinates move while someone is watching a video. For our research, we took sports and video games videos eye movement data files which include eye movement trace for each video clip of five sports videos and nine video games videos which were taken at a frequency of 240 Hz (i.e., 240 samples/s). They include the following information that we analyze in this paper about the eye movement:

- (x, y): instantaneous eye position coordinates while watching the screen.
- Pdiam: pupil diameter; it has defaulted to 0.
- Status:

0 Fixation excluding eye blinking.
1 In saccade excluding eye blinking.
2 fixation during an eye blink.
3 Saccade during an eye blink.
4 Smooth pursuit excluding eye blinking.
5 Unknown (e.g., loss of tracking).

More detailed descriptions of the data can be found in [24].

4 Data Analyses and Results

4.1 For Hypothesis1

To test H1, we perform a one-tailed F-test between the x- (and y-) coordinates of eye movement data when watching sports clips and those when watching video game clips. Table 1 shows the result.

From the result, it is noticed that for both coordinates, the results are significant, and the null hypothesis is rejected. That means when watching non-interactive video clips (sports), people's eyes do wander around more, both horizontally and vertically, compared to when watching more interactive video clips (video games). One possible explanation for this phenomenon is the sense and practice of control initiated by video game playing, which we believe carry over to just watching video game clips.

Table 1 The F-test result for comparing x- and y-coordinates of watching VG and SP videos

Pair	Num df	Den df	F	*p*-value	C.I	Ratio of var.
VG x–SP x	128,730	40,197	0.51	<2.2e−16	0.00–0.52	0.51
VG y–SP y	128,730	40,197	0.98	0.0213	0.00–0.99	0.98

VG video game data, SP sports data

Table 2 The Chi-square result for comparing eye status data of watching VG and SP videos

Pair	X-squared	df	Significance
VG–SP	296.25	25	<2.2e−16

VG video game data, SP sports data

4.2 *For Hypothesis2*

To investigate whether the distribution of the eye statuses is independent of the types of the videos people are watching, we performed a Chi-square test for the independence. Table 2 shows the results.

From the result, it is noticed that the distribution of the eye statuses is not independent of the types of the videos people are watching. Whether people are watching interactive videos or non-interactive videos will affect the distribution of their eye statuses.

5 Managerial Implications and Future Directions

Eye movement information has been one focal point of interests from researchers who want to investigate people's cognitive processes. Through the data collection and analysis of eye movement data while people engage in various activities, researchers have attempted to shed new lights on various cognitive processes. On the other hand, researchers have studied video game playing/viewing and their impacts on cognitive processes, especially when compared with regular video viewing [8]. However, there is limited literature that studies the eye movements during watching video game clips versus watching regular videos. In this paper, we attempt to fill this void. Based on [8]'s theory of sense and practice of control due to the interactivity nature of video game playing/viewing, we hypothesized about the relationships of watching different types of video clips (video game clips vs. regular sports clips) and their corresponding eye movement patterns. Based on the public eye tracking data available from Collaborative Research in Computational Neuroscience's (CRCNS) data sharing project, we show that when watching non-interactive video clips (regular sports), people's eye wander around more than when watching more interactive video clips (video games). We also show that the type of videos people are watching significantly affect the distribution of their eye status.

Our results have significant implications for a wide variety of IS practices and research directions. For example, in the learning technology design for children, researchers have realized that eye movement data capture and analyses is an ideal approach to infer the physiological behavior of children [25], as children may not be able to reliably express their conditions verbally. Our results may inform the learning technology designers to implement more interactive features on the components to which they want more focused visual attention from the children, in order to obtain betterment of learning process in children.

Another example is the design of IT artifact interface, such as website, which is an essential interface for HCI. By the eye tracking movements measure, we can observe the patterns of users surfing website with different levels of complexity [26]. Based on our results, we can improve the design of the website by enhancing interactive components in places where we prefer higher user visual attention and diminishing interactive features present in the places where we require less user's attention. This may lead to maximizing the likelihood of user satisfaction and return visits to the website.

Furthermore, in digital marketing, we can measure the effects of an advertisement stimulus on consumer attitudes [27]. Eye tracking can be used as one of the cognitive approaches, where we can track the eye movements and can relate the consumer psychology for better advertising methods. Based on our results, we can introduce higher interactive components in strategically located places in the advertisement in order to draw user's visual attention and develop an interest in the product.

In all of the above cases, designing experiments to verify the validity of those identified correlations would be of interest to many researchers. Once those correlations are validated, further research can be conducted to design the approached to actually implement the identified correlations.

References

1. Kurzhals, K., Fisher, B., Burch, M., Weiskopf, D.: Evaluating visual analytics with eye tracking. In: Proceedings of the Fifth Workshop on Beyond Time and Errors. Novel Evaluation Methods for Visualization, pp. 61–69, ACM (2014)
2. Kim, S.H., Dong, Z., Xian, H., Upatising, B., Yi, J.S.: Does an eye tracker tell the truth about visualizations?: findings while investigating visualizations for decision making. IEEE Trans. Visual Comput. Graph. **18**(12), 2421–2430 (2012)
3. Kurzhals, K., Burch, M., Pfeiffer, T., Weiskopf, D.: Eye tracking in computer-based visualization. Comput. Sci. Eng. **17**(5), 64–71 (2015). https://doi.org/10.1109/mcse.2015.93, Retrieved from https://ieeexplore.ieee.org/document/7208761
4. Foulsham, T., Farley, J., Kingstone, A.: Mind wandering in sentence reading. Can. J. Exp. Psychol./Rev. Can. De Psychologie Expérimentale **67**(1), 51–59 (2013)
5. Saint-Aubin, J., Klein, R.M., Landry, T.: Age changes in the missing-letter effect revisited. J. Exp. Child Psychol. **91**(2), 158–182 (2005)
6. McLeod, P., Land, M.F.: From eye movements to actions: how batsmen hit the ball. Nat. Neurosci. **3**(12), 1340–1345 (2000)
7. Kaminska, O., Foulsham, T.: Eye-tracking social desirability bias. Bulletin DE Méthodologieciologique **130**(1), 73–89 (2016)

8. Grodal, T.: Video games and the pleasures of control. In: *Media entertainment: The psychology of its appeal*, pp. 197–213 (2000)
9. Downward, L., Booth, C.H., Lukens, W.W., Bridges, F.: A variation of the F-test for determining statistical relevance of particular parameters in EXAFS fits. (2009) http://www.osti.gov/scitech/biblio/953921
10. Wong, M., Castro-Alonso, J.C., Ayres, P., Paas, F.: Gender effects when learning manipulative tasks from instructional animations and static presentations. J. Educ. Tech. Soc. **18**(4), 37 (2015)
11. Konig, P., Wilming, N., Kietzmann, T.C., Ossandon, J.P., Onat, S., Ehinger, B.V., Kaspar, K.: Eye movements as a window to cognitive processes. J. Eye Mov. Res **9**, 1–16 (2016)
12. Nikolaev, A.R., Pannasch, S., Belopolskiy, A.V.: Eye movement-related brain activity during perceptual and cognitive processing. Front. Syst. Neurosci. (2014)
13. Reichle, E.D., Reineberg, A.E., Schooler, J.W.: Eye movements during mindless reading. Psychol. Sci. **21**(9), 1300–1310 (2010)
14. Franklin, M., Smallwood, J., Schooler, J.: Catching the mind in flight: using behavioral indices to detect mindless reading in real time. Psychon. Bull. Rev. **18**(5), 992–997 (2011)
15. Jian, YC.: Fourth graders' cognitive processes and learning strategies for reading illustrated biology texts: eye movement measurements. Education Letter Retrieved from https://search.proquest.com/docview/1762495026 (2016)
16. Spence, I., Feng, J.: Video games and spatial cognition. Rev. Gen. Psychol. **14**(2), 92 (2010)
17. Cowley, B., Charles, D., Black, M., Hickey, R.: Toward an understanding of flow in video games. Comput. Entertainment (CIE) **6**, 1–27 (2008) https://doi.org/10.1145/1371216.1371223, Retrieved from http://dl.acm.org/citation.cfm?id=1371223
18. Smith, J., Graham, T.: Use of eye movements for video game control. Paper presented at the es. (2006). https://doi.org/10.1145/1178823.1178847, Retrieved from http://dl.acm.org/citation.cfm?id=1178847
19. Alkan, S., Cagiltay, K.: Studying computer game learning experience through eye tracking. Br. J. Educ. Tech. (2007)
20. El-Nasr, M., Yan, S.: Visual attention in 3D video games. Paper presented at the es. (2006) https://doi.org/10.1145/1178823.1178849, Retrieved from http://dl.acm.org/citation.cfm?id=1178849
21. Almeida, S., et al.: The eyes and games: a survey of visual attention and eye tracking input in video games. In: Proceedings of the SBGames (2011)
22. Goldstein, R.B., Woods, R.L., Peli, E.: Where people look when watching movies: do all viewers look at the same place? Comput. Biol. Med. **37**, 957–964 (2007)
23. Smuc, M., Mayr, E., Windhager, F.: The game lies in the eye of the beholder: the influence of expertise on watching soccer. In: Proceedings of the Annual Meeting of the Cognitive Science Society (2010)
24. Itti, L., Carmi, R.: Eye-tracking data from human volunteers watching complex video stimuli. CRCNS.org (2009)
25. El-Sheikh, M., Keller, P.S., Erath, S.A.: Marital conflict and risk for child maladjustment overtime: skin conductance level reactivity as a vulnerability factor. J. Abnorm. Child Psychol. **35**(5), 715–727 (2007). https://doi.org/10.1007/s10802-007-9127-2. PMID:17503176
26. Wang, Q., Yang, S., Liu, M., Cao, Z., Ma, Q.: An eye-tracking study of website complexity from cognitive load perspective. Decis. Support Syst. **62**, 1–10 (2014). https://doi.org/10.1016/j.dss.2014.02.007, Retrieved from https://search.proquest.com/docview/1528511672
27. Machouche, H., Gharbi, A., Elfidha, C.: Implicit effects of online advertising on consumer cognitive processes. Acad. Mark. Stud. J. **21**(2), 1–17 (2017). Retrieved from https://search.proquest.com/docview/1987371530

The Impact of Using a Gamified Interface on Engagement in a Warehousing Management Task: A NeuroIS Research Proposal

Mario Passalacqua, Pierre-Majorique Léger, Sylvain Sénécal, Marc Fredette, Lennart E. Nacke, Xinli Lin, Karine Grande, Nicolas Robitaille, Liza Ziemer and Tony Caprioli

Abstract Engagement, or rather lack thereof has become a major issue because of its negative impact on productivity. Recently, gamification has successfully been implemented into corporate technological interfaces to increase engagement

---

M. Passalacqua (✉) · P.-M. Léger · S. Sénécal · M. Fredette
Department of Management, HEC Montreal, Montreal, Canada
e-mail: mario.passalacqua@hec.ca

P.-M. Léger
e-mail: pierre-majorique.leger@hec.ca

S. Sénécal
e-mail: sylvain.senecal@hec.ca

M. Fredette
e-mail: marc.fredette@hec.ca

L. E. Nacke
University of Waterloo, Stratford School of Interaction Design and Business, Waterloo, Canada
e-mail: lennart.nacke@acm.org

X. Lin · K. Grande · N. Robitaille · L. Ziemer · T. Caprioli
JDA Software, Waukesha, USA
e-mail: xinli.lin@jda.com

K. Grande
e-mail: karine.grande@jda.com

N. Robitaille
e-mail: nicolas.robitaille@jda.com

L. Ziemer
e-mail: liza.ziemer@jda.com

T. Caprioli
e-mail: tony.caprioli@jda.com

X. Lin · K. Grande · N. Robitaille · L. Ziemer · T. Caprioli
JDA Software, Montreal, Canada

F. D. Davis et al. (eds.), *Information Systems and Neuroscience*,
Lecture Notes in Information Systems and Organisation 29,
https://doi.org/10.1007/978-3-030-01087-4_23

of employees. This paper proposes a theory-driven experiment that examines the impact a gamified interface has on engagement and performance of workers in a warehouse-management task. Specifically, the experiment proposed in this paper compares how the integration of two different types of goal-setting (self-set goals or assigned goals) into a warehouse-employee interface will affect engagement and performance.

Keywords Engagement · Performance · Gamification · Information systems Electroencephalography

1 Introduction

Nearly two-thirds of warehouse employees are not engaged in their work [1]. This leads to a lack of employee productivity, a high turnover rate, more errors and less profitability; all factors greatly affect organisational efficiency [2]. In recent years, gamification of employee interfaces has been employed to combat this issue. Gamification is defined as the "use of game design elements in non-game contexts" [3]. In other words, gamification employs the engaging nature of elements used in video games to create engagement in another context. Some of the common elements used in gamified interfaces are points, levels, goal-setting, feedback, badges and leaderboards. Building upon Tondello et al.'s framework [4], the current study will focus on two of these: goal-setting and feedback. There have been very few attempts at integrating gamification into an employee user interface for technology used within a warehouse setting [4, 5]. As noted by Coffey [6], optimization within this setting has mostly focused on the task itself, rather than on the human performing it. Small [7] adeptly proposes that the lack of focus on the human provides a great opportunity to increase employee engagement through gamification.

The objective of this paper is to propose an experiment that can determine how the gamification of a warehouse employee interface affects employee engagement and performance. The experiment will also allow for the examination of the physiological mechanisms by which gamification affects performance. First, employee engagement and gamification literature will be reviewed. Hypotheses will then be presented, followed by the experimental methodology.

2 Literature Review

Literature on engagement shows that engagement is a multifaceted concept. It is comprised of behavioral, emotional and cognitive engagement. Behavioral engagement relates to participation and involvement. Emotional engagement comprises positive and negative reactions. Cognitive engagement relates to investment, thoughtfulness and willingness to put in effort towards the task [8]. Intuitively, it is easy to understand how an engaged workforce performs better. Empirically, Harter et al. [9] performed a meta-analysis using 339 research studies and found that employee engagement is related to nine performance outcomes: profitability, productivity, turnover, absenteeism, customer loyalty, safety incidents, shrinkage, patient safety incidents and quality (defects).

Self-determination theory (SDT), a psychological theory of human motivation, has emerged as the leading theory with regards to explaining human motivation. SDT distinguishes between two types of motivation: intrinsic, which refers to motivation that comes from within, and extrinsic, which refers to motivation that results from assigned outcomes or reward. Research shows that intrinsic motivation is the main type that is used to explain underlying motivational effects of game design elements [10]. SDT states that satisfying three basic psychological needs will lead to increased intrinsic motivation: (1) competence, described as an employee feeling they can efficiently and competently deal with a challenge; (2) autonomy, defined as the sense of freedom and will when performing a task; (3) relatedness, which is the feeling of connection to others [11].

So how exactly does intrinsic motivation from a gamified interface increase employee engagement? This can be explained through the lens of the Job Demands-Resource (JD-R) model. Basically, this model proposes that the intrinsic motivation generated through the satisfaction of SDT's three basic psychological needs by the implementation game design elements results in a greater availability of motivational resources. JD-R states that when employees have enough resources to deal with job demands, engagement is greatly increased [12]. For example, integrating a self-set goal mechanism into an employee interface can increase intrinsic motivation and available resources through the autonomy of the competence aspect of SDT. In other words, allowing employees to set their own goals may give them a certain sense of autonomy.

Complementary to SDT, goal-setting theory, another well-established theory of human-motivation, provides further insight into how game elements can increase engagement, specifically, the goal-setting game element. This theory states that people are generally motivated to achieve goals. This motivation is because of self-regulation, which is the modification of thought, affect, and behavior [13–15]. In fact, decades of psychological research exist documenting how goal setting increases engagement and performance [16]. However, there is much debate on whether self-set goals or assigned goals produce greater engagement and performance. As is noted in a meta-analysis by Harkins and Lowe [17], most of the previous studies comparing self-set versus assigned goals did not take into account necessary factors for a valid comparison. Other research into this comparison has shown that goal commitment is higher when goals are self-set [18]. Because goal commitment is a strong moderator of the relationship between goals and performance [19], it can be argued that self-set goals may lead to better performance and possibly more engagement. Based on the reviewed literature, we have developed two hypotheses:

H1 The use of a gamified interface where goals are either self-set or assigned and feedback is received will lead to higher engagement and performance when compared to no gamification.

H2 The use of a gamified interface where goal-setting is self-set will lead to higher engagement and performance when compared to assigned goal.

3 Methods

3.1 Experimental Design

This study uses a within-subject design. Twenty subjects aged between 18 and 25 will participate in this study. They will be taken from our institution's participant pool. The current experiment was approved by our institution's research ethics board.

Building upon recommendations by Liu et al. [20] our experiment was designed bearing two types of outcomes in mind: experiential and instrumental. The following experiment will examine the impact of using a gamified interface on an experiential outcome (engagement) and an instrumental outcome (task performance) during a warehousing management task. In this case, a warehouse management task refers to picking specific items from various shelves and placing them into a bin. The implemented elements are goal setting (self-set vs. assigned) and feedback. Goal-setting and feedback have been integrated together because research has consistently shown that the motivational effects of goal-setting are most effective when the participant knows how he/she is progressing towards that goal, via some sort of feedback [17].

Three experiment conditions were developed to answer the research questions.

Condition 1: In this condition, participants will go through the picking task (see Sect. 3.2 for details about the task) without any set goal, without any feedback. This serves as a control condition.

Condition 2: In this condition, participants will be able to set their own goals at the beginning of the condition (e.g. The average time to complete the following task is five minutes. Today, I want to beat the average by 45 s). When participants are done, they will receive on-screen feedback about their performance (e.g. "Good job, you have reached your goal").

Condition 3: In this condition, participants will be assigned a goal (average completion time). All 20 participants will be assigned the same goal. They will also receive on-screen feedback about their performance.

We have chosen to always present condition 1 first based on what has been found in the literature. It is clear within the literature that having a task with a goal followed by a task without a goal will lead to lower engagement and performance in the latter task [21]. The order of the conditions 2 and 3 will be counterbalanced to reduce a possible ordering effect.

Fig. 1 Panasonic FZ-N1

3.2 *Experimental Setup and Stimuli*

A simulated warehouse was set up at the institution's research facilities, the room is 11 × 17 feet and has five metal bookshelves lined up on a wall. Also, there are four cameras set up around the room, so the participant can be seen at all times. The bookshelves were divided into three columns and four rows. Each compartment having its own unique identifier (e.g. A01001). The picking device used is the Panasonic FZ-N1, a fully rugged device with the Android operating system (version 6.0.1) (see Fig. 1). This device is about the same size as an average smartphone. This device will be strapped to the participant's arm.

3.3 *Experimental Tasks*

Participants will have to complete 12 picking tasks in each condition. A single picking task consists of taking a certain quantity of the same item from a compartment (e.g. pick five blue pens from A03002). Not all picks are equal in complexity (e.g. two erasers vs. five small white paper clips in small box with about 100 paper clips in various colours). Pick complexity therefore had to be operationalized to assure equal complexity in all conditions. An order picking complexity matrix was created based on research by Frazelle [22] and Errasti [23]. Simply put, pick complexity was determined by the quantity of the picked item, and its number of characteristics that add complexity (e.g. size, colour, brand, type). Because each of the 12 picks had a score, we are able to make sure pick complexity is constant across all conditions.

3.4 *Measurements*

As mentioned above this study will look at engagement and performance as outcome variables. Physiological measures were used to be able to capture the task engagement without interfering in the task itself, therefore maximizing the ecological validity. All physiological data will be synchronized to allow for the best possible quantification of engagement elements, as is recommended by Leger et al. [24] and Charland et al. [25]. In this case, two of three facets of engagement can be measured physiologically. Emotional engagement can be inferred by measuring emotional valence (positive or negative), as well as emotional arousal (calm/aroused). Electrodermal activity, which is the variance in electrical conductivity in response to sweat secretions, has been shown to be a valid measure of arousal. Electrocardiography, which measures the heart electrical activity is another valid measure of arousal [26]. As for emotional valence, it can be with electroencephalography (EEG) [4]. Cognitive engagement is measured using electroencephalography (EEG), which is the measurement of neuron synchronization in the brain. To properly measure cognitive engagement, Pope et al. [27] created a validated engagement index which measures the power spectral density of three bands (beta/ (alpha + theta)) [28, 29]. This index is more complex than the one suggested in the NeuroIS literature (e.g. [30]). For more information about the physiological tools in this study, refer to the book "Fundamentals of NeuroIS", written by Riedl and Léger [31]. Goal commitment and the emotional facet of engagement will be measured with questionnaires. They will be answered on a tablet at the end of each condition, therefore they will not interfere with the task. As mentioned above, the emotional facet of engagement can be inferred by measuring valence and arousal. The Affective Slider [32], which composed of a valence slider and an arousal slider, is one of the most reliable ways to measure self-report valence and arousal. The Affective Slider is composed of two sliders. To measure goal-commitment, a five-item questionnaire recommended by Klein et al. [33] was used. As for picking performance, it will be based on two factors: time taken to complete the task compared to the average (calculated during pretests) and task errors (wrong item or quantity).

3.5 Procedure

Firstly, the physiological measures will be installed on the participant. Participants then fill out a demographic questionnaire. Participants will then be explained the picking tasks and they will have the opportunity to practice with a training task. Participants then complete the conditions. After each of the 3 conditions, participants will answer the Affective Slider, as well as the goal-commitment questionnaire on a tablet. A post-experiment interview will then be administered to gain further insight into device and interface usability, as well as condition preference.

4 Next Step and Conclusion

We believe that the proposed experiment addresses the need for theory-driven gamification research that allows practitioners to understand the underlying mechanisms behind the integration of game-design elements within a technological interface. Moreover, this study will contribute theoretically and practically to the current body of knowledge. Theoretically, this study will allow for the direct comparison of self-set versus assigned goals, a topic that is still under debate. Practically, this study tests game-elements that can be implemented into a variety of interfaces in diverse contexts, making it of interest to practitioners.

References

1. Jackson-Martin, J.: Strategies for catalyzing workforce engagement in warehouse operations. Doctoral dissertation, Walden University (2017)
2. Mehta, D., Mehta, N.K.: Employee engagement: a literature review. Economia. Seria Manag. **16**(2), 208–215 (2013)
3. Deterding, D.D., Khaled, R., Nacke, L.: Gamification: toward a definition. CHI 2011 gamification workshop (2011)
4. Small, A.A.: Gamification as a means of improving performance in human operator processes. Doctoral dissertation, Massachusetts Institute of Technology (2017)
5. Klevers, M., Sailer, M., Günthner, W.A.: Implementation model for the gamification of business processes: a study from the field of material handling. In: Simulation and Gaming in the Network Society, pp. 173–184. Singapore: Springer (2016)
6. Coffey, D.: Zero in on Picking. Logistics Transp. Focus (1999)
7. JRA Limited.: Employee engagement: driving organisation performance (2007)
8. Fredricks, J.A., Blumenfeld, P.C., Paris, A.H.: School engagement: potential of the concept, state of the evidence. Rev. Educ. Res. **74**(1), 59–109 (2004)
9. Harter, J.K., Schmidt, F.L., Agrawal, S., Plowman, S.K., Blue, A.: The relationship between engagement at work and organizational outcomes. Gallup Poll Consulting University Press, Washington (2016)
10. Ryan, R.M., Rigby, C.S., Przybylski, A.K.: Motivation pull of video games: a self-determination theory approach. Motiv. Emot. **30**, 347–365 (2006)
11. Deci, E.L., Ryan, R.M.: Self-determination theory: when mind mediates behavior. J. Mind Behav. **1**, 33–43 (1980)
12. Demerouti, E., Bakker, A.B., Nachreiner, F., Schaufeli, W.B.: The job demands-resources model of burnout. J. Appl. Psychol. **86**(3), 499 (2001)
13. Kanfer, R., Ackerman, P.L.: Motivation and cognitive abilities: an integrative/aptitude-treatment interaction approach to skill acquisition. J. Appl. Psychol. **74**(4), 657e690 (1989) http://dx.doi.org/10.1037/0021-9010.74.4.657
14. Karoly, P.: Mechanisms of self-regulation: a systems view. Annu. Rev. Psychol. **44**, 23e52 (1993) http://dx.doi.org/10.1146/annurev.ps.44.020193.000323
15. Latham, G.P., Locke, E.A.: Self-regulation through goal setting. Organ. Behav. Hum. Decis. Process. **50**(2), 212e247 (1991) http://dx.doi.org/10.1016/0749-5978(91)90021-k
16. Landers, R.N., Bauer, K.N., Callan, R.C.: Gamification of task performance with leaderboards: a goal setting experiment. Comput. Hum. Behav. (2015) https://doi.org/10.1016/j.chb.2015.08.008
17. Harkins, S.G., Lowe, M.D.: The effects of self-set goals on task performance. J. Appl. Soc. Psychol. **30**(1), 1–40 (2000)
18. Locke, E., Latham, G., Erez, M.: The determinants of goal commitment. Acad. Manag. Rev. **13**(1), 23–39 (1988)
19. Locke, E.A., Latham, G.P.: Building a practically useful theory of goal setting and task motivation: a 35-year odyssey. Am. Psychol. **57**(9), 705 (2002)
20. Liu, D., Santhanam, R., Webster, J.: Towards meaningful engagement: a framework for design and research of gamified information systems. MIS Q. **41**(4) (2017)
21. Locke, E., Latham, G.: New directions in goal-setting theory. Current Dir. Psychological Sci. **15** (2006) https://doi.org/10.1111/j.1467-8721.2006.00449.x

22. Frazelle, E., Frazelle, E.: World-class Warehousing and Material Handling (Vol. 1). New York: McGraw-Hill (2002)
23. Errasti, A.: Logística de almacenaje: diseño y gestión de almacenes y plataformas logísticas. World Class Warehousing. Ediciones Pirámide, España (2011)
24. Léger, P.M., Sénecal, S., Courtemanche, F., de Guinea, A.O., Titah, R., Fredette, M., Labonte-LeMoyne, É.: Precision is in the eye of the beholder: application of eye fixation-related potentials to information systems research. J. Assoc. Inf. Syst. **15**(10), 651 (2014)
25. Charland, P., Léger, P.M., Sénécal, S., Courtemanche, F., Mercier, J., Skelling, Y., Labonté-Lemoyne, E.: Assessing the multiple dimensions of engagement to characterize learning: a neurophysiological perspective. J. Vis. Exp. **101**, e52627 (2015). https://doi.org/10.3791/52627
26. Cacioppo, J.T., Tassinary, L.G., Berntson, G.: Handbook of Psychophysiology. Cambridge University Press, Cambridge (2007)
27. Pope, A.T., Bogart, E.H., Bartolome, D.S.: Biocybernetic system evaluates indices of operator engagement in automated task. Biol. Psychol. **40**(1–2), 187–195 (1995)
28. Chaouachi, M., Chalfoun, P., Jraidi, I., Frasson, C.: Affect and mental engagement: towards adaptability for intelligent systems. In Proceedings of the 23rd International FLAIRS Conference, AAAI Press (2010)
29. Freeman, F.G., Mikulka, P.J., Scerbo, M.W., Scott, L.: An evaluation of an adaptive automation system using a cognitive vigilance task. Biol. Psychol. **67**(3), 283–297 (2004)
30. Müller-Putz, G.R., Riedl, R., Wriessnegger, S.C.: Electroencephalography (EEG) as a research tool in the information systems discipline: foundations, measurement, and applications. CAIS **37**, 46 (2015)
31. Riedl, R., Léger, P. M.: Fundamentals of NeuroIS. Studies in Neuroscience, Psychology and Behavioral Economics. Springer, Berlin, Heidelberg (2016)
32. Betella, A., Verschure, P.F.: The affective slider: a digital self-assessment scale for the measurement of human emotions. Plos One **11**(2), e0148037 (2016)
33. Klein, H.J., Wesson, M.J., Hollenbeck, J.R., Wright, P.M., DeShon, R.P.: The assessment of goal commitment: a measurement model meta-analysis. Organ. Behav. Hum. Decis. Process. **85**(1), 32–55 (2001)

Effect of Emotion on Content Engagement in Social Media Communication: A Short Review of Current Methods and a Call for Neurophysiological Methods

Melanie Schreiner and René Riedl

Abstract Engagement with content is vital for companies to achieve overall marketing goals (e.g., sales). Emotional content has the potential to grab attention and evoke the desired engagement. Our goal is to review the research methods used in the extant literature on the emotional effect on content engagement in social media communication. The findings show an unbalanced use of methods. Content analysis and emotion coding procedures are the dominant methods, while other methods have hardly been used. Based on this finding, we argue that future research needs to deploy neurophysiological methods to capture the complex emotion construct. Because neurophysiological methods are often applied in experimental settings, an increasing use of these methods would also imply a more advanced discovery of causal effects, thereby better clarifying the role of emotion in the content engagement process.

Keywords Content engagement effect · Emotion · Social media communication

1 Introduction

The global dissemination of social media platforms makes them indispensable channels in marketing communication. Companies can contact social media users and communicate with them. However, intense usage produces a vast amount of data from users and companies. Companies need to create content in order to gain users' attention and evoke engagement on a regularly basis [1]. Emotional content could

M. Schreiner (✉) · R. Riedl
University of Applied Sciences, Steyr, Upper Austria, Austria
e-mail: melanie.schreiner@fh-steyr.at

R. Riedl
e-mail: rene.riedl@fh-steyr.at

R. Riedl
Johannes Kepler University, Linz, Austria

F. D. Davis et al. (eds.), *Information Systems and Neuroscience*,
Lecture Notes in Information Systems and Organisation 29,
https://doi.org/10.1007/978-3-030-01087-4_24

grab attention in order to evoke engagement, but companies struggle in the challenging process of content creation [2, 3].

Due to the nature of social media, engagement indicates the efficiency in marketing communication. Engagement appears in an interaction experience and can be expressed in three dimensions: behavior, cognition, and emotion. The emotional component of the definition indicates an emotional effect to the social media content, thereby affecting users' engagement with the content (e.g., a like for a posting). Emotion is well known for its influence on communication processes in marketing and information systems research [e.g., 4, 5]. Importantly, studies demonstrate an effect of emotion on content engagement in social media communication [6]. Although social media have significant relevance in the daily routine of millions of users, it has been researched with low intensity [e.g., 3].

Our goal is to review and assess prior research from a methodological perspective about the effect of emotion on content engagement in social media. Specifically, we address the following research question: *Which methods have been used and which methods could be used to study the emotional impact on content engagement in social media marketing communication?* The following section contains a theoretical background of the relations between engagement, emotion, and content. The next sections delineate the literature analysis method and findings. A discussion, conclusions and future research are described in the last section.

2 Engagement in Social Media Marketing

Engagement has frequently been stated as a common goal in social media marketing communication [7]. Customer engagement is defined as "[…] a *psychological state* that occurs by virtue of *interactive, cocreative customer experiences* with a *focal agent/object* (e.g., a brand) in focal service relationships. It occurs under a specific set of context-dependent conditions generating differing CE levels; and exists as a *dynamic, iterative process* within service relationships that *cocreate value*. […] It is a *multidimensional concept* subject to a context- and/or stakeholder-specific expression of relevant cognitive, emotional and/or behavioral dimensions." [8, p. 260, italics in original]. Engagement with content comprises the user's interaction experience with a company's content on social media platforms, where the response can be expressed on cognitive, emotional, and/or behavioral levels. This expression (e.g., share of a post) is dependent on the context and can occur within a dynamic, iterative process (e.g., brand excitement). Previous studies provide evidence of content engagement's positive effect on sales performance [9] and branding goals [10]. Therefore, content engagement affects the success of social media marketing.

The stated customer engagement definition already reveals the effect of emotion on content engagement. The term *emotion* involves various definitions and interpretations. Schmidt-Atzert et al. [11] define emotion as „[…] a qualitative, descriptive state, which occurs with changes on one or more levels: feeling, physical state and expression" [11, p. 25, translated by the authors]. Scherer (2005) extends the term as "episode of interrelated, synchronized changes in the states of all or most of the five organismic subsystems in response to the evaluation of an external or internal stimulus event as relevant to major concerns of the organism" [12, p. 697]; thus, triggers of emotion can be internal or external stimuli such as marketer-generated content. Importantly, this definition highlights the importance of neurophysiology, because changes in an organism as a response to stimuli are inherently biological. It follows that studying the influence of emotion on content engagement implies a multimethod approach. Specifically, a research approach that does not consider neurophysiology is incomplete. From a measurement perspective, the emotion construct can be captured via the dimensions valence and arousal [e.g., 13], or categorically (e.g., happiness or anger) [e.g., 14].

Emotions have been proven to have essential effects in marketing communication [e.g., 5]. Previous studies have already shown that emotions may influence the advertising effectiveness (e.g., attitude) of content [5]. Marketers need to create branded content with affective messages in order to elicit emotions [15] and subsequently to provoke engagement in social media. Therefore, emotions (indirectly via content engagement) will have an effect on the success in social media marketing [6, 16].

3 Literature Review Method

To answer the research question, we conducted a structured literature review [17]. First, we carried out a literature search with keywords and used various sources, such as leading journals and conference proceedings. Additionally, we conducted forward and backward research. Based on the results of a first search on Google Scholar, we generated a keyword list.[1] The keywords were paired to produce search terms for a systematic search. We conducted a literature research and applied our keyword list within information systems research[2] based on the basket of 8[3] and marketing

[1]Keyword list: social media engagement, popularity brand post, popularity brand content, interaction brand post, interaction brand content, engagement brand post, engagement brand post, social media content, Facebook like, customer brand engagement + social media, content engagement, online engagement, user engagement, content strategy, viral online content, social media content emotion, emotional content, emotional text, emotional communication, emotional video, emotional picture, emotional image content, affective content, emotional virality, emotional engagement, emotional participation, emotional interaction, emotional popularity, emotional reaction, emotional endorsement, emotional rebroadcasting, affective engagement, affective virality.

[2]Information systems journals*: INFORM SYST RES, MIS QUART, J MANAGE INFORM SYST, J ASSOC INF SYST, J INF TECHNOL, INFORM SYST J, J STRATEGIC INF SYST, EUR J INFORM SYST.

[3]http://aisnet.org/?SeniorScholarBasket.

publications[4] based on the German Academic Association for Business Research.[5] We removed publications where we had no match of topic and content or due to a non-empirical research approach. Finally, we were left with 50 papers[6] focusing on content engagement in social media. We investigated the constructs of the research models and the applied research method characteristics.

4 Results

Our results show that the interest in the topic started in 2011 (see Fig. 1). Most of the studies (33) apply a mixed methods approach (qualitative and quantitative methods). Further nine publications use a pure qualitative and eight publications a pure quantitative research approach. A total of 42 studies investigated the engagement construct in the context of Facebook, whereas a limited number considered Twitter (3), YouTube (3), Instagram (2), or other platforms (5; Kaixin, MySpace, Weibo, Groupon, Renren).

Next, we analyzed the research models. Overall, 32 publications focus on the content-engagement (C-EG) relation. This type of research model is investigated intensely and dominates the research domain. Further 15 publications focus on the content-emotion-engagement (C-EM-EG) research model. Another three publications investigate the content-emotion (C-EM) relation, and no publication focuses on the emotion-engagement (EM-EG) relation. These findings reveal a lack of investigations in the emotional effect on content engagement. Specifically, publications which conceptualize emotion as mediator between content and engagement are rare (i.e., low number of C-EM-EG).

So far, most researchers performed content analyses (40), which frequently were implemented in a case study design; we identified this method combination 20 times. Additionally, the case study method is used 22 times overall. The survey method is mainly applied within field experiments; both research methods are used with mod-

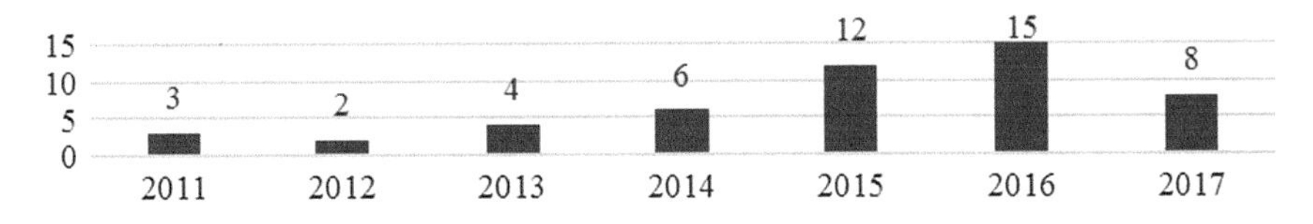

Fig. 1 Publications by year (absolute numbers)

[4]Marketing journals*: J MARKETING RES, J MARKETING, J CONSUM RES, MARKET SCI, INT J RES MARK, J ACAD MARKET SCI, J RETAILING, J SERV RES – US, J INTERACT MARK

* Abbreviations are based on Web of Science (https://images.webofknowledge.com/WOK46P9/help/WOS/A_abrvjt.html).

[5]http://vhbonline.org/vhb4you/jourqual/vhb–jourqual–3/teilrating–mark.

[6]The complete list of papers is available upon request.

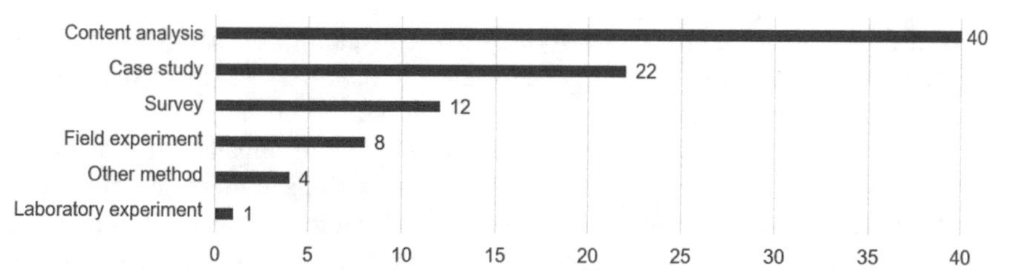

Fig. 2 Research methods used in previous studies (absolute numbers)

erate frequency. Previous studies were hardly conducted in the form of laboratory experiments (we identified only one study). Other research methods such as development of algorithms (e.g., for prediction models) are identified in four papers. As preliminary studies have almost totally neglected experimental research, our findings do not show a high degree of methodological plurality (see Fig. 2.).

A limited number of literature (15 papers) have investigated in the impact of emotion on content engagement. Considering the operationalization of the emotion construct, mainly three categories can be identified: (i) emotion dimensions, (ii) emotion categories, and (iii) general emotional message appeal. Result on (i): Eight studies operationalize the emotion construct dimensional whereby eight studies measure valence, one study measures arousal and no one considers dominance for emotion measurement. Result on (ii): Seven publications apply categories to capture emotion. The number of categories varies from 7 to 32, some of them did not mention the number. Result on (iii): Four studies exert a message appeal approach where they rate if emotions occur or not [e.g., 18]. Overall, our results indicate that most of the studies operationalize the emotion construct by using highly simplified measurement approaches, such as a pure valence-based approach without measurement of arousal.

5 Discussion, Conclusion and Future Research

The first publications, which targeted the topic of this paper, have been published in 2011. Our findings show that the emotion construct is hardly considered in the content engagement process in marketing and information systems research. Methodologically, we found that the emotion construct is predominantly measured by coding procedures based on valence only. Although the literature provides different perspectives and definitions of the emotion construct, all agree on the complexity of emotions [e.g., 19, 20]. Therefore they recommend to capture emotional responses based on multiple methods [21]. The Use of neurophysiological measures could overcome the observed shortcoming and provide more holistical insights on emotional impact [22]. In essence, physiological methods and tools support a deeper understanding of why and how emotions influence content engagement in social media, as outlined for IS research in a book by Riedl and Léger [23]. We propose to extend the self-report measurement framework by neurophysiological methods and apply, for example, startle reflex and skin conductance to measure emotional valence and arousal in social media communication (see Table 1).

The variety in research methods is currently low and unbalanced. For example, experiments, in particular those using neurophysiological measures, are hardly conducted. Because laboratory experiments are critical to establish causal relationships (here the nomological network is content → emotion → content engagement), future research should overcome the current methodological deficits in order to make possible a better understanding of the role of emotions in social media marketing. Neurophysiological methods will—most likely—reveal novel insights into this nomological network. Moreover, it is critical to mention that startle reflex and skin conductance are not the only neurophysiological methods which can be used in the present study context. In addition to these methods that relate to autonomic nervous system measurement, also methods related to central nervous system measurement could be applied, such as functional magnetic resonance imaging (fMRI) [e.g., 29] and electroencephalography (EEG) [e.g., 30]. Further methods are described in the NeuroIS literature [e.g., 23, 31, 32] and in other disciplines such as psychophysiology [e.g., 27].

A limitation of our work concerns the process of the literature review which involved not all possibly relevant keywords, and the keywords were not applied in all possibly relevant research fields. Further research, therefore, should extend the keyword list and consider research in other disciplines, such as psychology. Finally, it is also critical to mention that emotion in content engagement processes is related to other concepts. For example, one major concept in psychology is flow. The flow construct is considered as a cognitive state in an interaction process (e.g., during navigating a website) [33–35]. It follows that future research should consider the existing insights on flow (and the knowledge on its neurophysiological foundations) [e.g., 36], as well as the knowledge on related constructs, in order to advance research on the effects of emotion on content engagement in social media communication.

Table 1 Proposed emotion measurement in social media communication

Emotion	Measurement method		Reference (example)
Category	Self-report	Semantic differential	[13]
Valence, arousal, and dominance	Self-report	Self-Assessment Manikin (SAM)	[24]
Valence	Neurophysiological method	Startle reflex	[25–27]
Arousal	Neurophysiological method	Skin conductance	[26–28]

References

1. Luarn, P., Lin, Y.-F., Chiu, Y.-P.: Influence of Facebook brand-page posts on online engagement. Online Inf. Rev. **39**, 505–519 (2015)
2. Tafesse, W.: Content strategies and audience response on Facebook brand pages. Market. Intell. Plann. **33**, 927–943 (2015)
3. Swani, K., Milne, G.R., Brown, B.P., Assaf, A.G., Donthu, N.: What messages to post? evaluating the popularity of social media communications in business versus consumer markets. Ind. Mark. Manag. **62**, 77–87 (2017)
4. Powell, T.E., Boomgaarden, H.G., de Swert, K., de Vreese, C.H.: A clearer picture. The contribution of visuals and text to framing effects. J. Commun. **65**, 997–1017 (2015)
5. Holbrook, M.B., Batra, R.: Assessing the role of emotions as mediators of consumer responses to advertising. J. Consum. Res. **14**, 404–420 (1987)
6. Yu, J.: We look for social, not promotion: brand post strategy, consumer emotions, and engagement - a case study of the Facebook brand pages. GSTF J. Media Commun. **1**, 32–41 (2014)
7. de Vries, L., Gensler, S., Leeflang, P.S.H.: Popularity of brand posts on brand fan pages: an investigation of the effects of social media marketing. J. Interact. Mark. **26**, 83–91 (2012)
8. Brodie, R.J., Hollebeek, L.D., Juric, B., Ilic, A.: Customer engagement: conceptual domain, fundamental propositions, and implications for research. Conceptual domain, fundamental propositions, and implications for research. J. Serv. Res. **14**, 252–271 (2011)
9. Ha, S., Kankanhalli, A., Kishan, J.S., Huang, K.-W.: Does social media marketing really work for online SMEs? an empirical study. In: AIS (ed.) ICIS 2016 - International Conference on Information Systems, AIS (2016)
10. Hudson, S., Huang, L., Roth, M.S., Madden, T.J.: The influence of social media interactions on consumer–brand relationships. A three-country study of brand perceptions and marketing behaviors. Int. J. Res. Mark. **33**, 27–41 (2016)
11. Schmidt-Atzert, L., Peper, M., Stemmler, G.: Emotionspsychologie. Ein Lehrbuch. Kohlhammer, Stuttgart (2014)
12. Scherer, K.R.: What are emotions? And how can they be measured? Soc. Sci. Inf. **44**, 695–729 (2005)
13. Mehrabian, A., Russell, J.A.: An Approach to Environmental Psychology. The MIT Press, Cambridge (1974)
14. Izard, C.E.: Human Emotions. Plenum, NY (1977)
15. Chen, K.-J., Kim, J., Lin, J.-S.: The effects of affective and cognitive elaborations from Facebook posts on consumer attitude formation. J. Consum. Behav. **14**, 208–218 (2015)
16. Jaakonmäki, R., Müller, O., vom Brocke, J.: The impact of content, context, and creator on user engagement in social media marketing. In: HICSS (ed.) HICSS 2017 - Hawaii International Conference on System Sciences 50th Anniversary (2017)
17. vom Brocke, J., Simons, A., Niehaves, B., Riemer, K., Plattfaut, R., Cleven, A., others: Reconstructing the giant: on the importance of rigour in documenting the literature search process. In: ECIS (ed.) ECIS 2009 - 17th European Conference on Information Systems, 9, pp. 2206–2217 (2009)
18. Ashley, C., Tuten, T.: Creative strategies in social media marketing: an exploratory study of branded social content and consumer engagement. Psychol. Mark. **32**, 15–27 (2015)
19. Walla, P.: Affective processing guides behavior and emotions communicate feelings: towards a guideline for the NeuroIS community. In: Davis, F.D., Riedl, R., vom Brocke, J., Léger, P.-M., Randolph, A.B., Walla, P. (eds.) Information Systems and Neuroscience. Lecture Notes in Information Systems and Organisation, pp. 141–150. Springer, Berlin (2018)
20. Gregor, S., Lin, A.C.H., Gedeon, T., Riaz, A., Zhu, D.: Neuroscience and a nomological network for the understanding and assessment of emotions in information systems research. J. Manag. Inf. Syst. **30**, 13–48 (2014)
21. Micu, A.C., Plummer, J.T.: Measurable emotions. How television ads really work. J. Advertising Res. **50**, 137–153 (2010)

22. Dimoka, A., Banker, R.D., Benbasat, I., Davis, F.D., Dennis, A.R., Gefen, D., Gupta, A., Ischebeck, A., Kenning, P., Pavlou, P.A., et al.: On the use of neurophysiological tools in IS research. Developing a research agenda for NeuroIS. MIS Q. **36**, 679–702 (2012)
23. Riedl, R., Léger, P.-M.: Fundamentals of NeuroIS. Information systems and the brain. Springer, Berlin (2016)
24. Bradley, M.M., Lang, P.J.: Measuring emotion: the self-assessment manikin and the semantic differential. J. Behav. Ther. Exp. Psychiatry **25**, 49–59 (1994)
25. Walla, P., Koller, M.: Emotion is not what you think it is: startle reflex modulation (SRM) as a measure of affective processing in NeuroIS. In: Davis, F.D., Riedl, R., vom Brocke, J., Léger, P.-M., Randolph, A.B., Walla, P., Koller, M. (eds.) Information Systems and Neuroscience. Lecture Notes in Information Systems and Organisation, pp. 181–186. Springer, Berlin (2015)
26. Walla, P., Brenner, G., Koller, M.: Objective measures of emotion related to brand attitude. A new way to quantify emotion-related aspects relevant to marketing. PLoS ONE **6**, e26782 (2011)
27. Amrhein, C., Mühlberger, A., Pauli, P., Wiedemann, G.: Modulation of event-related brain potentials during affective picture processing: A complement to startle reflex and skin conductance response?. Int. J. Psychophysiol. Official J. Int. Organ. Psychophysiol **54**, 231–240 (2004)
28. Bradley, M.M., Miccoli, L., Escrig, M.A., Lang, P.J.: The pupil as a measure of emotional arousal and autonomic activation. Psychophysiology **45**, 602–607 (2008)
29. Dimoka, A.: How to conduct a functional magnetic resonance (fMRI) study in social science research. MIS Q. **36**, 811–840 (2012)
30. Müller-Putz, G., Riedl, R., Wriessnegger, C.S.: Electroencephalography (EEG) as a research tool in the information systems discipline: foundations, measurement, and applications. Commun. Assoc. Inf. Syst. **37**, 911–948 (2015)
31. Riedl, R., Banker, R.D., Benbasat, I., Davis, F.D., Dennis, A.R., Dimoka, A., Gefen, D., Gupta, A., Ischebeck, A., Kenning, P., et al.: On the foundations of NeuroIS. Reflections on the Gmunden retreat 2009. Commun. Assoc. Inf. Syst. **27**, 243–264 (2010)
32. Riedl, R., Davis, F., Hevner, A.: Towards a NeuroIS research methodology. Intensifying the discussion on methods, tools, and measurement. J. Ass. Inf. Syst. **15**, i–xxxv (2014)
33. Ghani, J.A., Deshpande, S.P.: Task characteristics and the experience of optimal flow in human—computer interaction. J. Psychol. **128**, 381–391 (1994)
34. Shin, N.: Online learner's 'flow' experience: an empirical study. Br. J. Edu. Technol. **37**, 705–720 (2006)
35. Novak, T.P., Hoffman, D.L., Yung, Y.-F.: Measuring the customer experience in online environments. A structural modeling approach. Mark. Sci. **19**, 22–42 (2000)
36. Mauri, M., Cipresso, P., Balgera, A., Villamira, M., Riva, G.: Why is Facebook so successful? Psychophysiological measures describe a core flow state while using Facebook. Cyberpsychol. Behav. Soc. Network. **14**, 723–731 (2011)

To like or Not to Like in the World of Instagram: An Eye-Tracking Investigation of Instagram Users' Evaluation Process for Liking an Image

Yu-feng Huang, Feng-yang Kuo and Chia-wen Chen

Abstract For image-based social media (e.g., Instagram or Snapchat), understanding people's decision behind their liking of photos is critical to researchers and practitioners. The liking decision toward an image, however seemingly simple and effortless for browsers, involves an interplay among evaluation dimension (hedonic vs. utilitarian), social influence (pre-existing number of likes), user characteristics, and underlying cognitive activities (effort and attention). The preliminary results from our eye-tracking studies show that the utilitarian evaluation of an image is negatively associated with its liking probability, effort (pupil dilation), and attention (fixation time). Social influence is shown to affect long-term social media users by increasing their hedonic rating and liking intention. The results suggest that using eye movements to predict the liking intention in social media requires the understanding of products' prominent evaluation dimension and users' characteristics. Discussions and future work are also presented.

Keywords Social media · Social recommendation · Eye-tracking Image evaluation

1 Introduction

Recently, picture-based social media such as Instagram have been widely used not only by individual users but by sellers as a channel to broadcast their products. Gaining likes for picture are important for both categories of posters because likes seem

Y. Huang (✉) · F. Kuo · C. Chen
National Sun Yat-Sen University, Kaohsiung, Taiwan
e-mail: evanhuang@mis.nsysu.edu.tw

F. Kuo
e-mail: bkuo@mis.nsysu.edu.tw

C. Chen
e-mail: m44020038@student.nsysu.edu.tw

F. D. Davis et al. (eds.), *Information Systems and Neuroscience*,
Lecture Notes in Information Systems and Organisation 29,
https://doi.org/10.1007/978-3-030-01087-4_25

to generate more likes, and in turn, due to the algorithms applied by social media, increase exposure of both pictures and their posters [1]. However, while pressing a "like" button seems easy to browsers, making this decision actually involves complex interaction between preference and social influence that are built toward a picture. In addition, the preference constructed toward a picture might involve multiple evaluative dimensions, especially the hedonic (enjoyment) utilitarian (usefulness) ones. Specifically, consumers can be motivated to focus either on the hedonic or the utilitarian dimension, and this focus biases subsequent buying and evaluation behavior [2]. Consumers motivated by a utilitarian goal pay more attention to utilitarian aspects of a product, and attend more to the hedonic aspects when a hedonic goal is activated [3]. While the two evaluative dimensions have long been identified and studied, how social influence can interact with the two dimensions remains largely unknown. Besides, the dichotomous outcome of a liking decision (to-like vs. not-to-like) is a surrogate result of the interplay between preference (can have more than one evaluative dimension) and social influence, masking the specific contributing effect of each individual factor.

Given all these complexities it has been difficult for practitioners and researchers, who usually merely examine behavioral outcomes, to disentangle the contributing factors. A good way to unpack these complexities is to examine the attentional process that underlies the interaction between preference and social influence. Attention, captured by eye trackers, is strongly related to people's cognitive activities, including even those complex ones such as decision making and problem solving, and can be used to infer peoples' emotional states and cognitive load [4]. Those data are useful to compensate behavioral outcomes by providing additional information and even opportunity to test competing theories [5]. With the eye-tracking technology, we further explore the mechanisms that contribute to a liking decision. Here we raise the following research questions: can we predict liking decision from eye movements? How may social influence affect this prediction?

Attention is strongly coupled with preference. Specifically, decision making studies that involve comparison among pictures have shown that attention, measured by eye fixations, is biased toward those chosen pictures. That is, the probability that people fixate at the eventually chosen picture increases over time. This phenomenon is called "gaze cascade" effect [6], which might be explained with mere exposure and preferential looking. However, this line of study examines mostly the preference construction process and has not considered social influence. Because of its usefulness in predicting purchase and subsequent user ratings, the importance of social influence in modern shopping scenarios cannot be overestimated [7]. Besides, gaze cascade studies involve a comparison among pictures, while in picture-based social network pictures are usually displayed in a single evaluation situation. To examine this problem, a study has applied the drift-diffusion model, which asserts that a choice is made when attention process has accumulated enough evidence toward it, has shown that pictures of products are more likely to be chosen when positive reviews are fixated and negative reviews are not fixated and vice versa [8].

Nevertheless, previous studies have not considered the fact that a product can be evaluated with the two critical dimensions: hedonic and utilitarian. These two dimensions have different effects on satisfaction [9], emotion [10], decision making [11], and intention to use [12]; besides, compared with utilitarian evaluation, hedonic evaluation might be performed with less effort [13, 14]. In a similar vein, eye tracking studies of online shopping have shown that people using emotional decision process, compared with those using calculative decision process, tend to pay less attention toward choices [15]. It is, therefore, likely that the two evaluative dimensions can result in distinct types of cognitive activities, which can be observed with attention (fixation time of image) and effort (pupil dilation). Furthermore, while eye-tracking studies have examined the effect of customer reviews on product evaluation, it remains uncertain how social influence can affect the two evaluative dimensions. One possibility is that hedonic evaluation is more susceptible to social influence because users spend less time for this dimension [14] and hence social influence might gain more weight on the final liking decision. In short, this study examines (1) how the two dimensions affect attention and liking intention, and (2) how social influence affect the evaluation of the two dimensions.

In our two studies, we have adopted cake images to answer the questions. Cake images are selected because they can be evaluated simultaneously from hedonic and utilitarian dimensions [16] and are common content that users share on picture-based social media. Average pupil dilations are collected as index for processing effort and fixation time of image for attention. Social influence is manipulated using different levels of pre-existing numbers of likes associated with the images. Behavioral results include liking intention and ratings of hedonic and utilitarian evaluation. We also include users' experience using Instagram to examine its effect on image evaluation.

2 Method

2.1 Study 1 (Behavior Study)

The purpose of study 1 is to establish an image pool, in which each image has its properties measured. We collected 164 cake images from Instagram. A total number of 124 participants were recruited in multiple sessions to rate these images on PCs in a computer room. Taste rating and convenience rating were used as the hedonic and utilitarian dimensions for the subsequent eye-tracking study. Forty cakes images were selected so that the scores of the two dimensions were uncorrelated.

2.2 Study 2 (Eye-Tracking Study)

To examine the effect of evaluation dimensions (hedonic vs. utilitarian) and social influence (pre-existing number of likes), the stimulus of study 2 mimicked the look of Instagram but with only two pieces of information: an image centered on the screen and a number of pre-existing likes presented on the left bottom (Fig. 1). Note that this manipulation allows us to exclude other contributing variables (e.g., posters' comments).

A group of 99 participants joined our eye-tracking study. They were instructed to imagine that they were viewing images on Instagram at their own pace. Participants were assigned to either experimental group (n = 56) or control group (n = 43). In the experimental group, participants were provided with 20 levels of pre-existing numbers of likes (ranging from 0 to 2,754,229). The presentation order of cake images and their combination with numbers of likes were randomized. In the control group the social influence was printed as "number of likes unknown." After viewing each image, they were asked to indicate whether to "like" the image and rate their taste and convenience scores on a 1 to 5 Likert scale (the higher the score, the tastier and more convenient the image is perceived). Specifically, for the taste score, participants were asked "I think that the cake is tasty" (not agree—agree). For the convenience score, participants were asked "I think the cake is available in nearby supermarket or convenience store." Then participants completed a demographic questionnaire to indicate their Instagram experience and were compensated and debriefed.

The fixation and pupil data were collected using the Eye Link 1000 system (SR Research Ltd., Canada) at 250 Hz sampling rate. Participants' heads were stabilized with a chin rest, and a nine-point calibration was executed before the experiment. During the experiment, a drift correction was performed between pictures to increase eye movement measurement accuracy. Pictures (visual angle was 13.87° width and 21.98° height) were centered on a 19″ LCD and at a levelled distance of 60 cm to participants' eyes. Fixating time was quantified with the sum of fixation duration of all fixations that landed on pictures. Average pupil dilation, which is a relative measure, was retrieved from the Data Viewer software.

Fig. 1 Examples of stimuli

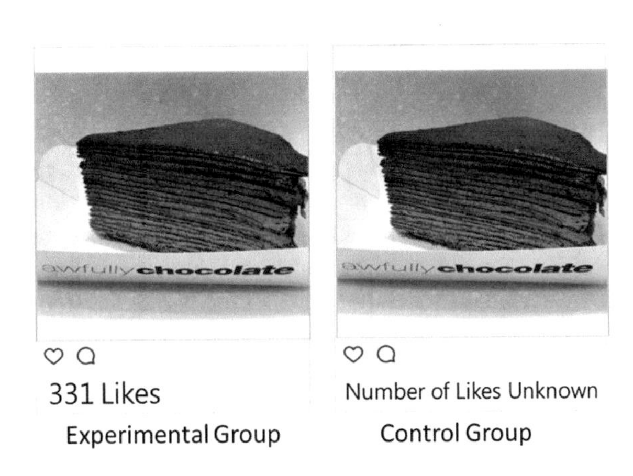

3 Results

3.1 Manipulation Check

The ratings of taste and convenience dimensions remained uncorrelated in study 2 (r = 0.02, $p = 0.88$). This suggested that in study 2 the utilitarian and the hedonic dimensions were independently rated.

3.2 Attention, Ratings and Liking Intention

To examine factors affecting liking decision toward an image, we conducted an image based analysis. Data were aggregated on the basis of individual image, resulting 40 analytical units. Then we conducted a path analysis (Fig. 2). In this model eye movements (pupil dilation and fixation time of cake image) predict taste and convenience ratings, which in turn predict liking intention. The results showed that effort (pupil dilation) and attention (fixation time) were negatively correlated with the convenience score. Conversely, pupil dilation and attention were positively correlated with to taste rating but were not statistically significant. Furthermore, liking intention could be positively predicted by taste score and negatively predicted by convenience score.

3.3 Social Influence

To examine the effect of social influence (pre-existing numbers of likes) on evaluation and liking intention, we aggregated data on the basis of individual pre-existing number of likes, resulting in 20 analytical units. The control group was aggregated

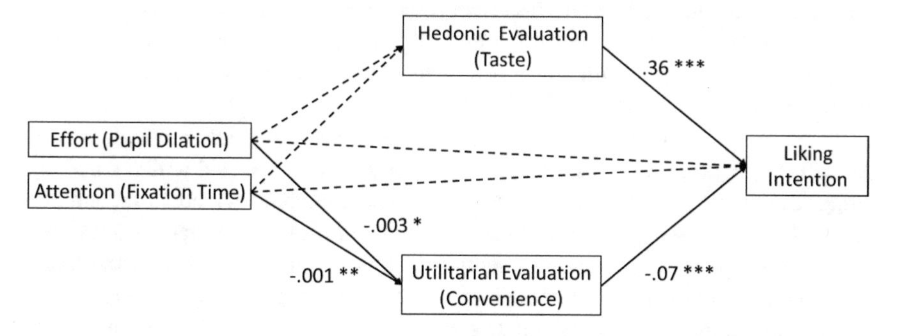

Fig. 2 Path analysis results of image evaluation (*$p < 0.05$, **$p < 0.01$, ***$p < 0.001$)

into one unit as a comparison to the experimental group. Furthermore, the analysis was conducted separately between the more experienced users, who had used Instagram for more than 3 years (the median score), and the less experienced users, who had used it for no more than 3 years. Participants who never used Instagram were excluded from analysis (n = 12, 5 in the experimental group, 7 in the control group). Correlations were conducted between log-transformed pre-existing number of likes (0 was excluded) and convenience rating, taste rating, and liking intention.

For the less experienced user group, the social influence had negative but unreliable effect on convenience rating, taste rating, and liking intention (Fig. 3, left column). For the experienced user group, the social influence had no significant effect on convenience ratings, but had positive effect on taste ratings and liking probability (Fig. 3, right column). The average ratings of the control group (also separated by Instagram experience) were plotted with horizontal dotted lines. Finally, we did not find social influence's effect on pupil dilation or attention in either of the user groups.

4 Discussion

Previous attentional studies on the interaction between preference and social influence have not considered that preference can be constructed with hedonic and/or utilitarian dimensions. Our preliminary results might inform this line of studies in three aspects.

First, our results show that utilitarian evaluation negatively predicts people's liking decision of pictures, while hedonic evaluation positively predicts this decision. This result supports the notion that a product image can be simultaneously evaluated in multiple evaluative dimensions [9, 11–14], and that the dimensions can independently contribute to the dichotomous and surrogate liking decision. This suggests that, for both practitioners and researchers, to understand preference to a product picture it is important to consider the prominent evaluation dimension of this product.

Second, our study also shows that social influence (pre-existing number of likes) can increase hedonic rating, but this effect is only evident in the long-term Instagram user group. While this finding remains preliminary, it nevertheless suggests that the perception of social influence is affected by evaluative dimensions and user characteristics.

Third, we showed that effort and attention are negatively correlated with the utilitarian rating (statistically significant) but are positively correlated with the hedonic rating (though not significant). The negative association between attention and utilitarian dimension suggest that utilitarian dimension of a picture acts as a piece of negative information [8]; paying attention to this dimension decreases the likelihood that a picture is liked. However, we find no evidence of a reliable relationship between eye movements (pupil dilation and fixation time) and the eventual liking intention or evidence showing that social influence can directly affect eye move-

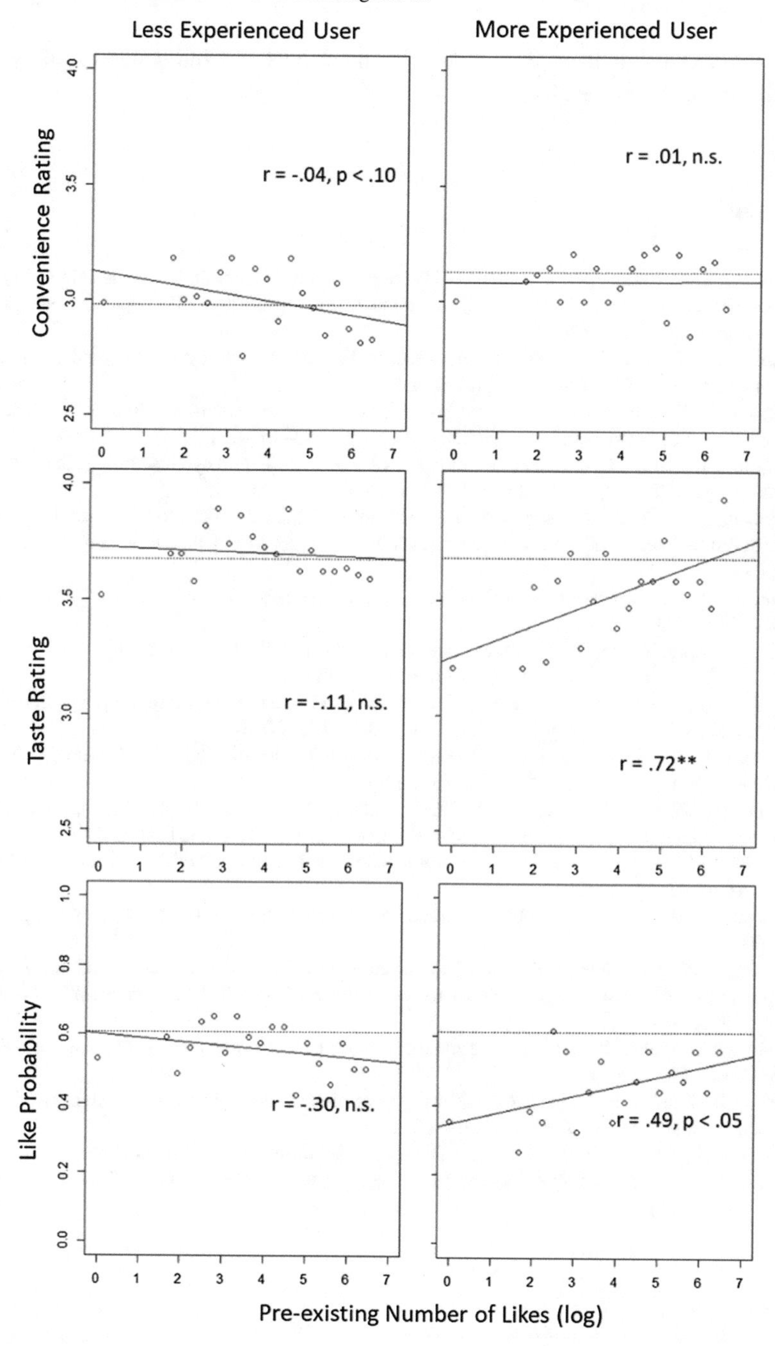

Fig. 3 Results of social influence (pre-existing numbers of likes), separately tested in more and less experienced Instagram user group

ments. Future work is needed to understand the robustness and generalizability of current eye movement results.

References

1. Lua, A.: Understanding the Instagram algorithm: 7 key factors and why the algorithm is great for marketers [cited 2018 Apr 30]; Available from: https://blog.bufferapp.com/instagram-algorithm#engagement (2017)
2. Babin, B.J., Darden, W.R., Griffin, M.: Work and/or fun: measuring hedonic and utilitarian shopping value. J. Consum. Res. **20**(4), 644–656 (1994)
3. Adaval, R.: How good gets better and bad gets worse: understanding the impact of affect on evaluations of known brands. J. Consum. Res. **30**(3), 352–367 (2003)
4. Orquin, J.L., Mueller, S.: Loose, attention and choice: a review on eye movements in decision making. Acta Physiol. (Oxf) **144**(1), 190–206 (2013)
5. Johnson, E.J., Schulte-Mecklenbeck, M., Willemsen, M.C.: Process models deserve process data: comment on Brandstatter, Gigerenzer, and Hertwig (2006). Psychol. Rev. **115**(1), 263–273 (2008)
6. Shimojo, S., et al.: Gaze Bias both reflects and influences preference. Nat. Neurosci. **6**(12), 1317–1322 (2003)
7. Zhu, F., Zhang, X.: Impact of online consumer reviews on sales: the moderating role of product and consumer characteristics. J. Mark. **74**(2), 133–148 (2010)
8. Ashby, N.J., Walasek, L., Glöckner, A.: The effect of consumer ratings and attentional allocation on product valuations. Judgment Decis. Mak. **10**(2), 172 (2015)
9. Chitturi, R., Raghunathan, R., Mahajan, V.: Delight by design: the role of hedonic versus utilitarian benefits. J. Mark. **72**(3), 48–63 (2008)
10. Mano, H., Oliver, R.L.: Assessing the dimensionality and structure of the consumption experience: evaluation, feeling, and satisfaction. J. Consum. Res. **20**(3), 451–466 (1993)
11. Okada, E.M.: Justification effects on consumer choice of hedonic and utilitarian goods. J. Mark. Res. **42**(1), 43–53 (2005)
12. Van der Heijden, H.: User acceptance of hedonic information systems. MIS Q. **28**(4), 695–704 (2004)
13. Huang, Y.-F., et al.: Hedonic evaluation can be automatically performed: an electroencephalography study of website impression across two cultures. Comput. Hum. Behav. **49**, 138–146 (2015, August)
14. Pham, M.T., et al.: Affect monitoring and the primacy of feelings in judgment. J. Consum. Res. **28**(2), 167–188 (2001)
15. Huang, Y.-F., Kuo, F.-Y.: An eye-tracking investigation of internet consumers' decision deliberateness. Internet Res. **21**(5), 541–561 (2011)
16. Vu, T.M.H., Tu, V.P., Duerrschmid, K.: Design factors influence consumers' gazing behaviour and decision time in an eye-tracking test: a study on food images. Food Qual. Prefer. **47**, 130–138 (2016)

Cognitive Work Protection—A New Approach for Occupational Safety in Human-Machine Interaction

Christian Neu, Elsa Andrea Kirchner, Su-Kyoung Kim, Marc Tabie, Christian Linn and Dirk Werth

Abstract Previous occupational safety concepts in human-machine interaction scenarios are based on the principle of spatial separation, reduction of collision force or distance monitoring between humans and robots. Collaborative robot systems and semi-automated machines are working closely with people in more and more areas, both spatially and functionally. Therefor a new approach for occupational safety in close human-machine collaboration scenarios is presented. It relies on a real-time EEG measurement of human workers with brain computer interfaces and a subsequent adjustment of the robot system based on the detected cognitive states.

Keywords Occupational safety · Human machine interaction
Brain computer interfaces

C. Neu (✉) · C. Linn · D. Werth
August-Wilhelm Scheer Institut, Saarbruecken, Germany
e-mail: christian.neu@aws-insitut.de

C. Linn
e-mail: christian.linn@aws-insitut.de

D. Werth
e-mail: dirk.werth@aws-insitut.de

E. A. Kirchner · S.-K. Kim · M. Tabie
Deutsches Forschungszentrum Für Künstliche Intelligenz,
Robotics Innovation Center, Bremen, Germany
e-mail: elsa.kirchner@dfki.de

S.-K. Kim
e-mail: su-kyoung.kim@dfki.de

M. Tabie
e-mail: marc.tabie@dfki.de

E. A. Kirchner
Research Group Robotics, Mathematics and Computer Science,
University of Bremen, Bremen, Germany

F. D. Davis et al. (eds.), *Information Systems and Neuroscience*,
Lecture Notes in Information Systems and Organisation 29,
https://doi.org/10.1007/978-3-030-01087-4_26

1 Introduction

According to a study by the International Federation of Robotics, the number of industrial robots sold in 2016 was 16% higher than in the previous year. Annual sales growth of 10% on average is expected until 2020 [1]. In addition to the possible increases in efficiency and productivity made possible by the growing number of robots in working environments across countries, the protection and safety of the employees involved must always come first. Accidents at work not only have serious consequences for those affected but are also a cost factor that should not be neglected for the companies involved and the economy in general. The main causes of accidents at work are human behavior errors based on carelessness, stress or hecticness [2]. Especially in close cooperation with industrial robots, this is triggered by complex motion sequences, unpredictable changes in position and speed or unexpected starting of the robots [3]. Current safety strategies to avoid work accidents in human machine interactions are based on a strict spatial separation between the robot and the work area of the human operator. However, collaborative work scenarios that are characterized by a very close human-machine interaction are increasingly coming into focus. Collaborative robot systems and semi-automated machines will work closely with people in more and more areas, both spatially and functionally. Particularly in these situations, special attention must be paid to the protection and safety of people. Classical safety strategies can no longer be applied for these collaborative work scenarios. The close physical cooperation between man and machine requires adapted and reliable occupational safety concepts.

Another aspect is the problem of under- or overstraining at the workplace. According to stress reports of recent studies, already in 2012 about 18% of employees felt that they were either professionally or quantitatively underchallenged at their work. About 23% of those questioned suffered from work overload [4]. However, mental health problems can arise from permanent under- or overchallenging. In order to master the unavoidable change in work requirements in a socially acceptable manner, new methods are therefore necessary to optimally design the working conditions for the individual person.

In this paper we present a new approach to increase occupational safety in collaborative working scenarios, including the potential to optimize the individual working conditions in a rapidly changing environment. It is based on cognitive measurements of human workers which are analyzed and used to optimize the co-working situations with robots with the aim to avoid work accidents. Following a design science methodology [5], we first demonstrate the relevance of such an approach followed by a conceptual solution which acts as the artifact and a subsequent discussion and evaluation.

2 Related Work

Especially in the field of human-robot interaction, the aspect of occupational safety and security is of great importance. Current approaches for collaborative robot systems are mainly focused on aspects of force and power limitation as well as speed and distance monitoring [3, 6]. On the one hand, there are approaches to minimize the collision force in a potential human-robot contact. On the other hand, technologies were developed to detect persons entering the safety area of the robot to initiate appropriate safety measures of the robot [7, 8]. The analysis and use of cognitive states to control and influence physical objects (such as robots) is an approach that has been less considered so far. Most of the current research relates to medical applications [9–11]. Especially in the industrial sector, i.e. in production or manufacturing, other requirements arise which have not yet been addressed to this extent. The electroencephalogram ("EEG") is a representation of electrical brain activity measured at the head surface and detected by means of metal electrodes and a conductive medium [12]. The EEG has so far mainly been used for the detection of neurological diseases in medicine, for the investigation of brain functions in research as well as in the context of therapy and rehabilitation. Further applications can be found in the field of brain-computer interfaces, i.e., the use of EEG signals to decipher mental states (fatigue, stress, etc.) and the improvement of human-machine interaction based on them [13–15]. The mental states can be analyzed in the EEG using the P300 or changes in the frequency bands. At a high workload, changes in the alpha and theta frequency bands are observed, e.g. increases in theta activity in the frontal brain area and reduction of alpha activity in the parietal area [16–22]. However, individual studies report increases in alpha activity [23, 24], which can be attributed to a general large variance of individual differences [25]. With onset of fatigue, however, activity in both frequency bands is increased [26–31]. In the case of the P300, a reduction in amplitude can be observed both with a high workload and with onset of fatigue [14, 32]. Errors in robot behavior or faulty human-robot interaction can be measured in the EEG as error-related potentials and detected in real time using machine learning methods [33–35]. The error-related potential can also be used as implicit feedback from humans (e.g. encouraging learning) in robot learning approaches [36].

Based on recent research a next step in the direction of implementing functional solutions within realistic industrial applications is needed to prove that cognitive work protection using physiological data is applicable and a strong contribution to worker's safety.

3 Technical Developments and Feasibility of the Approach

Our approach for occupational safety in close human-machine collaboration scenarios is feasible due to technical developments of the used core components that have come up over the last few years. This applies in particular to the fields of **EEG data**

analysis methods, **digital platform technologies and robotic control**, as well as **smart devices** (e.g. in terms of wearables).

By means of the introduction of machine learning methods [37] and advanced signal processing techniques (e.g. [38]) **EEG analysis** is now possible in almost real time and more importantly cognitive states and intentions of humans can be detected or inferred in single trial [39]. This development is the basis for applying brain computer interfaces (BCIs) in real world settings, such as done by embedded Brain Reading (eBR) [39]. Further, the improvements in EEG recording and analysis techniques that allow EEG analysis and analysis of other physiological data recorded under non-static conditions such as walking and running [40] or cycling [41, 42] did also enabled the integration of psychophysiological data into the control of robots such as exoskeletons [15]. Thus, due to the combination of both developments it is no longer required that persons wearing EEG recording equipment are not allowed to move at all while recording the data in often shielded lab environments [43]. Instead, free movements in cooperation with robotic systems in real world settings is now possible [36].

On the other hand, the development of embedded processing hardware and advanced embedded software solutions [44] does even enable the implementation of small recording as well as analysis techniques that already incorporate advanced real-time EEG processing [45] and even classifier training or adaptation on embedded devices [46].

Despite improvements in hardware and data analysis techniques the integration of physiological data into the control of robots or its usage for the improvement of human-machine interfaces required new concepts for deep integration and failure free usage of highly uncertain data such as EEG data. New concepts were developed that allow to infer on the context of interaction to make use of low level information from EEG data to infer on high level intentions of a user or to automatically detect markers in the EEG based on the current activity of the supported person that can be used to infer on the cognitive or mental state of the user [47]. The latter one is even possible without using a secondary task to measure workload [14].

Finally, advances in **EEG sensor systems** regarding usability and costs lead to a widespread use of EEG data in gaming and entertainment (e.g. [48–50]). Similar to the development of mobile cell phones this development is a major driver for solutions in the field of low cost and easy to use EEG sensor systems that are of need for our proposed application.

Digital platforms are new marketplaces by which the exchange of goods, services and other added value can be organized and realized using digital technologies. They enable interactions and transactions between interested participants and objects (e.g. machines, networks, institutions, etc.) [51, 52]. In this context "digital technologies" such as standardized interfaces, user and role administration, service catalogs, database technologies, intelligence and analytics software components serve as enabler and technological framework for the running of this connecting technology. These technological developments lead to a widespread use of digital platforms in the production as well as in the service sector [51]. Within the Industrial IOT scenario, platforms have the potential to realize the holistic framework of a "smart factory", in which **centralized robotic control scenarios** can be established as well [53]. In the field of data intelligence and analytics, big-data scenarios in particular can be implemented e.g. as predictive analyses and real-time data processing [54].

The use of **smart devices** such as **wearables** reflects the current development of the "quantify yourself" trend, which mainly focused on tracking and analyzing data from everyday activities such as sports, weight control, sleeping activities and other habits [55]. Technologically, applications can be deployed on the wearable that establish an interface to other decentralized services thus analyzing data in real-time. Besides sports and clinical approaches [56], new approaches arise in field of occupational safety e.g. for personalized construction safety environment using techniques like physiological monitoring; environmental sensing; proximity detection; and location tracking [57].

Regarding those technical developments, the combination of the components leads to the conception of the following holistic approach.

4 Cognitive Work Protection

The first goal of our approach (Fig. 1 shows the holistic design) is to measure the cognitive condition, for example the stress level or the workers' ability to concentrate, and thus to optimize interaction with robots and machines in real time with a view to increased safety. The second objective is to reduce accidents at work that occur in cooperation between humans and robots and to promote the physical and mental health of employees.

Starting point of our approach are EEG measurements of employees during the operation of machines or in cooperation with robots. The electrical activity of the

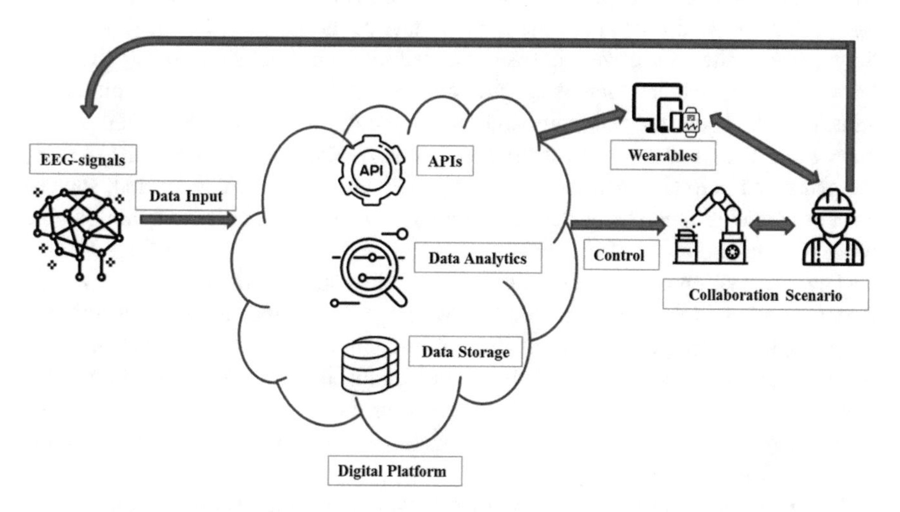

Fig. 1 Design representation of the holistic framework

brain is determined by electrodes that measure the voltage fluctuations at the head surface. The focus is on the electrode design and the ergonomic design of the sensor system. The system is optimized for a minimum number of sensors that are able to detect the desired cognitive states. For this purpose, the brain and thus head regions are identified which have the highest information content when determining the various cognitive states.

The electrodes based on flexible polymer substrates are installed at the identified head- and electrode positions in safety goggles and thus integrated into the safety equipment that is required in any case. Based on the identified electrode positions, optimal electrode flexibility and shape is identified by various prototypes and tests on person. These optimal electrode shapes should ensure reliable hair penetration and head contact as well as high long-term wearing comfort.

The measured brain waves are recorded in real time and transmitted via a wireless interface to a central digital platform. The data is processed there by means of automated analysis methods. Based on data preparation, characteristics are generated and patterns are classified using methods of machine learning and artificial intelligence. Thus, online measured EEG patterns can be assigned to certain cognitive states. EEG analyses can be performed in regards to time and frequency. Regarding the timing sequence, changes in natural potentials generated by the brain can be detected. An example of this is the potential P300 attributed to the attention of a test person; with an increasing workload, the amplitude of the signal decreases. A higher latency of the potential can also occur. To do this, the data is first filtered in time and then a spatial filter (e.g. xDAWN) [38] is trained, which reduces the amount of data and extracts important characteristics. Then the extracted characteristics are classified with a classifier like a support vector machine [58]. If the target signal changes, the processing chain can no longer correctly classify this signal, which can then be an indication of an overload or distraction of the subject. Furthermore, there are different frequency bands in the EEG. The activity in the respective frequency bands allows conclusions to be drawn about the mental state of the subject. To use this online, the energy of the respective band is measured under normal stress or no stress on the subject. This can be done by spectral analysis, for example. The determined rest values are then continuously compared with the current values in the application in order to be able to evaluate any differences directly. Error related potentials [33–36] are generated in the brain when a subject perceives something unexpected, such as a robot behaving incorrectly. These potentials can be classified with a comparable processing chain to that of the P300.

The information about the employee's cognitive states is then used for two different functions: to optimize interaction with robots and machines and to provide feedback for the employee himself. Based on the determined cognitive state of the employee, a rule-based control of the robot is realized. For example, the higher the measured stress level of the employee, the lower the speed of the robot. The robot is also controlled in real time and is connected via wireless communication to the digital platform [51, 53].

Within the feedback system, the real-time results of the EEG analysis, i.e. the information about the cognitive state, are visualized and made available to the employee via mobile devices. This ensures that the employee is permanently informed about his own data. This is realized via various visualization components (pie charts, alerts etc.) and haptic signals (e.g. vibration alarm). On the other hand, algorithms are included to identify patterns in continuous EEG measurements and to generate recommendations for optimizing working conditions. For example, with the help of a time series or classification analysis, which are usually found in Business Intelligence and Analytics components, mental state developments such as fatigue in time can be anticipated and thus recommendations can be displayed early via text windows and alarms (e.g. via wearables [55–57].

In order to ensure adequate data protection, the design of the system has to ensure that only the employee himself receives information about his EEG measurements. The resulting recommendations for improving working conditions are also only made available to the employee in a first step.

5 Discussion and Future Research

The Cognitive Work Protection approach has the potential to change the way occupational safety is realized in human-machine interaction scenarios. Compared to classical approaches it allows a close collaboration and real collaborative work between humans and robots. It is the first approach to directly detect the main reasons for work accidents—namely stress, fatigue, inattention—and to trigger countermeasures in a proactive way.

Besides occupational safety, the presented approach offers the possibility to improve the general working conditions in particular with respect to work overload and boredom. As discussed, this can be realized by providing recommendations for improving the working conditions to the employees themselves. In a next step however, the method can also be used for a company-wide management of human resources. Employees could be assigned varying tasks depending on their skills and level of satisfaction. For example, employees that typically feel unchallenged performing the same tasks every day could be dynamically assigned to frequently changing tasks, while the workload of employees feeling overstrained could be reduced. In this scenario, the presented approach therefore not only reduces work accidents and improves the working conditions of employees, but also has the potential to increase a company's productivity through managing human resources in an optimal way.

Despite the great potential of the approach, there are also challenges that need to be overcome for practical applications. From a technical point-of-view the main challenge is to construct a reliable and comfortable EEG measurements sensor system that does not interfere employees in their work. Current advances in EEG sensor systems show promising results in this regard and hint to a realistic possibility for a practical realization. The second and most important challenge comes with the fact that EEG sensors determine highly sensitive data in the human working environment within which the employee is in a relationship of dependence. In this regard, urgent questions arise which are of an ethical, social and legal nature and which must be sufficiently addressed before a practical realization.

In our future research, we plan to address these challenges and drive the development of a Cognitive Work Protection system, that is accepted by employees and employers alike. Therefor we will develop a prototype that can be tested and evaluated regarding usability, security and privacy issues.

References

1. International Federation of Robotics, World Robotics 2017, https://ifr.org/downloads/press/Executive_Summary_WR_2017_Industrial_Robots.pdf
2. Hofmann, D.A., Burke, M.J., Zohar, D.: 100 years of occupational safety research: from basic protections and work analysis to a multilevel view of workplace safety and risk. J. Appl. Psychol. **102**(3), 375–388 (2017)
3. Zhang, D., Wei, B., Rosen, M.: Overview of an engineering teaching module on robotics safety. In: Zhang, D., Wei, B. (eds.) Mechatronics and Robotics Engineering for Advanced and Intelligent Manufacturing. Lecture Notes in Mechanical Engineering. Springer, Cham (2017)
4. Hassard, J., Teoh, K.R.H., Visockaite, G., Dewe, P., Cox, T.: The cost of work-related stress to society: A systematic review. J. Occup. Health Psychol. **23**(1), 1–17 (2018)
5. Hevner, A.R., March, S.T., Park, J., Ram, S.: Design science in information systems research. MIS Q. **28**(1) (2004)
6. Kirchner, E.A., de Gea Fernandez, J., Kampmann, P., Schröer, M., Metzen, J.H., Kirchner, F.: Intuitive Interaction with Robots - Technical Approaches and Challenges, pp. 224–248. Springer, Heidelberg (2015)
7. Kulic, D.: Safety for human robot interaction, https://ece.uwaterloo.ca/~dkulic/pubs/DKulicThesisFinal.pdf
8. de Gea Fernandez, J., Mronga, D., Gnther. M., Knobloch, T., Wirkus, M., Schrer, M., Trampler, M., Stiene, S., Kirchner, E., Bargsten, V., Bnziger, T., Teiwes, J., Krger, T., Kirchner, F.: Multimodal sensor-based whole-body control for humanrobot collaboration in industrial settings. Robot. Auton. Syst. **94**, 102–119 (2017). ISSN: 0921-8890, https://doi.org/10.1016/j.robot.2017.04.007, URL http://www.sciencedirect.com/science/article/pii/S0921889016305127
9. Sunny, T.D., Aparna, T., Neethu, P., Venkateswaran, J., Vishnupriya, V., Vyas, P.S.: Robotic arm with brain – computer interfacing. Procedia Technol. **24**, 1089–1095 (2016)
10. Latif, M.Y. et al.: Brain computer interface based robotic arm control. In: 2017 International Smart Cities Conference (ISC2), Wuxi, pp. 1–5 (2017)
11. Wang, J., Liu, Y, Tang, J.: Fast robot arm control based on brain-computer interface. In: 2016 IEEE Information Technology, Networking, Electronic and Automation Control Conference, Chongqing, pp. 571–575 (2016)
12. Niedermeyer, E., Lopes da Silva, F.H.: Electroencephalography: Basic Principles, Clinical Applications and Related Fields, 3rd edition, Lippincott. Williams & Wilkins, Philadelphia (1993)
13. Chaudhary, U., Birbaumer, N., Ramos-Murguialday, A.: Brain–computer interfaces for communication and rehabilitation. Nat. Rev. Neurol. **12**, 513–525 (2016)
14. Kirchner, E.A., Kim, S.K., Wöhrle, H., Tabie, M., Maurus, M., Kirchner, F.: An intelligent man-machine interface - multi-robot control adapted for task engagement based on single-trial detectability of P300. Front. Hum. Neurosci. **10**, 291 (2016)
15. Wöhrle, H., Kirchner, E.A.: Online classifier adaptation for the detection of p 300 target recognition processes in a complex teleoperation scenario. In: da Silva, H.P., Holzinger, A., Fairclough, S., Majoe, D. (eds.) Physiological Computing Systems, Vol. 8908 of Lecture Notes in Computer Science, pp 105–118. Berlin, Heidelberg: Springer (2014)

16. Gundel, A., Wilson, G.F.: Topographical changes in the ongoing EEG related to the difficulty of mental tasks. Brain Topogr. **5**(1), 17–25 (1992)
17. Scerbo, M.W., Freeman, F.G., Mikulka, P.J.: A brain-based system for adaptive automation. Theor. Issues Ergon. Sci. **4**(1–2), 200–219 (2003)
18. Postma, M.A., Schellekens, J.M.H., Hanson, E.K.S., Hoogeboom, P.J.: Fz theta divided by Pz alpha as an index of task load during a PC-based air traffic control simulation. In: De Waard, D., Brookhuis, K.A., van Egmond, R., Boersema, T. (eds.) Human Factors in Design, Safety, and Management, pp. 465–470 (2005)
19. Berka, C., Levendowski, D.J.: EEG correlates of task engagement and mental workload in vigilance, learning, and memory tasks. Aviat. Space Environ. Med. **78**(5), B231–B244 (2007)
20. Holm, A., Lukander, K., Korpela, J., Sallinen, M., Müller, K.M.: Estimating brain load from the EEG. Sci. World J. **9**, 639–651 (2009)
21. Kamzanova, A.T., Kustubayeva, A.M.: Use of EEG workload indices for diagnostic monitoring of vigilance decrement. Hum. Factors **56**(6), 136–1149 (2014)
22. Dasari, D., Shou, G., Ding, L.: ICA-derived EEG correlates to mental fatigue, effort, and workload in a realistically simulated air traffic control task. Front. Neurosci. **11**, 297 (2017)
23. Boksem, M.A.S., Meijman, T.F., Lorist, M.M.: Effects of mental fatigue on attention: an ERP study. Cogn. Brain. Res. **25**, 107–116 (2005)
24. Käthner, I., Wriessnegger, S.C., Müller-Putz, G.R., Kübler, A., Haldera, S.: Effects of mental workload and fatigue on the P300, alpha and theta band power during operation of an ERP (P300) brain–computer interface. Biol. Psychol. **102**, 118–129 (2014)
25. Pigeau, R., Hoffmann, R. Purcell, S., Moffitt A.: The effect of endogenous alpha on hemispheric asymmetries and the relationship of frontal theta to sustained attention. Defense Technical Information Center (1987)
26. Akerstedt, T., Gillberg, T.: Subjective and objective sleepiness in the active individual. Int. J. Neurosci. **52**, 29–37 (1999)
27. Lal, S.K.L., Craig, A.: Driver fatigue: electroencephalography and psychological assessment. Psychophysiology **39**, 313–321 (2002)
28. Marcora, S.M., Staiano, W., Manning, V.: Mental fatigue impairs physical performance in humans. J. Appl. Physiol. **106**, 857–864 (2009)
29. Tanakal, M., Shigihara, Y., Ishii, A., Funakura, M., Kanai, E., Watanabe, Y.: Effect of mental fatigue on the central nervous system: an electroencephalography study. Behav. Brain Funct. **8**, 48 (2012)
30. Barwick, F., Arnett, P., Slobounov, S.: EEG correlates of fatigue during administration of a neuropsychological test battery. Clin. Neurophysiol. **123**(2), 278–284 (2012)
31. Zhaoa, C., Zhaoa, M., Liu, J., Zhengb, C.: Electroencephalogram and electrocardiograph assessment of mental fatigue in a driving simulator. Accid. Anal. Prev. **45**, 83–90 (2012)
32. Kok, A.: On the utility of P3 amplitude as a measure of processing capacity. Psychophysiology **38**(3), 557–577 (2001)
33. Kim, S.K., Kirchner, E.A.: Classifier transferability in the detection of error related potentials from observation to interaction. In: Proceedings of IEEE international conference of system, man, cybernetics, pp. 3360–3365 (2013)
34. Chavarriaga, R., Sobolewski, A.: Millán, J.d.R.: Errare machinale est: the use of error-related potentials in brain-machine interfaces. Front. Neurosci. **8**, 208 (2014)
35. Kim, S.K., Kirchner, E.A.: Handling few training data: classifier transfer between different types of error-related potentials. IEEE Trans. Neural Syst. Rehabil. Eng. **24**(3), 320–332 (2016)
36. Kim, S.K., Kirchner, E.A., Stefes, A., Kirchner, F.: Intrinsic interactive reinforcement learning – using error-related potentials for real world human-robot interaction. Sci. Rep. **7**, 17562 (2017)
37. Lotte, F., Bougrain, L., Cichocki, A., Clerc, M., Congedo, M., Rakotomamonjy, A., Yger, F: A review of classification algorithms for EEG-based brain–computer interfaces: a 10 year update. J. Neural Eng. **15**(3) (2018)

38. Roy, R.N., Bonnet, S., Charbonnier, S., Jallon, P., Campagne, A.: A comparison of ERP spatial filtering methods for optimal mental workload estimation. In: 37th Annual International Conference of the IEEE Engineering in Medicine and Biology Society (EMBC), pp. 7254–7257, Milan (2015)
39. Kirchner, E.A., Kim, S.K., Straube S., Seeland, A., Wöhrle, H., Krell, M. M., Tabie, M. Fahle, M.: On the applicability of brain reading for predictive human-machine interfaces in robotics. PLoS ONE **8**(12), e81732, 12 (2013)
40. Gwin, J.T., Gramann, K., Makeig, S., Ferris, D.P.: Removal of movement artifact from high-density eeg recorded during walking and running. J. Neurophysiol. **103**(6), pp. 3526–3534 (2010, June)
41. Kohli, S., Casson, A.J.: Towards out-of-the-lab EEG in uncontrolled environments: feasibility study of dry EEG recordings during exercise bike riding. Conf. Proc. IEEE Eng. Med. Biol. Soc., pp. 1025–1028 (2015)
42. Bailey, S.P., Hall, E.E., Folger, S.E., Miller, P.C.: Changes in EEG during graded exercise on a recumbent cycle ergometer. J. Sports Sci. Med. **7**(4), 505–511 (2008)
43. Reis, P.M.R., Hebenstreit, F., Gabsteiger, F., von Tscharner, V., Lochmann, M.: Methodological aspects of EEG and body dynamics measurements during motion. Front. Hum. Neurosci. **8**, 156 (2014)
44. Wöhrle, H., Teiwes, J., Krell, M.M., Seeland, A., Kirchner, E.A., Kirchner, F.: Reconfigurable dataflow hardware accelerators for machine learning and robotics. In: Proceedings of European Conference on Machine Learning and Principles and Practice of Knowledge Discovery in Databases, (ECML PKDD-2014), 15.9.2014–19.9.2014, Nancy, Springer, pp. 129–138 (2014)
45. Wöhrle, H., Tabie, M., Kim, S.K., Kirchner, E., Kirchner, F. (2017). A hybrid FPGA-based system for EEG- and EMG-based online movement prediction. Sensors 17 (2017)
46. Wöhrle, H., Teiwes, J., Krell, M.M., Kirchner, E.A., Kirchner, F.: A dataflow-based mobile brain reading system on chip with supervised online calibration. In: Congress Proceedings International Congress on Neurotechnology, Electronics and Informatics (NEUROTECHNIX-2013), Vilamoura, Portugal, SCITEPRESS Digital Library, 18–20 September 2013
47. Kirchner, E.A., Drechsler, R.: A formal model for embedded brain reading. Ind. Robot Int. J. **40**(6), 530–540 (2013)
48. https://www.emotiv.com, 2018/04/30
49. http://www.choosemuse.com/, 2018/04/30
50. https://www.slashgear.com/portable-eeg-machine-shows-how-music-affects-the-brain-during-exercise-19519957/, 2018/04/30
51. Jaekel, M.: Die Macht der digitalen Plattformen. Wegweiser im Zeitalter einer expandierenden Digitalshpäre und künstlicher Intelligenz. Springer, Wiesbaden (2017)
52. Tiwana, A.: Platform Ecosystems. Aligning Architecture, Governance, and Strategy. Morgan Kaufmann, Waltham (2014)
53. Shariatzadeh, N., Lundholm, T., Lindberg, L., Sivard, G.: Integration of digital factory with smart factory based on Internet Of Things. Procedia CIRP **50**, 512–517 (2016)
54. Lee, J., Bagheri, B., Jin, C.: Introduction to cyber manufacturing. Manuf. Lett. **8**, 11–15 (2016)
55. Klasnja, P., Hekler, E.B.: Wearable technology and long-term weight loss. JAMA. **317**(3), 317–318 (2017)
56. Pevnick, J.M., Birkeland, K., Zimmer, R., Elad, Y., Kedan, I.: Wearable technology for cardiology: an update and framework for the future. Trends Cardiovasc. Med. **28**(2), 144–150 (2018)
57. Awolusi, I., Marks, E., Hallowell, M.: Wearable technology for personalized construction safety monitoring and trending: review of applicable devices. Autom. Constr. **85**, 96–106 (2018)
58. Chang, C.-C., Lin, C.-J.: LIBSVM: a library for support vector machines. ACM Trans. Intell. 852 Sys. Technol. (TIST) **2**(27), 1–27 (2011)

Analysis of Heart Rate Variability (HRV) Feature Robustness for Measuring Technostress

David Baumgartner, Thomas Fischer, René Riedl and Stephan Dreiseitl

Abstract Technostress has become an important topic in the scientific literature, particularly in Information Systems (IS) research. Heart rate variability (HRV) has been proposed as a measure of (techno)stress and is widely used in scientific investigations. The objective of the pilot study reported in this paper is to showcase how the preprocessing/cleaning of captured data can influence the results and their interpretation, when compared to self-report data. The evidence reported in this paper supports the notion that NeuroIS scholars have to deliberately make methodological decisions such as those related to preprocessing of physiological data. It is therefore crucial that methodological details are presented in NeuroIS papers in order to create a better understanding of the study results and their implications.

Keywords Data preprocessing · Heart rate variability (HRV)
NeuroIS research methodology · Technostress · Signal feature · Stress

1 Introduction/Motivation/Related Work

In recent years, technostress has become an important concept in NeuroIS research (e.g., [1, 2]). It has been argued and demonstrated that mixed methods research,

D. Baumgartner · S. Dreiseitl
University of Applied Sciences Upper Austria, Hagenberg, Austria
e-mail: david.baumgartner@fh-hagenberg.at

S. Dreiseitl
e-mail: stephan.dreiseitl@fh-hagenberg.at

T. Fischer (✉) · R. Riedl
University of Applied Sciences Upper Austria, Steyr, Austria
e-mail: thomas.fischer@fh-steyr.at

R. Riedl
e-mail: rene.riedl@fh-steyr.at

R. Riedl
Johannes Kepler University, Linz, Austria

F. D. Davis et al. (eds.), *Information Systems and Neuroscience*,
Lecture Notes in Information Systems and Organisation 29,
https://doi.org/10.1007/978-3-030-01087-4_27

particularly involving physiological measures, is crucial for this research domain [2–6]. Amongst other methods, the collection of heart rate data and the calculation of heart rate variability (HRV) as a measure of stress have been proposed as viable additions, particularly to field studies investigating technostress [7, 8].

As part of a larger project, we are currently investigating the potential of several data collection methods for technostress research in a longitudinal field study. This includes heart rate data, which we collect using consumer level devices. In this paper, we report on a pilot study in which we tested the feasibility of letting individuals track their own heart rate during their working hours using a chest belt with a heart rate sensor. The objective of this pilot study is to assess the quality of the data generated in this manner. We showcase how the preprocessing/cleaning of captured data can influence the results and their interpretation when compared to self-report data, which we obtained by having study participants fill out a technostress questionnaire. By showing how data collected using consumer-grade devices can be useful to assess individual stress levels, we also specifically seek to support the call for further technostress research in field settings [2, 9].

In particular, we are interested in how data cleaning influences the correlation coefficient between HRV data and self-report data. While the HRV analysis methods investigated here are in no way particular to technostress research, they are nevertheless a necessary first step in every data processing endeavor.

In Sect. 2, we present an overview of the data collection procedures that were applied, the heart rate features that were extracted and the how the features were preprocessed. Then, in Sect. 3, we present our results regarding the influence of preprocessing methods on the results and their interpretation. Finally, in Sect. 4, we discuss our findings and present some recommendations for future research utilizing heart rate data in technostress research.

2 Materials and Methods

2.1 Data Collection

We collected our data in the week of 11/27/2017–12/01/2017 using a Polar H7 chest belt in combination with a smartphone app[1] that collected the captured data. Fifteen employees (12 female, 3 male) of a publishing company with its headquarters close to Salzburg, Austria, participated in the study. They were instructed to put on the chest belt after they had arrived at work and to start the data collection on the smartphone app. They stopped the data collection upon leaving the workplace. After removing unusable data, we retained 44 samples from 11 study participants from one week of data collection. At the end of the workweek, which for most of the participants was Friday, the participants were also invited to take part in an online survey, which

[1] https://itunes.apple.com/at/app/heart-rate-variability-logger/id683984776?mt=8 [03/05/2017].

included questions on their technostress level using a German version [10] of the "Technostress Creators" questionnaire by Ragu-Nathan et al. [11]. The questionnaire was deliberately handed out after the workweek as we wanted to avoid any changes in perception (e.g., situations that are pointed out as stressful in the questionnaire are then seen as more stressful due to heightened awareness) that could bias our results.

2.2 Heart Rate Features

The literature reports on a variety of indicators extracted from electrocardiography (ECG) recordings that can help in assessing HRV [12, 13]. In general, most of these indicators are calculated either in the time-domain or the frequency-domain of the signal. When analyzing the signal in the time-domain, the most relevant aspect is the time interval between subsequent peaks in the QRS complex, i.e., the time duration of one heartbeat (for details, see for example [12]). This interval is known as the RR interval; RR intervals of normal signals are known as NN intervals. In this work, we focus on the following five indicators, which we will call *features* of the signal. The first three are from the time domain, measured in milliseconds, the last two from the frequency domain, measured in squared milliseconds [13]:

SDNN: The standard deviation of NN intervals in the signal.
SDANN: The standard deviation of the averages (taken over five minute segments) of the NN intervals in the signal.
RMSSD: This feature depends on the differences of subsequent NN intervals. Square these differences, then take the square root of the arithmetic mean of these squares.
LF: The power of the signal in the low-frequency spectrum (0.04–0.15 Hz).
HF: The power of the signal in the high-frequency spectrum (0.15–0.5 Hz).

2.3 Data Preprocessing/Cleaning

Preprocessing and cleaning of data is a necessary initial step in most data analysis tasks, particularly in analyses of physiological data. We chose the Kubios HRV software, a state of the art tool for studying the variability of heart-rate intervals, mainly for its data-cleaning functionality, ease of use, and the wide range of HRV features it calculates. Kubios thus fits our requirements, and can also be considered the standard software in this application domain (e.g., [14, 15]). It provides tools for artifact removal (missed or spurious beats), analysis methods in the time and frequency domains, as well as the ability to calculate a number of less frequently used features (such as entropy measures, or measures calculated from a recurrence plot).

Obtaining artifact-free raw data samples from a consumer-grade device is almost impossible in a real-world setting. In this study, the quality of the data obtained from the Polar H7 chest belt depends mostly on its correct placement. Most HRV features and metrics are highly influenced by noisy data [16]; the degree of preprocessing and noise removal thus has a direct influence on the features reported by HRV software tools.

It has to be noted, though, that using consumer-grade devices in this context is already a very important deliberate choice, with several associated benefits and challenges (e.g., low intrusiveness, but particular need for data cleaning). Nonetheless, despite the particular obstacles, Schellhammer and colleagues [7] have demonstrated that consumer-grade devices are a valuable addition to technostress research in the field, particularly if one is interested in heart rate measures. Yet, just as technostress studies in the field are still scarce [2], so are technostress studies that have applied heart rate measures and reported their data cleaning procedures, aside from simply referencing the tool they used (e.g., AcqKnowledge in [17]). We therefore focus on this particular step in the research process, but want to highlight that there are several other challenges generated by the selection of a specific research design and measurement method in NeuroIS [1].

Kubios HRV allows threshold-based artifact correction, with automatic correction available in some versions [18]. In this preprocessing step, every RR interval value is compared to a simple moving median over a time window. An RR interval value is considered an artifact if it differs more than a specified threshold from this local median. Five equidistant threshold values are available in Kubios HRV, corresponding to increasingly lenient requirements for values to be considered artifacts: On the lowest level, only those values further than 0.45 s from the median are removed; on the highest level, all values further than 0.05 s from the median are removed (these values, given here for 60 bpm, are adjusted for heart rate). The correction uses cubic spline interpolation, which can lead to unusual beats if too many beats are corrected.

3 Results

Below, we present two sets of numerical results: one for the effects of data preprocessing on the HRV features, and one for the effects of data preprocessing on the correlation of HRV feature values with self-reported technostress levels.

The effect of five artifact correction thresholds on the features described in Sect. 2.2 is measured by average percentage change over all data sets. For each feature, its baseline value was calculated on the raw unchanged data samples. Compared to this baseline, the percentage change turned out to be quite high for most features. The results of applying five different artifact correction levels to five features is shown in Fig. 1. In this figure, we can observe that the correction at the threshold level "very low" already resulted in a percentage difference ranging from 2.6% to 20.9% on the different features. At this level, the average percentage of corrected RR intervals is only 0.46%. It is surprising that there is such a noticeable

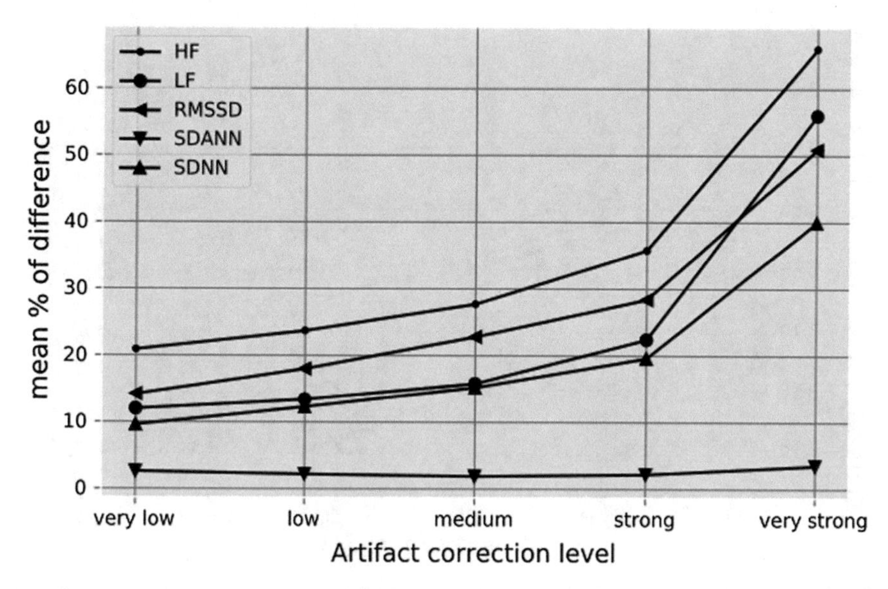

Fig. 1 Average percentage difference between baseline (raw feature values) and feature values obtained at different artifact correction levels

difference, given this small amount of correction. The percentages of corrected RR intervals were, in increasing order of artifact correction levels, 0.79%, 1.26%, 3.08% and 25.03%, respectively. The LF, HF, RMSSD and SDNN features especially show large percentage changes, and thus appear to be highly influenced by the degree of artifact correction.

It can be observed that the SDANN feature is much more robust than the other features. This can be attributed to the fact that the SDANN feature is defined as the standard deviation of the average RR interval calculated over 5 min periods [13]; this additional averaging smooths out the effect of the artifact corrections.

Figure 2 illustrates the influence of the artifact correction level on the correlation coefficient between feature values and the self-reported technostress levels. We can observe that changing the correction level results, for two features, in a change of sign of the correlation coefficient between raw and corrected data. As in Fig. 1, the SDANN feature is the most robust feature, with the other features showing small negative correlations with self-reported technostress levels. This concurs with the literature, where high self-reported stress levels are reported to be correlated with low HRV standard deviations and low LF/HF ratios [8].

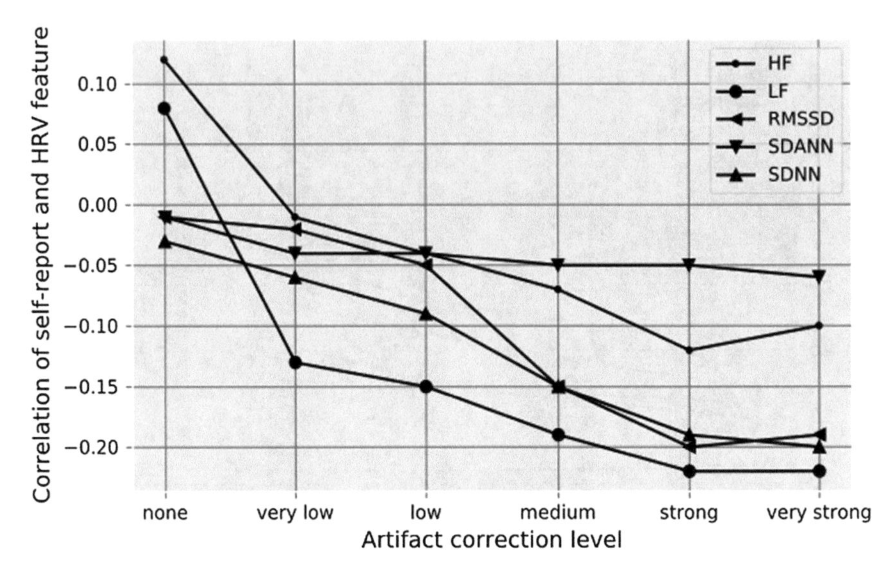

Fig. 2 Correlation of the questionnaire technostress level results and feature values obtained at different artifact correction levels

4 Discussion

Our calculations suggest that analyzing an entire data sample may detect too many artifacts, and thus result in a biased overcorrection. According to the state of the art in HRV artifact removal tools [18, 19], there is currently no possibility to fully automate the detection, correction and analysis process. Therefore, a human investigator is necessary to select threshold levels appropriate for the particular data analysis task. Hence, in the context of the research reported here, determining how well self-report data correlates with HRV data requires human intervention in the data-preprocessing stage.

Previous reviews of the literature have shown that there is still a need for field studies applying neurophysiological measures in technostress research [2, 9]. Yet, current developments in the area of consumer-grade devices may allow for investigations using this approach to be conducted more frequently in the future [7]. In order to support individual researchers interested in such settings, who then have to preprocess the generated data, in this paper we present a concrete example based on real world data. We show that in NeuroIS research, scholars have to make multiple decisions with respect to data preprocessing and analysis and present some potential effects of these decisions (e.g., on the correlation between measures).

As indicated by Riedl and colleagues in a paper on NeuroIS research methodology ([1], specifically refer to Section "Objectivity" starting on page xix), such decisions “may affect the corroboration and/or rejection of the research hypothesis [and therefore] it is important that authors report details related to study design, data collection, preprocessing, and analysis in their papers" (p. 26). Transparency in this regard may also help to foster the creation of more robust results which can then, for example, be building blocks for automated approaches to the analysis and use of data (e.g., when creating stress-sensitive adaptive enterprise systems [20]).

Based on the evidence reported in this paper, we make a renewed call for deliberately making methodological decisions (such as those related to preprocessing of physiological data) and presenting methodological details in NeuroIS papers.

Acknowledgements This research was funded by the Upper Austrian Government as part of the Ph.D. program "Digital Business International", a joint initiative between the University of Applied Sciences Upper Austria and the University of Linz, and as part of the project "Digitaler Stress in Unternehmen" (Basisfinanzierungsprojekt) at the University of Applied Sciences Upper Austria.

References

1. Riedl, R., Davis, F.D., Hevner, A.R.: Towards a NeuroIS research methodology: intensifying the discussion on methods, tools, and measurement. J. Assoc. Inf. Systems **15**, i–xxxv (2014)
2. Fischer, T., Riedl, R.: Technostress research: a nurturing ground for measurement pluralism? Commun. Assoc. Inf. Systems **40**, 375–401 (2017)
3. Tams, S., Hill, K., de Guinea, A.O., Thatcher, J., Grover, V.: NeuroIS—alternative or complement to existing methods? illustrating the holistic effects of neuroscience and self-reported data in the context of technostress research. J. Assoc. Inf. Systems **15**, 723–753 (2014)
4. Riedl, R.: On the biology of technostress: literature review and research agenda. DATA BASE Adv. Inf. Systems **44**, 18–55 (2013)
5. Riedl, R., Kindermann, H., Auinger, A., Javor, A.: Technostress from a neurobiological perspective—system breakdown increases the stress hormone cortisol in computer users. Bus. Inf. Systems Eng. **4**, 61–69 (2012)
6. Riedl, R., Kindermann, H., Auinger, A., Javor, A.: Computer breakdown as a stress factor during task completion under time pressure: identifying gender differences based on skin conductance. Adv. Hum. Comput. Interact. 1–8 (2013)
7. Schellhammer, S., Haines, R., Klein, S.: Investigating technostress in situ: understanding the day and the life of a knowledge worker using heart rate variability. In: IEEE Proceedings of HICSS 2013, pp. 430–439 (2013)
8. Hjortskov, N., Rissén, D., Blangsted, A.K., Fallentin, N., Lundberg, U., Søgaard, K.: The effect of mental stress on heart rate variability and blood pressure during computer work. Eur. J. Appl. Physiol. **92**, 84–89 (2004)
9. Fischer, T., Riedl, R.: Theorizing technostress in organizations: a cybernetic approach. In: Thomas, O., Teuteberg, F. (eds.) Proceedings of the 12th International Conference on Wirtschaftsinformatik, pp. 1453–1467 (2015)
10. Maier, C., Laumer, S., Weinert, C., Weitzel, T.: The effects of technostress and switching stress on discontinued use of social networking services: a study of facebook use. Inf. Systems J. **25**, 275–308 (2015)
11. Ragu-Nathan, T.S., Tarafdar, M., Ragu-Nathan, B.S., Tu, Q.: The consequences of technostress for end users in organizations: conceptual development and empirical validation. Inf. Systems Res. **19**, 417–433 (2008)
12. Xhyheri, B., Manfrini, O., Mazzolini, M., Pizzi, C., Bugiardini, R.: Heart rate variability today. Prog. Cardiovasc. Dis. **55**, 321–331 (2012)
13. Task Force of the European Society of Cardiology, and the North American Society of Pacing and Electrophysiology: Heart rate variability: Standards of measurement, physiological interpretation, and clinical use. Eur. Heart J. **17**, 354–381 (1996)

14. Giles, D., Draper, N., Neil, W.: Validity of the polar V800 heart rate monitor to measure RR intervals at rest. Eur. J. Appl. Physiol. **116**, 563–571 (2016)
15. Wijaya, A.I., Prihatmanto, A.S., Wijaya, R.: Shesop Healthcare: Android Application to Monitor Heart Rate Variance, Display Influenza and Stress Condition Using Polar H7. Unpublished (2016)
16. Shaffer, F., Ginsberg, J.P.: An overview of heart rate variability metrics and norms. Frontiers Public Health **5**, 1–17 (2017)
17. Neben, T., Schneider, C.: Ad intrusiveness, loss of control, and stress: a psychophysiological study. In: AIS Proceedings of ICIS 2015 (2015)
18. Tarvainen, M.P., Niskanen, J.-P., Lipponen, J.A., Ranta-Aho, P.O., Karjalainen, P.A.: Kubios HRV—heart rate variability analysis software. Comput. Methods Programs Biomed. **113**, 210–220 (2014)
19. Kaufmann, T., Sütterlin, S., Schulz, S.M., Vögele, C.: ARTiiFACT—a tool for heart rate artifact processing and heart rate variability analysis. Behav. Res. Methods **43**, 1161–1170 (2011)
20. Adam, M.T.P., Gimpel, H., Maedche, A., Riedl, R.: Design blueprint for stress-sensitive adaptive enterprise systems. Bus. Inf. Systems Eng. **59**, 277–291 (2017)

Wearable Devices: A Physiological and Self-regulatory Intervention for Increasing Attention in the Workplace

Monica Fallon, Kai Spohrer and Armin Heinzl

Abstract Despite stress associated with work overload, employees are still expected to maintain attentional focus and generate new knowledge. However, attention in the work environment is a scarce resource making completing tasks under stress increasingly difficult. There are few technological interventions used in the IS literature targeted at both decreasing stress and increasing attention. Wearable device technologies may facilitate such processes due to their ability to collect real-time physiological measures and cue individuals at moments when they should take action. Self-regulation theories consider attentional resources and cognitive processes used to consciously control performance, thoughts, and the recognition of emotions. However, stressors reduce the availability of attentional resources, where maximum attention only occurs during moderate levels of physiological arousal. We examine both cognitive and physiological paths affecting attentional processes and propose a technology-mediated intervention to study these effects.

Keywords Wearable devices · Stress · Haptic vibrations · Self-regulation
Attention · Work overload

1 Introduction and Background

Maintaining attention is a key resource for knowledge workers. However, stress may be impeding attentional focus. Work overload is a significant workplace stressor and results in employee turnover intentions and work exhaustion [1, 2]. Work overload is the perception that assigned work exceeds an individual's capability or skill level [3,

M. Fallon (✉) · K. Spohrer · A. Heinzl
University of Mannheim, Mannheim, Germany
e-mail: fallon@uni-mannheim.de

K. Spohrer
e-mail: spohrer@uni-mannheim.de

A. Heinzl
e-mail: heinzl@uni-mannheim.de

F. D. Davis et al. (eds.), *Information Systems and Neuroscience*,
Lecture Notes in Information Systems and Organisation 29,
https://doi.org/10.1007/978-3-030-01087-4_28

4]. Increases in stress in a work environment can negatively affect employee performance [5–10]. Despite perceptions of work overload, employees are still expected to generate useful information and knowledge. In order to do so, they are required to maintain their attentional focus. However, reduced barriers to interruptions and increased expectations of availability seem to be creating an environment where demands exceed individuals' abilities [11–13]. Despite what is known about the negative effects of stress on performance and the struggle to maintain attention in the work environment, there are few interventions targeted at both decreasing stress and increasing attention.

Well-designed technology could help knowledge workers cope with stress, maintain attention, and keep up their performance. Technology may provide an effective means to promote individual self-awareness of stress [14, 15]. More specifically, wearable device technologies may facilitate such processes due to their ability to collect real-time physiological measures as an indicator of stress and cue individuals at moments when they should take action [16]. By using elementary features of wearable devices (i.e. heart rate sensors, notifications, and haptic vibrations) in combination with cognitive and physiological processes, we aim to reduce stress and increase attention.

Although IS research has focused on taking breaks to reduce strain from overload [17, 18], less research has examined the subsequent cognitive effects, despite calls to study these effects [14]. Taking a break may be important for maintaining attention, especially when performing resource intensive tasks [19]. However, simply giving employees the opportunity to take breaks may not always be effective. Motivated individuals may set high norms, making other workers feel pressured to continue working [11]. This pressure may be relieved through a short and discreet intervention facilitated by a wearable device. A short physiological and cognitive intervention may provide the stress-relieving benefits to sustain attention.

Resource allocation theories consider attentional resources in terms of *cognitive processes* and how they are allocated to direct and maintain attentional focus [20]. Such cognitive processes include self-regulation, which enables the ability to monitor performance and maintain attentional control [19, 21]. Self-regulation theories describe a cognitive process for the ability to consciously control task performance, thoughts, and the recognition of emotions [22]. For example, individuals who are more aware of their current mental activity, including those who have been trained in mindfulness techniques, are better able to regulate their attention [23]. Furthermore, stressors reduce the availability of attentional resources, where maximum attention only occurs during moderate levels of physiological arousal [24–26]. This illustrates a *physiological process* for decrements in attention. For example, stressed individuals tend to lose attentional focus and have more off-task thoughts [19, 24, 25]. Off-task thoughts are thoughts unrelated to the task at hand and detracts from performance on attentional tasks. Thus, we examine cognitive and physiological paths affecting attentional processes and propose a technology-mediated intervention. Against this backdrop, we aim to answer the following research questions:

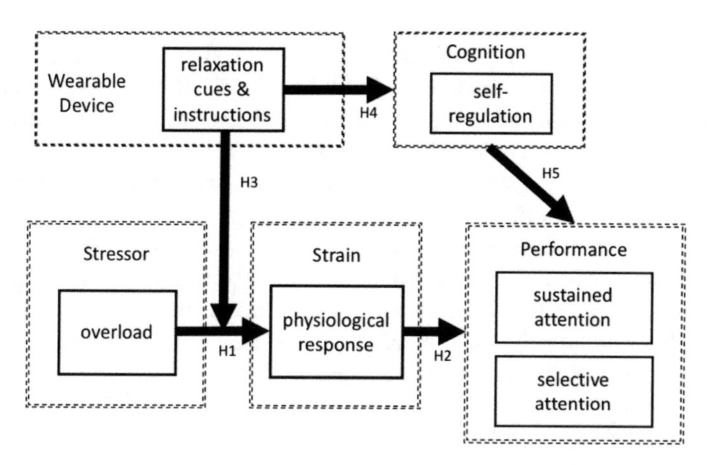

Fig. 1 Research model

(1) How can self-regulatory theories be used to explain the effects of stress on attention?
(2) Can cues on a wearable device facilitate relaxation techniques and decrease stress through physiological processes?
(3) Can relaxation cues on a wearable device increase attention through self-regulatory cognitive processes?

2 Hypotheses Development

Lazarus and Folkman [27] describe stress as "a particular relationship between the person and the environment that is appraised by the person as taxing or exceeding his or her resources" (p. 19). By this description, overload—the perception that assigned work exceeds an individual's capability or skill level [3, 4]—would also induce stress. Overload is experienced when the requirements of the task are too high and there are too many demands for the individual to fill [28]. Strain is the resulting psychological and physiological response to these environmental demands [29]. An increase in heart rate has been shown to accompany perceptions of a high load and compares favorably with identifiable stress during the work day [30, 31]. Thus, we propose that when facing overload, individuals will have a physiological response that can be detected by heart rate sensors on a wearable device (Fig. 1).

H1: *Overload will positively affect the physiological response to stress.*

Stressors reduce the availability of attentional resources, where maximum attention only occurs during moderate levels of physiological arousal [24–26]. Participants facing overload may have more work to complete, thus, providing more opportunities to seem productive. However, an increased physiological response due to overload results in more errors [32]. We propose that more errors are the result of attentional lapses due to strain. When experiencing stress, individuals tend to have more off-task thoughts and decrements in attention [18, 20, 24, 33, 34]. Thus, we expect that when individuals face overload and are strained, they will also have a reduced availability of attentional resources. Therefore, we propose:

H2: *The physiological response to overload has a negative effect on sustained attention and selective attention.*

We propose that the features implemented in the wearable device will facilitate deep breathing (see Sect. 3.1 for features). Deep breathing contributes to a physiological response characterized by decreased heart rate and increased parasympathetic activity [35]. Additionally, deep breathing can aid in decreasing perceived stress [36–39]. We propose that the features implemented in the wearable device will facilitate these processes.

H3: *Relaxation cues and instructions will negatively moderate the relationship between overload and the physiological response to overload.*

Breathing techniques are an essential part of mindfulness [40], which promotes a balance between a relaxed and attentive state of mind [41]. Self-regulation is an essential component of mindful exercises where one must bring awareness and attention to the current moment and focus on the breath [40]. We propose that the features implemented in the wearable device will allow the facilitation of these self-regulatory processes.

H4: *Relaxation cues and instructions will positively affect self-regulation.*

The ability to self-regulate has been found to be a key component in enhancing cognition and attentional processes [42]. Additionally, self-regulatory processes include self-monitoring and self-evaluation. They are responsible for directing and maintaining attentional control [20]. Theories of self-regulation illustrate that people can make themselves more aware of their current mental state and can regulate their attention to suppress and prevent the onset of unwanted thoughts [21, 23]. Thus, we propose that self-regulation can be used as a way to refocus and maintain attentional control.

H5: *Self-regulation positively affects sustained attention and selective attention.*

3 Methods

3.1 Features of Wearable Device

3.2 Experimental Design

To empirically validate our research model, we plan to perform a controlled experiment that allows us to measure physiological, behavioral, and self-reported data. We will use heart rate as a physiological measure to capture real-time responses to stress. Behavioral measures will be used to measure objective performance on an attention task. Self-report measures will be used to capture self-regulatory processes

Feature	Purpose	Function
Push notification	Notifies participant of increased heart rate	Awareness of physiological response
Screen	Displays instructions telling individuals to inhale deeply while the device vibrates and exhale when it stops vibrating	Informational instructions to facilitate deep breathing
Haptic vibrations	Rhythmic and kinesthetic vibrations creating the sense of touch—vibrates slowly 6 times per minute	Sensory instructions to facilitate deep breathing

and perceptions of work overload stress. We use a multi-method approach because it can achieve better explanatory power and helps to avoid method-bias [10, 43]. Additionally, physiological measures of stress have been shown to explain and predict variance in performance over and above the prediction afforded by a self-reported stress measure [9]. We propose an independent group design with repeated measures. Participants will be randomized to one of two groups (wearable device vs control). The Apple Watch Breathe Feature has capabilities to detect heart rate and the accuracy has been tested [44]. Additionally, the application is used in combination with haptic vibrations and its goal is to relieve stress [45]. We will use this software and the Apple Watch to test our hypotheses.

Baseline Attention Task: The d2 Test of Attention will be used to measure baseline attention and changes in attention [46, 47]. The d2 Test of Attention is a paper-and pencil cancelation task measuring *sustained and selective attention*. The test was chosen because these abilities have been predicted to be positively affected by mindfulness training and brief mindfulness interventions such as deep breathing [36, 40, 48]. Such attentional processes are also important for knowledge workers [11]. The psychometric properties of the test have been well supported [49]. The d2 sheet contains 14 lines of letters, and the task is to cross out "*d*s" with two dashes, which are interspaced with distractors. The time limit for each line is 20 s.

Pre-Test for Overload: To account for individual differences in perceptions of overload, participants will complete a baseline *overload* task [32, 50]. The test involves completing anagrams, or rearranging letters to form a word. Because one participant might perceive ten anagrams in five minutes as overload, while another participant may perceive the same set of anagrams as a moderate workload, we concluded that it is necessary to get baseline measures. Participants will be given a test trial where they will be asked to fill out as many anagrams as they can while working at a normal pace. They will not explicitly be given a time limit, so they experience time pressure in the experimental trial, however, they will have five minutes to complete as many anagrams as possible. At the end of the test trial, the experimenter will collect the anagrams from participants and use the results to calculate the number of anagrams that each participant should receive in the experimental trial.

Heart Rate Baseline: Next, participants will be instructed to rest for 10 min while sitting still to measure baseline levels of physiological activity. 10 min is the resting period used in other standardized stress tasks [51]. During this time average *resting heart rate* will be measured. Heart rate is used as a measure of the sympathetic division of the autonomic nervous system and has been used in IS literature to measure physiological stress [8]. Heart rate will be continuously measured for the rest of the experiment.

Stress Overload Task: All participants will perform an experimental overload task [32, 50]. Participants will be told, "You will be given anagrams to decode and you should decode as many as possible in 5 min. The amount of anagrams you will be given is a standard rate, designed to be appropriate for college-ability students. After 5 min, the anagrams will be removed and you will be given more." Unbeknownst to participants, they will begin with 200% more anagrams compared to the pre-test trial and a timer will be displayed. Anagrams will be taken away after 5 min and participants will receive new ones. The task will go on for 15 min maximum or until the wearable device detects a heart rate above 20% of their resting heart rate. That is the average increase in heart rate after five minutes in a standardized laboratory stress task [51].

Intervention: *Wearable Device group*: When the wearable device group's heart rate exceeds 20% of their resting heart rate, the anagrams will be taken away and they will receive a push notification on the wearable device saying, "Your heart rate appears higher than normal. Take two minutes to relax and follow the instructions." Two minutes is chosen so that the intervention can be short and discreet, but still enough time to physiologically recover from stress [52]. The participant will have to click 'continue' at which point instructions will be shown, i.e. "Inhale deeply while the watch vibrates and exhale when it stops." The vibrations will be formed by haptic vibrations. Haptic vibrations are kinesthetic vibrations that create the sense of touch. These vibrations will rhythmically vibrate slowly six times per minute for two minutes. Breathing at a rate of around six breaths per minute results in increases in heart rate variability [53] reflecting a reduction of stress levels. And heart rate is strongly correlated with heart rate variability measures [54]. *Control Group*: When the control group's heart rate exceeds 20% of their resting heart rate, the anagrams will be taken away and they will wait 2 min before the next task. They will not be instructed to alter their breathing in any way and although they will be wearing a wearable device, it will not be activated.

Attention Task: All participants will perform the d2 Test of Attention for the second time in order to measure changes from baseline sustained and selective attention. We choose three outcomes hypothesized to be the most sensitive based on the literature mentioned above (1) the total error rate (E; commissions and omissions); (2) the error percentage (E%, calculated as $E/TN \times 100$, where TN represents the total number of processed items); and, following the d2 manual, (3) the error distribution (ED), defined as the error sums for three test sections (lines 1–5, lines 5–10, and lines 11–14).

Self-Report measures: Before and after the experiment participants will complete the following questionnaires: perceived stress scale [55] on-task/off-task scales [20, 56] and self-regulatory processes scales [20].

4 Expected Contributions

This study would yield several practical contributions. It will illustrate how the physiological response associated with work overload can be used as real-time information to provide appropriate timing of cues for relaxation. The integration of bio signals into IS is an advantageous design aspect [14] and we incorporate this aspect into the wearable device. Additionally, this research will contribute insights into how elementary features of a wearable device, such as cues and haptic vibrations, can be used to facilitate relaxation techniques. Such a device could be used in a work environment where employees need a high level of attention and have a high workload, for example financial traders or investors [5]. Additionally, the implications of such a device for agile software development teams may be promising, where individuals may work better as a team when under less stress.

This study will also provide theoretical insights for IS research by illustrating cognitive and physiological mechanisms for how a wearable device can influence both strain and performance. We explain these processes through resource allocation and self-regulatory theories. Self-regulation and physiological arousal appear to be important factors influencing the maintenance of attention [20, 22, 26]. We propose a technology-mediated intervention using these mechanisms to positively affect attention. Our comprehensive model illustrating the proposed effects will bring together research on resource allocation theories, self-regulation theories, and how information technology can facilitate these processes.

The results from this study will provide opportunities for future research. For example, these ideas may be extended to examine automatic use processes and how they can be facilitated through wearable devices. The effects on attention may be amplified through extended learning with the wearable device. After learning has occurred, a task becomes more automatic. Thus, secondary tasks can be completed with little cognitive effort [19, 57]. Therefore, individuals may be able to complete relaxation tasks while performing other primary tasks and maintaining attention. Using the elementary features described above may potentially facilitate these processes with little cognitive effort. Future research should consider how learning affects these processes.

References

1. Moore, J.: One road to turnover: an examination of work exhaustion in technology professionals. MIS Q. **24**(1), 141–168 (2000)

2. Ayyagari, R., Grover, V., Purvis, R.: Technostress: technological antecedents and implications. MIS Q. **35**(4), 831–858 (2011)
3. Kahn, R.: Organizational Stress: Studies in Role Conflict and Ambiguity. Wiley, New York, NY (1964)
4. McGrath, J.: Stress and Behavior in Organizations. Rand McNally College Publishing, Chicago (1976)
5. Lo, A.W., Repin, D.V.: The psychophysiology of real-time financial risk processing. J. Cogn. Neurosci. **14**(3), 323–339 (2002)
6. Ellis, A.P.J.: System breakdown: the role of mental models and transactive memory in the relationship between acute stress and team performance. Acad. Manag. J. **49**(3), 576–589 (2006)
7. Tarafdar, M., Pullins, E.B., Ragu-Nathan, T.S.: Technostress: negative effect on performance and possible mitigations. Inf. Syst. J. **25**(2), 103–132 (2015)
8. Riedl, R.: On the biology of technostress: literature review and research agenda. Data Base Adv. Inf. Syst. **44**(1), 18–55 (2013)
9. Tams, S., Thatcher, J., Hill, K., Ortiz de Guinea, A., Grover, V.: NeuroIS—alternative or complement to existing methods? Illustrating the holistic effects of neuroscience and self-reported data in the context of technostress research. J. Assoc. Inf. Syst. **15**, 723–753 (2014)
10. Moody, G.D., Galletta, D.F.: Lost in cyberspace: the impact of information scent and time constraints on stress, performance, and attitudes online. J. Manag. Inf. Syst. **32**(1), 192–224 (2015)
11. Davis, G.B.: Anytime/anyplace computing and the future of knowledge work. Commun. ACM **45**(12), 67–73 (2002)
12. Davenport, T., Beck, J.: The Attention Economy: Understanding the New Currency of Business. Harvard Business School Press, Cambridge, MA (2001)
13. Tu, Q., Wang, K., Shu, Q.: Computer-related technostress in China. Commun. ACM **48**(4), 77–81 (2005)
14. Astor, P.J., Adam, M.T.P., Jerčić, P., Schaaff, K., Weinhardt, C.: Integrating biosignals into information systems: a NeuroIS tool for improving emotion regulation. J. Manag. Inf. Syst. **30**(3), 247–278 (2013)
15. Adam, M.T., Gimpel, H., Maedche, A., Riedl, R.: Design blueprint for stress-sensitive adaptive enterprise systems. Bus. Inf. Syst. Eng. **59**(4), 277–291 (2017)
16. Patel, M., Asch, D., Volpp, K.: Wearable devices as facilitators, not drivers of health behavior change. *JAMA* (2015)
17. Galluch, P.P.S., Grover, V., Thatcher, J.J.B.: Interrupting the workplace: examining stressors in an information technology context. J. Assoc. Inf. Syst. **16**(1), 1–47 (2015)
18. Brillhart, P.E.: Technostress in the workplace: managing stress in the electronic workplace. J. Am. Acad. Bus. **5**, 302–307 (2004)
19. Randall, J.G., Oswald, F.L., Beier, M.E.: Mind-wandering, cognition, and performance: a theory-driven meta-analysis of attention regulation. *Psychol. Bull.* 1–21, (2014)
20. Kanfer, R., Ackerman, P.L.: Motivation and cognitive abilities: an integrative/aptitude-treatment interaction approach to skill acquisition. J. Appl. Psychol. **74**(4), 657–690 (1989)
21. Smallwood, J., Schooler, J.W.: The restless mind. Psychol. Bull. **132**(6), 946–958 (2006)
22. Baumeister, R., Heatherton, T., Tice, D.: Losing Control: How and Why People Fail at Self-Regulation. Academic Press (1994)
23. Mrazek, M.D., Franklin, M.S., Phillips, D.T., Baird, B., Schooler, J.W.: Mindfulness training improves working memory capacity and GRE performance while reducing mind wandering. Psychol. Sci. **24**(5), 776–781 (2013)
24. Humphreys, M.S., Revelle, W.: Personality, motivation, and performance: a theory of the relationship between individual differences and information processing. Psychol. Rev. **91**(2), 153–184 (1984)
25. Hancock, P.A., Warm, J.: A dynamic model of stress and sustained attention. Hum. Factors **31**(5), 519–537 (1989)

26. Hockey, G.: Compensatory control in the regulation of human performance under stress and high workload: a cognitive-energetical frameowrk. Biol. Psychol. **45**, 73–93 (1997)
27. Lazarus, R., Folkman, S.: Stress, Appraisal, and Coping. Springer, New York (1984)
28. Tarafdar, M., Tu, Q., Ragu-Nathan, B., Ragu-Nathan, T.: The impact of technostress on role stress and productivity. J. Manag. Inf. Syst. **24**(1), 301–328 (2007)
29. Selye, H.: Handbook of stress: Theoretical and clinical aspects, pp. 7–17. Free Press, New York (1993)
30. Trimmel, M., Meixner-Pendleton, M., Haring, S.: Stress response caused by system response time when searching for information on the Internet. Hum. Factors **45**(4), 615–621 (2003)
31. Hexnigan, J.K., VVortham, A.W.: Analysis of workday stresses on industrial managers using heart rate as a criterion. *Ergonomics* **18**(6), 675–681 (1975)
32. Sales, S.: Some effects of role overload and role underload. Organ. Behav. Hum. Perform. **5**(6), 592–608 (1970)
33. Smallwood, J., et al.: Subjective experience and the attentional lapse: task engagement and disengagement during sustained attention. Conscious. Cogn. **13**, 657–690 (2004)
34. Hansen, A., Johnsen, B., Thayer, J.: Vagal influence on working memory and attention. Int. J. Psychophysiol. **48**, 263–274 (2003)
35. Varvogli, L., Darviri, C.: Stress management techniques: evidence-based procedures that reduce stress and promote health. Heal. Sci. J. **5**(2), 74–89 (2011)
36. Ma, X., et al.: The effect of diaphragmatic breathing on attention, negative affect and stress in healthy adults. Front. Psychol. **8**, 1–12 (2017)
37. Arsenio, W.F., Loria, S.: Coping with negative emotions: connections with adolescents' academic performance and stress. J. Genet. Psychol. **175**(1), 76–90 (2014)
38. Prinsloo, G.E., Derman, W.E., Lambert, M.I., Rauch, L.H.: The effect of a single session of short duration biofeedback- induced deep breathing on measures of heart rate variability during laboratory-induced cognitive stress: a pilot study. Appl Psychophysiol Biofeedback **38**, 81–90 (2013)
39. Wang, S.-Z., et al.: Effect of slow abdominal breathing combined with biofeedback on blood pressure and heart rate variability in prehypertension. J. Altern. Complement. Med. **16**(10), 1039–1045 (2010)
40. Bishop, S.R., et al.: Mindfulness: a proposed operational definition. Clin. Psychol. Sci. Pract. **11**(3), 230–241 (2004)
41. Wallace, A.: The Attention Revolution: Unlocking the Power of the Focused Mind. Wisdom Publications, Boston, MA (2006)
42. Moore, A., Malinowski, P.: Meditation, mindfulness and cognitive flexibility. Conscious. Cogn. **18**(1), 176–186 (2009)
43. Fischer, T., Riedl, R.: Technostress research: a nurturing ground for measurement pluralism? Commun. Assoc. Inf. Syst. **40**(17), 375–401 (2017)
44. Wang, R., et al.: Accuracy of wrist-worn heart rate monitors. Jama Cardiol. **2**(1), 104–106 (2017)
45. Apple: Use the Breathe App (2017) [Online]. Available: https://support.apple.com/en-gu/HT206999
46. Brickenkamp, R.: Afmerksamkeits-Belastungs-Test (Test d2) [The D2 Test of Attention], 9th edn. Hogrefe, Göttingen, Germany (2002)
47. Brickenkamp, R., Zillmer, E.: The D2 Test of Attention. Hogrefe and Huber, Seattle, WA (1998)
48. Zeidan, F., Johnson, S.K., Diamond, B.J., David, Z., Goolkasian, P.: Mindfulness meditation improves cognition: evidence of brief mental training. Conscious. Cogn. **19**(2), 597–605 (2010)
49. Bates, M.E., Lemay, E.P.: The d2 test of attention: construct validity and extensions in scoring techniques. J. Int. Neuropsychol. Soc. **10**(3), 392–400 (2004)
50. Froggatt, K., Cotton, J.: The impact of type A behavior pattern on role overload-induced stress and performance attributions. J. Manage. **13**(1), 87–98 (1987)
51. Kirschbaum, C., Pirke, K.-M., Hellhammer, D.H.: The 'trier social stress test'—a tool for investigating psychobiological stress responses in a laboratory setting. Neuropsychobiology **28**, 76–81 (1993)

52. Kudielka, B., Buske-Kirschbaum, A., Hellhammer, D., Kirschbaum, C.: Differential heart rate reactivity and recovery after psychosocial stress (TSST) in healthy children, younger adults, and elderly adults: the impact of age and gender. Int. J. Behav. Med. **11**(2), 116–121 (2004)
53. Lehrer, P.M., et al.: Heart rate variability biofeedback increases baroreflex gain and peak expiratory flow. Psychosom. Med. **805**, 796–805 (2003)
54. Hart, J.: Association between heart rate variability and novel pulse rate variability methods. J. Can Chiropr. Assoc. **57**(3), 243–250 (2012)
55. Cohen, S., Kamarck, T., Mermelstein, R.: A global measure of perceived stress. J. Health Soc. Behav. **24**(4), 385–396 (1983)
56. Matthews, G., Joyner, L., Gilliland, K., Campbell, S.E., Falconer, S., Huggins, J.: Validation of a comprehensive stress state questionnaire: towards a state big three. Personal. Psychol. Eur. **7**, 335–350 (1999)
57. Ferratt, T.W., Prasad, J., Dunne, E.J.: Fast and slow processes underlying theories of information technology use. J. Assoc. Inf. Syst. **19**(1), 1–22 (2018)

Exploring Flow Psychophysiology in Knowledge Work

Michael T. Knierim, Raphael Rissler, Anuja Hariharan, Mario Nadj and Christof Weinhardt

Abstract We report on a first exploration of a new paradigm to study flow physiology in knowledge work that we call controlled experience sampling (cESM) in order to build a bridge for flow physiology research to more unstructured tasks. Results show that the approach elicits a consistent flow experience with intensities as least as high as in an established difficulty-manipulated math task. Yet, significantly lower stress perceptions and heart rate variability (HRV) responses are found in the cESM approach which highlights gaps and consequences for the diagnostic potential of HRV features for the understanding of flow physiology and automated flow observation in bio-adaptive systems.

Keywords Flow · Psychophysiology · Knowledge work · Adaptive systems

1 Introduction

Flow, the experience of complete involvement in a challenging task where action occurs fluidly [1], describes a desirable state in the work environment as it is supposed to improve worker performance and well-being [2, 3]. Due to the complexity of knowledge work, flow has also been proposed as a metric to evaluate knowledge

M. T. Knierim (✉) · R. Rissler · A. Hariharan · M. Nadj · C. Weinhardt
Karlsruhe Institute of Technology (KIT), Karlsruhe, Germany
e-mail: michael.knierim@kit.edu

R. Rissler
e-mail: raphael.rissler@kit.edu

A. Hariharan
e-mail: anuja.hariharan@kit.edu

M. Nadj
e-mail: mario.nadj@kit.edu

C. Weinhardt
e-mail: christof.weinhardt@kit.edu

F. D. Davis et al. (eds.), *Information Systems and Neuroscience*,
Lecture Notes in Information Systems and Organisation 29,
https://doi.org/10.1007/978-3-030-01087-4_29

work outcomes and work environments [4]. While flow facilitation is still a major challenge [3, 5], due to the complex requirements (e.g. absence of distractions, structure of the task, challenge of the task, physiological and psychological state of the individual, etc.) [3, 5], especially the advancements on flow physiology [6, 7] propose interesting avenues for supportive bio-adaptive systems [8–10]. However, at present most of the physiological research is conducted in highly controlled game tasks, leaving gaps to understand flow physiology in more unstructured tasks that are typical in knowledge work [4, 11]. Especially as knowledge work demand is estimated to increase strongly due to artificial intelligence advancements [12], we aim to build a bridge towards increased external validity by adapting an original flow research method, the experience sampling method (ESM) [13], to a laboratory setting. This presents a controlled approach (cESM) in which individuals work on a knowledge work task whilst being observed using neurophysiological sensors and being interrupted multiple times to "catch flow in the act". By analyzing experience across interruptions, and by comparing them to a standard flow induction approach, we aim to answer the main research question of how well the cESM approach can elicit flow. Knowledge workers are a promising target population for flow physiology research as they are highly trained individuals that face challenging tasks requiring creative solutions. With this property, they fall perfectly in the theoretically described situation for flow experience [14]. Also, knowledge workers are often times working in sedentary positions in computer-mediated environments, which makes their natural work environment already similar to laboratory settings and ideal for physiological assessment. In sum, our work contributes significantly to the present literature by (1) advancing the understanding of flow elicitation in laboratory settings, by (2) extending flow physiology research to the knowledge work context, and by (3) delivering insights into flow physiology across tasks.

2 Background

Flow Theory. Flow is characterized by nine distinct dimensions in Csikszentmihalyi's theory: (1) challenge-skill balance, (2) clear goals, (3) unambiguous feedback, (4) autotelic experience, (5) action-awareness merging, (6) sense of control, (7) loss of self-consciousness, (8) transformation of time, and (9) concentration on the task at hand [15]. Dimensions 1–3 are deemed antecedents, 5–9 characteristics of the experience itself, and dimension 4 a consequence of flow experiences [16]. Flow has been studied in a wide range of different activities from sports [17] to artistic and musical performance [18, 19], computer gaming [11, 20], literary writing [1, 21] or reading [22], and been found to appear remarkably similar across tasks.

Research Paradigms. The flow characterizations have in the past primarily been developed from self-reports (first interviews, then surveys, especially by use of the ESM) [13, 23]. The ESM was developed early in flow research to overcome interview limitations and catch flow while it is occurring through repeated interruption over the course of a day [13]. Only more recently experimental flow induction has been developed with the primary paradigm of difficulty manipulation (DM) [23]. By providing a task in a low, balanced, or high difficulty condition, experiences of boredom, flow, and overload are to be elicited respectively. While the approach has been deemed useful to elicit contrasts, it has also been criticized as to whether real flow experiences are elicited (given the low involvement often present in experiment tasks) [11].

Physiology. Flow physiology research has extensively made use of the DM paradigm. Having surveyed 20 studies on the peripheral nervous system (PNS) [6] and two studies published since then [14, 24], we find that 12 of 22 studies used this paradigm. Furthermore, 16 of 22 studies used game tasks. This shows a focus with limitations to external validity, that has spawned calls for creative laboratory research [7]. In this research, we provide an approach adaption by transferring the ESM method to a controlled setting (cESM), with similarities to previous work with musicians [25, 26]. In doing so, we build on established psychometric instruments, like the Flow Short Scale (FKS) [27] to assess flow experience, together with biometric measures. Central hypothesis on PNS flow physiology are moderate [28–30] or high sympathetic activation in flow [25, 31, 32] (compared to boredom or overload experiences). Also, the parasympathetic modulation of sympathetic nervous system activity (maybe even non-reciprocal co-activation) [20, 24, 28] has been discussed. Therefore, especially HRV metrics have so far played a central role in understanding PNS activation levels during flow [6].

3 Method

Materials and Procedure. Our study was conducted in a laboratory setting with air-conditioned cabins, one participant at a time. Each participant worked on (1) writing a research project report, and on (2) solving math equations manipulated in difficulty both within the same session. Scientific writing represents a challenging and frequent task for scholars and students (typical knowledge workers). Furthermore, writing has in the past been related to flow and engaging experiences [1, 21, 33]. Participants brought their own thesis, were given time to review the state of their work and to define a challenging, yet achievable goal for a writing session of 20–25 min using the SMART mnemonic [34, 35]. The math task was chosen as reference to an established, DM flow induction paradigm [32, 36, 37]. This task, in which participants sum two or more numbers, was replicated from [36] with two adjustments as task difficulty was found to be too high in initial pre-tests. The adjustments are: In the boredom condition, subjects had to always solve randomly generated equations in one of three forms ($101 + 1, +2, \text{or} +3$). In the flow condition, difficulty was increased/decreased when three responses in a row were correct/incorrect. The order of math and writing task was randomized and so were the orders of the three math task conditions, resulting in a total number of 12 procedures (2 tasks * 3! math condition combinations) that were all executed once. The complete procedure is outlined in Fig. 1.

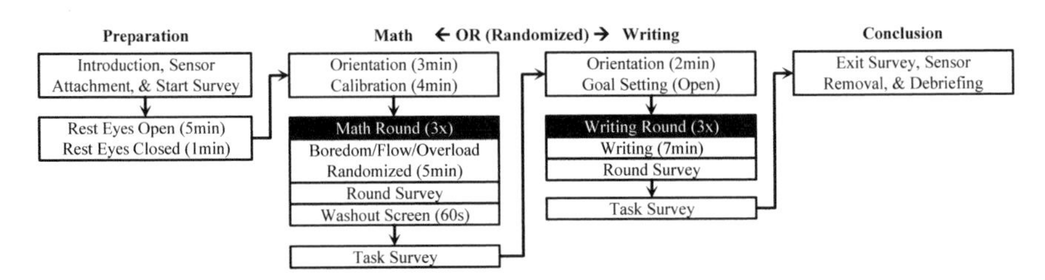

Fig. 1 The experiment procedure visualized

Measures. Round surveys included instruments on flow and task demand (the ten item Flow Short Scale by [27] and one additional task demand item also by [27]), stress (a five item construct by [38]), and affect (the single item SAM scales for arousal and valence by [39]), amongst others. Task surveys included instruments on task importance [27]. All reported survey instruments used 7-point Likert scales (with the exception of the SAM affect scales which used 9-point Likert scales). ECG data was collected using gelled chest electrodes in Lead II configuration. EDA data was collected from the left foot using gold cup electrodes. Both ECG and EDA were collected through a BiosignalsPlux hub at a sampling frequency of 1000 Hz. Additionally, participants wore a 14-channel Emotiv Epoc + EEG headset. In this report, we focus on the analyses of ECG data only.

Participants. 12 students (3 female) ages 21–30 participated voluntarily. In the recruitment survey participants reported average thesis challenge levels of 4.3 (SD: 0.98) and Wilcoxon comparisons showed no difference in preference for math or writing tasks as measured by three items from [36].

4 Results

Data Processing. In the scope of this article we report on four psychometric (flow, demand, stress, arousal) and two physiological (RMSSD, HF-HRV as two HRV features derived in processing ECG data) variables compared across six sampling points (three math conditions, three writing interruptions). For all variables, outliers were removed (>2 SD from the construct mean). Normal distribution (Shapiro-Wilk test) and variance homogeneity (Fligner-Killen test) were violated for many samples, prompting the use of non-parametric tests. One item was removed from the stress construct, improving Cronbach's Alpha in the math boredom condition. Given the high internal consistency across samples (see Table 1), and corroborating results from the self-assessment manikin arousal item [39], the stress construct was kept in the analysis. Friedman tests indicated the presence of main effects at significant levels ($p < .01$ for psychometric data, $p < .05$ for physiological data). Variable means and standard deviations are shown in Table 2. Post hoc pairwise Wilcoxon comparisons of group means are shown in Table 3.

Table 1 Cronbach's alpha levels across sampling points (M-F/B/O/All = math boredom/flow/overload/all math samples, W1-3/All = writing sample 1-3/all writing samples)

Variable	MB	MF	MO	M-All	W1	W2	W3	W-All
Flow	0.68	0.88	0.89	0.80	0.77	0.98	0.93	0.95
Stress	0.84	0.58	0.78	0.84	0.84	0.78	0.90	0.82

Table 2 Variable means and standard deviations (in parentheses) across sampling points

Variable	MB	MF	MO	W1	W2	W3
Flow	4.03 (0.80)	4.53 (1.16)	4.02 (1.21)	5.43 (0.59)	4.93 (1.55)	5.09 (1.14)
Stress	2.96 (1.32)	4.18 (0.74)	4.75 (1.18)	2.81 (1.26)	2.79 (1.02)	2.64 (1.07)
Arousal	2.73 (1.19)	6.18 (0.98)	6.27 (1.27)	3.33 (1.50)	2.82 (1.08)	3.91 (1.64)
Demand	1.45 (0.93)	5.18 (0.60)	6.09 (0.70)	4.42 (0.90)	4.17 (1.19)	3.73 (1.01)
ΔRMSSD	−5.55 (9.44)	−1.42 (9.12)	−4.54 (8.05)	−10.49 (7.36)	−8.91 (6.01)	−8.59 (10.21)
ΔHF-HRV	−151.30 (319.30)	−197.48 (189.06)	−232.65 (199.07)	−434.16 (286.05)	−453.47 (305.14)	−374.23 (376.30)

Demand. For a difficulty manipulation check, the demand variable was inspected (cf. [28, 31]). Significant differences were found between all math conditions, showing increasing demand from boredom to overload conditions, confirming manipulation success. The demand levels in the writing samples lay consistently between the math boredom and overload condition, and possibly even below the math flow condition, indicated at trend level ($p < .09$). Aside from one trend level difference between writing sample 1 and 3, no differences were found within the writing task. In all other variables, no differences were found across the writing samples and comparison data was therefore excluded from the report (Table 4).

Flow. Comparisons of the FKS reports indicate significant differences between the math flow and overload condition, and repeated, significant differences between the math boredom and overload conditions with the writing samples. Also, there is a trend level indication of higher flow in the first writing sample than in the math flow condition. Within the writing task, there were no significant differences. Therefore, flow was reported at least as high in writing consistently across all writing task interruptions. We also find support for this observation of consistency in the range of flow reports per participant (mean range = 1.13, SD = 0.62).

Stress and Arousal. Comparison of stress reports revealed significant differences between all three math task conditions, with increasing levels from boredom to overload. Within the writing task, the stress levels did not differ significantly. Stress levels were consistently lower in writing than in the math flow and overload conditions. Arousal report comparisons show almost the same pattern, with increasing arousal from math boredom to overload, although with only a significant difference between boredom condition and the other two. Arousal levels were consistently lower in writing than in math flow and overload conditions.

Table 3 Psychometric data wilcoxon mean comparison *p*-values

	Demand			Flow			Stress			Arousal		
	MB	MF	MO	MB	MF	MO	MB	MF	MO	MB	MF	MO
MF	**<.01**			>.1			**<.05**			**<.01**		
MO	**<.01**	**<.05**		>.1	**<.05**		**<.01**	**<.05**		**<.01**	>.1	
W1	**<.01**	**<.09**	**<.01**	**<.01**	**<.09**	**<.01**	>.1	**<.05**	**<.01**	>.1	**<.01**	**<.01**
W2	**<.01**	**<.09**	**<.01**	>.1	>.1	**<.05**	>.1	**<.01**	**<.01**	>.1	**<.01**	**<.01**
W3	**<.01**	**<.05**	**<.01**	**<.05**	>.1	**<.09**	>.1	**<.01**	**<.01**	**<.09**	**<.01**	**<.01**

Bold indicates significant results

Table 4 Physiological data wilcoxon mean comparison *p*-values

	ΔRMSSD			ΔHF-HRV		
	MB	MB	MB	MB	MF	MO
MF	>.1			>.1		
MO	>.1	**<.05**		>.1	>.1	
W1	>.1	**<.01**	**<.05**	**<.05**	**<.09**	**<.09**
W2	>.1	**<.05**	**<.05**	**<.05**	**<.05**	**<.05**
W3	>.1	**<.05**	>.1	**<.05**	**<.05**	**<.09**

HRV. We furthermore analyzed HRV metrics that have been central to previous flow physiology research [6]. In order to conduct this analysis, ECG data was processed using the Python toolbox NeuroKit [40] to derive time-series data of adjacent heartbeat intervals (RR-intervals). Afterwards, based on the RR-interval data, HRV features were computed in the same toolbox and cross-validated using the R toolbox RHRV [41]. Similar to this research (cf. [20, 28, 31]) change scores were used in the analysis ($\Delta HRV = HRV_{task} - HRV_{baseline}$) of five-minute window time-domain (SDNN, RMSSD) and frequency-domain (LF-HRV, HF-HRV) features preceding each survey. Friedman tests of main effects across the sampling points were only significant for RMSSD and HF-HRV, which is why only these two measures were investigated further. Both are similar in post hoc test results. Wilcoxon tests of mean differences revealed for the RMSSD feature a significantly higher level in the math flow condition than in the math overload condition. This was not corroborated by the HF-HRV feature, which indicated similar HRV levels across math task conditions, albeit with a decrease trend from boredom to overload condition, with the flow condition falling in-between. Similarity across sampling points was also found for all the writing task samples, in both features. However, most importantly, comparisons revealed significantly lower HRV levels in both RMSSD and HF-HRV in comparison to both math flow and math overload tasks. HF-HRV was also significantly lower in all writing task conditions compared to the math boredom condition, indicating consistently stronger parasympathetic withdrawal during the writing task.

Task Importance. After math and writing task, participants rated the importance of the task [27], with no significant differences (math mean = 3.82, writing mean = 4).

5 Discussion and Conclusion

Discussion. Within the math task, our results suggest a successful manipulation with comparable results to previous research, showing that flow experience (as indicated by self-reports) is experienced most strongly when task demands are optimally balanced with a participants skill levels [28, 29, 31]. Within the writing task, we find that all presented variables indicate a consistent experience, despite repeated interruption. This finding is important as interruptions are often considered a prime cause of lack of flow experiences [42]. Therefore, we initially anticipated more experiential variance due to repeated task interruption. It is possible that other factors in the writing task design (like the facilitation of goal setting early on) helped to mediate this interruption impact. Goal setting has been found to be an important step in the writing process that facilitates high quality work [34] and is theorized a prime conductor for flow experiences in the original theory [15]. We take the results here as first support that the cESM approach with a writing task can be used to elicit flow, at least at similar intensities that are elicited with a standard DM paradigm (the math task). However, even though tasks are reported as similarly important, writing is experienced as less stressful and demanding, and shows a stronger HRV reduction. A key reason for the stress difference could be the design-related, contrasted presence of multiple stress factors (demand overload, social-evaluative threat, lack of control) [43]. These factors have in the past purposefully been introduced to flow experiment designs in order to elicit motivated task performances [28, 36, 43] and have resulted in repeated sightings of psychometric reports that paint a picture of increased stress/arousal in balance and overload conditions compared to boredom conditions, even in contexts (e.g. gaming tasks) where threat experiences could be less likely than in for example math or public speaking tasks [20, 28, 29, 43]. Our results indicate, that a task that is naturally important to the individual [25], yet lacks these stressors, results in similar reported flow intensities without perceptions of strain. The critique on the aptitude of the difficulty manipulation paradigm to elicit real flow experience could therefore receive some support [11]. However, it should also be pointed out, that these results might indicate a central limitation to how flow is collected psychometrically (in this study, but also in general), as there could be experiential components to flow that are not captured by the FKS [27] that was employed here.

Within the math task, our HRV results are comparable with previous work showing increased PNS activation from flow to overload conditions (i.e. moderate activation in flow) [14, 24, 28, 29]. Within the writing task, the HRV similarity across the sampling points further supports the observation of a consistent experience. However, given this consistency it is hard to say if the reduced HRV (compared to the math task) is due to a qualitatively different flow experience, or due to other variables (e.g. task complexity, effort, etc.). Nevertheless, the finding that even though writing is perceived as less stressful, the corollary of lower HRV is interesting and could alternatively indicate that the proposition, that flow is actually a state of high physiological activation, is correct [25, 31, 32]. In any case, the comparison of the two tasks could explain these previously contrasting findings (i.e. why in some studies the physiological results point to moderate PNS activation and in some to high PNS activation) to be caused by experimental paradigms. Lastly, the results could also show a potential complication for future bio-adaptive systems work that uses physiological thresholds to infer experiential states.

Conclusion and Future Work. A central limitation of this study is the small sample size, which is why the results should be treated with care. Future work should increase variance in the writing task (e.g. by including a writing boredom phase, or temporally varied sampling), together with more psychometric scales (involvement, effort, effortlessness, etc.) to enable more nuanced insights on experiential and physiological processes. The indication of the aptitude of the writing task to elicit flow experiences in the lab is the first major contribution of this work. It highlights the potential of using the cESM approach as an alternative to study flow in laboratory settings. Secondly, the research contributes to the study of flow by extending it to the knowledge work domain. Thirdly, we contribute by studying the comparability and utility of physiological measures to observe flow. More work is needed to help us understand if our findings indicate increased or decreased diagnostic potential of PNS markers to observe flow experience in the context of knowledge work [9] and if CNS data, e.g. electroencephalographic (EEG) data, is additionally required for flow-supportive bio-adaptive systems. In this direction, most recent work highlights the potential of identifying EEG correlates of flow experiences in a DM paradigm (using the same math task that was used in this article) issuing calls for comparison with other tasks [44].

References

1. Csikszentmihalyi, M.: Flow and the Psychology of Discovery and Invention. Harper Collins, New York (1996)
2. Bakker, A.B., van Woerkom, M.: Flow at work: a self-determination perspective. Occup. Heal. Sci. **1**, 47–65 (2017)
3. Spurlin, S., Csikszentmihalyi, M.: Will work ever be fun again? In: Fullagar, C.J., Delle Fave, A. (eds.) Flow at Work: Measurement and Implications, pp. 176–187 (2017)
4. Quinn, R.W.: Flow in knowledge performance experience. Adm. Sci. Q. **50**, 610–641 (2005)
5. Ceja, L., Navarro, J.: "Suddenly I get into the zone": examining discontinuities and nonlinear changes in flow experiences at work. Hum. Relations. **65**, 1101–1127 (2012)
6. Knierim, M.T., Rissler, R., Dorner, V., Maedche, A., Weinhardt, C.: The psychophysiology of flow: a systematic review of peripheral nervous system features. Lect. Notes Inf. Syst. Organ. **25**, 109–120 (2017)
7. Harris, D.J., Vine, S.J., Wilson, M.R.: Neurocognitive mechanisms of the flow state. Prog. Brain Res. **237**, 221–243 (2017)
8. Adam, M.T.P., Gimpel, H., Maedche, A., Riedl, R.: Design blueprint for stress-sensitive adaptive enterprise systems. Bus. Inf. Syst. Eng. **59**, 277–291 (2016)
9. Rissler, R., Nadj, M., Li, M.X., Knierim, M.T., Maedche, A.: Got flow? Using machine learning on physiological data to classify flow. In: Proceedings of the Conference of Human Factors in Computing Systems (CHI) (2018)
10. Léger, P.M., Davis, F.D., Cronan, T.P., Perret, J.: Neurophysiological correlates of cognitive absorption in an enactive training context. Comput. Human Behav. **34**, 273–283 (2014)
11. Moller, A.C., Meier, B.P., Wall, R.D.: Developing an experimental induction of flow: effortless action in the lab. In: Bruya, B. (ed.) Effortless Attention, pp. 191–204 (2010)
12. Frey, C.B., Osborne, M.A.: The future of employment: how susceptible are jobs to computerisation? Technol. Forecast. Soc. Change. **114**, 254–280 (2017)
13. Csikszentmihalyi, M., Hunter, J.: Happiness in everyday life: the uses of experience sampling. J. Happiness Stud. **4**, 185–199 (2003)
14. Klarkowski, M.: The Psychophysiologial Evaluation of the Player Experience (2017)
15. Csikszentmihalyi, M.: Beyond Boredom and Anxiety. Jossey-Bass, San Francisco, CA (1975)
16. Nakamura, J., Csikszentmihalyi, M.: Flow theory and research. In: Lopez, S., Snyder, C.R. (eds.) Oxford Handbook of Positive Psychology, pp. 195–206. Oxford University Press, New York (2009)

17. Swann, C., Keegan, R.J., Piggott, D., Crust, L.: A systematic review of the experience, occurrence, and controllability of flow states in elite sport. Psychol. Sport Exerc. **13**, 807–819 (2012)
18. Gruzelier, J., Inoue, A., Smart, R., Steed, A., Steffert, T.: Acting performance and flow state enhanced with sensory-motor rhythm neurofeedback comparing ecologically valid immersive VR and training screen scenarios. Neurosci. Lett. **480**, 112–116 (2010)
19. de Manzano, O., Theorell, T., Harmat, L., Ullén, F.: The psychophysiology of flow during piano playing. Emotion **10**, 301–311 (2010)
20. Harmat, L., de Manzano, Ö., Theorell, T., Högman, L., Fischer, H., Ullén, F.: Physiological correlates of the flow experience during computer game playing. Int. J. Psychophysiol. **97**, 1–7 (2015)
21. Erhard, K., Kessler, F., Neumann, N., Ortheil, H.J., Lotze, M.: Professional training in creative writing is associated with enhanced fronto-striatal activity in a literary text continuation task. Neuroimage **100**, 15–23 (2014)
22. McQuillan, J., Conde, G.: The conditions of flow in reading: two studies of optimal experience. Read. Psychol. **17**, 109–135 (1996)
23. Moneta, G.B.: On the measurement and conceptualization of flow. In: Engeser, S. (ed.) Advances in Flow Research, pp. 23–50 (2012)
24. Tian, Y., Bian, Y., Han, P., Wang, P., Gao, F., Chen, Y.: Physiological signal analysis for evaluating flow during playing of computer games of varying difficulty. Front. Psychol. **8**, 1–10 (2017)
25. de Manzano, O., Theorell, T., Harmat, L., Ullén, F., de Manzano, Ö., Theorell, T., Harmat, L., Ullén, F.: The psychophysiology of flow during piano playing. Emotion **10**, 301–311 (2010)
26. Harmat, L., Ullen, F., de Manzano, O., Olsson, E., Elofsson, U., von Scheele, B., Theorell, T.: Heart rate variability during piano playing: a case study of three professional solo pianists playing a self-selected and a difficult prima vista piece. Music Med. **3**, 102–107 (2011)
27. Engeser, S., Rheinberg, F.: Flow, performance and moderators of challenge-skill balance. Motiv. Emot. **32**, 158–172 (2008)
28. Tozman, T., Magdas, E.S., MacDougall, H.G., Vollmeyer, R.: Understanding the psychophysiology of flow: a driving simulator experiment to investigate the relationship between flow and heart rate variability. Comput. Human Behav. **52**, 408–418 (2015)
29. Klarkowski, M., Johnson, D., Wyeth, P., Phillips, C., Smith, S.: Psychophysiology of challenge in play: EDA and self-reported arousal. In: Proceedings of the 2016 CHI Conference Extended Abstracts on Human Factors in Computing Systems, pp. 1930–1936. ACM Press, New York, New York, USA (2016)
30. Bian, Y., Yang, C., Gao, F., Li, H., Zhou, S., Li, H., Sun, X., Meng, X.: A framework for physiological indicators of flow in VR games: construction and preliminary evaluation. Pers. Ubiquitous Comput. **20**, 821–832 (2016)
31. Keller, J., Bless, H., Blomann, F., Kleinböhl, D.: Physiological aspects of flow experiences: skills-demand-compatibility effects on heart rate variability and salivary cortisol. J. Exp. Soc. Psychol. **47**, 849–852 (2011)
32. Ulrich, M., Keller, J., Gro, G., Grön, G.: Neural signatures of experimentally induced flow experiences identified in a typical fMRI block design with BOLD imaging. Soc. Cogn. Affect. Neurosci. **11**, 496–507 (2016)
33. Galluch, P.S., Grover, V., Thatcher, J.B.: Interrupting the workplace: examining stressors in an information technology context. J. Assoc. Inf. Syst. **16**, 1–47 (2015)
34. Flower, L., Hayes, J.R.: A cognitive process theory of writing. Coll. Compos. Commun. **32**, 365–387 (1981)
35. Doran, G.T.: There's a S.M.A.R.T. way to write managements's goals and objectives. Manage. Rev. **70**, 35–36 (1981)
36. Ulrich, M., Keller, J., Hoenig, K., Waller, C., Grön, G.: Neural correlates of experimentally induced flow experiences. Neuroimage **86**, 194–202 (2014)
37. Keller, J.: The flow experience revisited: the influence of skills-demands-compatibility on experiential and physiological indicators. In: Harmat, L., Andersen, F.O., Ullén, F., Wright, J., Sadlo, G. (eds.) Flow Experience, pp. 351–374 (2016)

38. Tams, S., Hill, K., Ortiz de Guinea, A., Thatcher, J., Grover, V.: NeuroIS—alternative or complement to existing methods? Illustrating the holistic effects of neuroscience and self-reported data in the context of technostress research. J. Assoc. Inf. Syst. **15**, 723–753 (2014)
39. Bradley, M.M., Lang, P.J.: Measuring emotion: the self-assessment manikin and the semantic differential. J. Behav. Ther. Exp. Psychiatry **25**, 49–59 (1994)
40. Makowski, D.: NeuroKit (2016). https://github.com/neuropsychology/NeuroKit.py
41. Martinez, C.A.G., Quintana, A.O., Vila, X.A., Touriño, M.J.L., Rodriguez-Liñares, L., Presedo, J.M.R., Penin, A.J.M.: Heart Rate Variability Analysis with the R Package RHRV. Springer (2017)
42. Rissler, R., Nadj, M., Adam, M.T.P., Mädche, A.: Towards an integrative theoretical framework of IT-mediated interruptions. In: Proceedings of the 25th European Conference on Information Systems (ECIS), Guimarães, Portugal, pp. 1950–1967 (2017)
43. Tozman, T., Zhang, Y.Y., Vollmeyer, R.: Inverted U-shaped function between flow and cortisol release during chess play. J. Happiness Stud. **18**, 1–22 (2017)
44. Katahira, K., Yamazaki, Y., Yamaoka, C., Ozaki, H., Nakagawa, S., Nagata, N.: EEG correlates of the flow state: a combination of increased frontal theta and moderate frontocentral alpha rhythm in the mental arithmetic task. Front. Psychol. **9**, 1–11 (2018)

Asking Both the User's Heart and Its Owner: Empirical Evidence for Substance Dualism

Ricardo Buettner, Lars Bachus, Larissa Konzmann and Sebastian Prohaska

Abstract Mind-body physicalism is the metaphysical view that all mental phenomena are ultimately physical phenomena, or are necessitated by physical phenomena. Mind-body dualism is the view that at least some mental phenomena are non-physical. While mind-related concepts are usually measured using questionnaires, body-related concepts are measured using physiological instruments. We breakdown the narrowed measuring approaches within the simplified mind-body discussion to all four possible substance-measuring pairs and evaluate the mind-body substance dualism theory versus the physicalism theory applying perceived and physiological measured stress data using a wearable long-term electrocardiogram recorder. As a result we derive empirical evidence and strong arguments against physicalism, and assess the overall strength of the benefits of NeuroIS instruments as complementary measures.

Keywords NeuroIS · Mind-body problem · Substance dualism · Physicalism
Stress · Health data

1 Introduction

The mind-body problem is one of the most fundamental problems in NeuroIS research. This problem refers to open philosophical questions concerning the existence of a human mind and if so, the (causal) interaction of the mind with its physical body [1]. While the mind is concerned with mental processes, thoughts and consciousness, the body is concerned with the physical aspects of the brain and how the brain is structured [2]. One major theoretical approach refers to the idea that the mind does not exist distinct from the brain (physicalism); the opposite approach comprises the idea that mind and body are different substances (substance dualism). If physicalism is true, human perceptions—which Information Systems scholars investigate

---

R. Buettner (✉) · L. Bachus · L. Konzmann · S. Prohaska
Aalen University, Aalen, Germany
e-mail: ricardo.buettner@hs-aalen.de

F. D. Davis et al. (eds.), *Information Systems and Neuroscience*,
Lecture Notes in Information Systems and Organisation 29,
https://doi.org/10.1007/978-3-030-01087-4_30

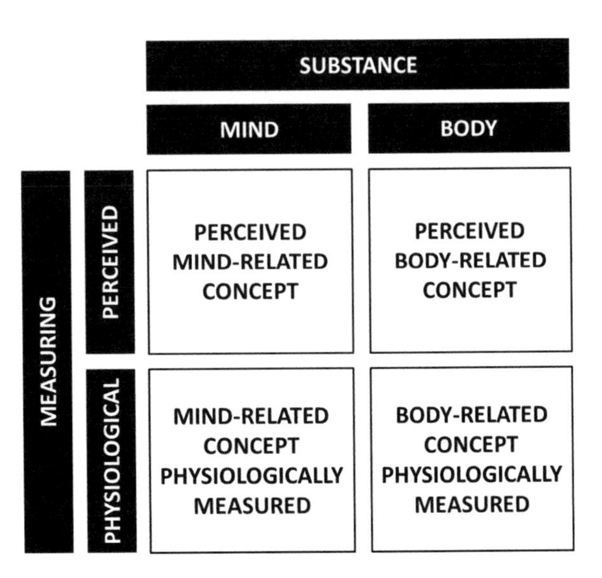

Fig. 1 The unfolded mind-body matrix

using self-rating instruments—simply reflect physical/brain activity, which could be directly analyzed using psychophysiological NeuroIS instruments. If substance dualism is true, it is interesting to see whether NeuroIS instruments can add complementary or contrary insights to subjective self-ratings.

While scholars have established arguments for both, physicalism and dualism (see [2] for an overview), we aim to derive more complete answers by systematically unpacking elements involved in the mind-body problem to all four possible substance-measuring pairs. More precisely, the stress concept can be measured using physiological instruments such as Heart Rate Variability [3–5] (body-related access) or using questionnaires such as the Perceived Stress Scale [6] (mind-related access). The discourse on the mind-body problem often narrows the discussion in the way that a concept is either measured physiologically (body-related concept physiologically measured) or mentally (perceived mind-related concept). This is not only true for the stress concept, but also for all other information systems concepts (e.g. mental workload [7–9]). To open the discussion to the two other possible measuring options (perceived body-related concept; mind-related concept physiologically measured), we propose the unfolded "mind-body matrix" as shown in Fig. 1.

To evaluate the possibility of gaining more insights into the mind-body problem using the unfolded mind-body matrix, this paper analyzes perceived stress data and body-related stress data, simultaneously captured from 851 participants.

While stress in general and technostress in particular are important concepts in the Information Systems (IS) and NeuroIS literatures [10–14], the stress context for the evaluation of the proposed mind-body matrix is an important one.

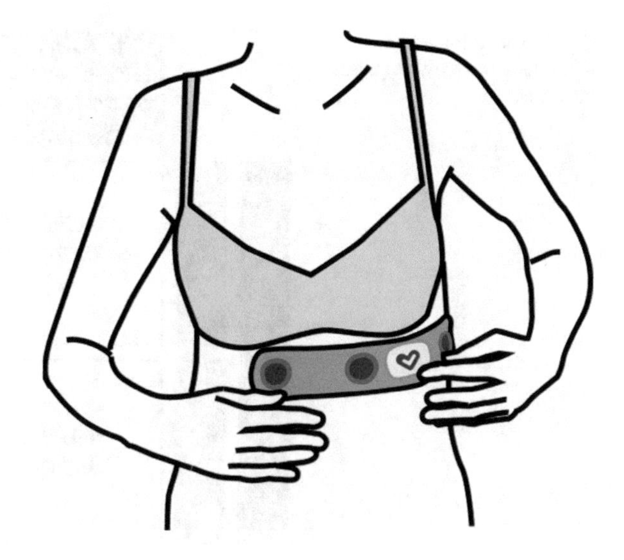

Fig. 2 The picture shows how the electrodes are attached to the body

2 Methodology

2.1 Data and Participants

Data used for this analysis were collected by corvolution GmbH, a spin-off company of the Karlsruhe Institute of Technology. 851 people (mean [±SD] age: 43.7 ± 11.9 years; 361 females, 490 males) participated in the study.

2.2 Instruments and Devices

Physically health data was recorded using the CM200 device, a wearable long-term electrocardiogram (ECG) recorder. The CM220 uses a patch-electrode-system with dry electrodes (Fig. 2). To improve ECG signal quality it is comprised of a 2-channel ECG redundancy and acquires additional context-data: barometer, thoracic-impedance, acceleration and temperature.

Based on this physiological health data, the Baevsky Stress Index [3] was calculated and used in our analysis. The Baevsky Stress Index is a heart-rate-variability derivative which is widely accepted in medicine to evaluate the state of the autonomic regulatory mechanisms of the cardiac rhythm under physical or psychological loads [4, 5]. Perceived stress as a mind-related concept was measured using the Perceived Stress Scale [6] with reliability according to Cronbach's α of 0.89. Perceived stress as a body-related concept was measured using the list of Thirty Stress Symptoms (e.g. 'I'm often tired over the day.', 'I often have tachycardia.', 'I often have digestive problems.') by Nathan and Rosch [15] with reliability measured by Cronbach's α of

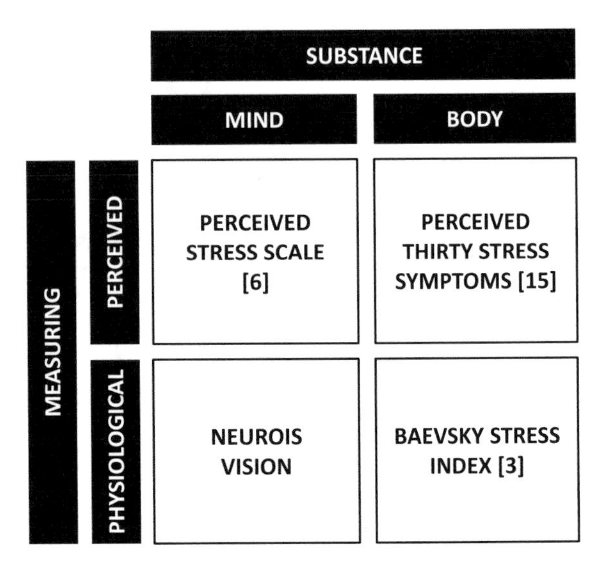

Fig. 3 The unfolded mind-body matrix with stress measures

0.94. Perceived stress as a mind-related concept directly measured by a physiological instrument is an open issue and an aim especially of NeuroIS scholars [16–19] (Fig. 3).

2.3 Procedure

Participants were asked to wear the CM200 electrocardiogram recorder over two days and to answer the questionnaires (Perceived Stress Scale and Perceived Thirty Stress Symptoms) after each day. All participants worked for at least one of the two days the measurement took place. They had no special task list to fulfil. The raw data from which the aggregated data were extracted comprise of 415 MB for each 48 h measurement. The Baevsky Stress Index was calculated using all data points during recording. A mean of the answers was derived from the questionnaires over two days.

3 Results

The relationships within the unfolded mind-body matrix can be found in Table 1.

Table 1 Relationships within the unfolded mind-body matrix (Spearman correlation)

Relation	Level of correlation	Level of significance
Perceived stress scale [6] and perceived thirty stress symptoms [15]	0.45	$p < 0.001$
Perceived thirty stress symptoms [15] and Baevsky stress index [3]	0.12	$p < 0.01$
Perceived stress scale [6] and Baevsky stress index [3]	0.037	n.s.

4 Discussion

On the pure perception level, our results show a substantial correlation of 0.45 between stress as a mind-related concept and stress as a body-related concept ($p < 0.001$). Participants perceived a substantial level of congruency between stress as a whole and stress symptoms.

But, from a pure body perspective we only found a very low correlation between the physiologically measured stress data and the perception of stress symptoms ($r = 0.12$, $p < 0.01$). In addition, from the traditional mind-body perspective we found no relationship between the perception of stress as a whole and the physiologically measured stress data (n.s.).

To summarize, we found (a) a substantial congruency on the perception level between mind and body, but (b) no or even a negligible relationship between mind and body data from a measurement perspective (physiologically measured versus perceived data). Our study offers empirical evidence against physicalism since stress perceptions simply did not reflect physiological stress data; neither for perceived stress as a whole (mind-related) nor for perceived stress symptoms (body-related). Using the unfolded mind-body matrix our study strengthens the arguments against physicalism, which overall strengthens the arguments for using NeuroIS instruments as complementary measures [10–14, 16–20].

5 Limitation and Future Work

The main limitation is related to the fact that the unfolded mind-body matrix is also based on the narrow conceptualizations of the concepts 'mind' and 'body/brain' as it is assumed in dualism and physicalism—assumptions that can also be challenged [2, 21].

In this work we report on body-related stress data physiologically measured only according to the Baevsky Stress Index. In the extended manuscript we will show the results for other ECG derivatives such as beat-to-beat measures, sinus arrhythmia, bradycardia or tachycardia [22].

In addition, future work could use other physiological stress measures such as skin conductance or stress hormones [12, 13, 20].

Acknowledgements We would like to thank the whole team of corvolution GmbH for collecting the data used for this analysis. In addition, we would like to thank the reviewers, who provided very helpful comments on the refinement of the paper.

References

1. Buettner, R.: Asking both the user's brain and its owner using subjective and objective psychophysiological neurois instruments. In: ICIS 2017 Proceedings: 38th International Conference on Information Systems, Seoul, South Korea, 10–13 December 2017 (2017)
2. Kim, J.: Mind in a Physical World. MA, U.S., MIT Press, Cambridge (2000)
3. Baevskii, R.M.: Analysis of heart rate variability in space medicine. Hum. Physiol. **28**(2), 202–213 (2002)
4. Zilov, V.G., Khadartsev, A.A., Eskov, V.V., Eskov, V.M.: Experimental study of statistical stability of cardiointerval samples. Bull. Exp. Biol. Med. **164**(2), 115–117 (2017)
5. Eskov, V.M., Eskov, V.V., Filatova, O.E.: Characteristic features of measurements and modeling for biosystems in phase spaces of states. Meas. Tech. **53**(12), 1404–1410 (2011)
6. Cohen, S., Kamarck, T., Mermelstein, R.: A global measure of perceived stress. J. Health Soc. Behav. **24**(4), 385–396 (1983)
7. R. Buettner, Cognitive Workload of Humans Using Artificial Intelligence Systems: Towards objective measurement applying eye-tracking technology. In: KI 2013 Proc., ser. LNAI, vol. 8077, pp. 37–48 (2013)
8. Buettner, R., Daxenberger, B., Eckhardt, A., Maier, C.: Cognitive workload induced by information systems: Introducing an objective way of measuring based on pupillary diameter responses. In: Pre-ICIS HCI/MIS 2013 Proceedings, 2013, Paper 20 (2013)
9. Buettner, R.: Analyzing mental workload states on the basis of the pupillary hippus. In: NeuroIS'14 Proceedings, p. 52 (2014)
10. Fischer, T., Riedl, R.: Technostress research: a nurturing ground for measurement pluralism? Commun. Assoc. Inf. Syst. **40**, 375–401 (2017)
11. Riedl, R., Kindermann, H., Auinger, A., Javor, A.: Computer breakdown as a stress factor during task completion under time pressure: Identifying gender differences based on skin conductance. Adv. Hum. Comp. Interact. 1–8 (2013)
12. Riedl, R.: On the biology of technostress: literature review and research agenda. DATA BASE. Adv. Inf. Syst. (ACM SIGMIS Database) **44**(1), 18–55 (2013)
13. Riedl, R., Kindermann, H., Auinger, A., Javor, A.: Technostress from a neurobiological perspective: system breakdown increases the stress hormone cortisol in computer users. Bus. Inf. Syst. Eng. **4**(2), 61–69 (2012)
14. Tams, S., Hill, K., de Guinea, A.O., Thatcher, J., Grover, V.: NeuroIS—alternative or complement to existing methods? illustrating the holistic effects of neuroscience and self-reported data in the context of technostress research. J. Assoc. Inf. Syst. **15**(10), 723–753 (2014)
15. Nathan, R.G., Rosch, P.J.: The Doctor's Guide to Instant Stress Relief. Random House Publishing Group, New York (1989)
16. Riedl, R., Banker, R.D., Benbasat, I., Davis, F.D., Dennis, A.R., Dimoka, A., Gefen, D., Gupta, A., Ischebeck, A., Kenning, P., Müller-Putz, G., Pavlou, P.A., Straub, D.W., vom Brocke, J., Weber, B.: On the foundations of NeuroIS: reflections on the gmunden retreat 2009. CAIS **27**, 243–264 (2010)

17. Riedl, R., Fischer, T., Léger, P.M.: A decade of neurois research: Status quo, challenges, and future directions. In: ICIS 2017 Proceedings: 38th International Conference on Information Systems, 10–13 December 2017, Seoul, South Korea (2017)
18. Dimoka, A., Banker, R.D., Benbasat, I., Davis, F.D., Dennis, A.R., Gefen, D., Gupta, A., Ischebeck, A., Kenning, P.H., Pavlou, P.A., Müller-Putz, G., Riedl, R., vom Brocke, J., Weber, B.: On the use of neurophysiological tools in is research: developing a research agenda for NeuroIS. MIS Quarterly **36**(3), 679–A19 (2012)
19. Riedl, R., Léger, P.-M.: Fundamentals of NeuroIS: Information Systems and the Brain. Springer, Berlin (2016)
20. Riedl, R., Davis, F.D., Hevner, A.R.: Towards a NeuroIS research methodology: Intensifying the discussion on methods, tools, and measurement. J. Assoc. Inf. Syst. **15**(10), 1–35 (2014)
21. Bennett, M.R., Hacker, P.M.S.: Philosophical Foundations of Neuroscience. Blackwell, Oxford (2003)
22. Thayer, J.F., Åhs, F., Fredrikson, M., Sollers, J.J., Wager, T.D.: A meta-analysis of heart rate variability and neuroimaging studies: Implications for heart rate variability as a marker of stress and health. Neurosci. Biobehav. Rev. **36**(2), 747–756 (2012)

Cardiovascular, Neurophysiological, and Biochemical Stress Indicators: A Short Review for Information Systems Researchers

Jürgen Vogel, Andreas Auinger and René Riedl

Abstract We conducted a systematic review of the scientific literature on indicators for stress measurement. The full texts of 128 articles (published in the period 1970–2017) were analyzed and we identified a total of 21 different stress indicators, including cardiovascular, neurophysiological, and biochemical measures. Moreover, we analyzed the frequency of use of the indicators. Glucocorticoids including the hormone cortisol (52 out of 128 articles), heart rate (HR) and heart rate variability (HRV) (50/128), as well as diastolic blood pressure (DBP) and systolic blood pressure (SBP) (40/128), are the dominant stress indicators. Also, we found that half of the articles (64/128) report about at least two different stress indicators, thus a combination of biological measurement approaches is relatively common in stress research. This review holds value for researchers in the Information Systems (IS) discipline and related interdisciplinary research fields such as technostress.

Keywords Blood pressure · Heart rate · Hormones · Measurement · Stress
Technostress

1 Introduction

Stress, in many respects, affects our social lives. By influencing our behavior individually, it has a distinct effect on communities, organizations, and society [1]. While in Eastern cultures stress is a sign of an unbalanced peace of mind, in Western countries

J. Vogel (✉) · R. Riedl
Johannes Kepler University, Linz, Austria
e-mail: k00621741@students.jku.at

R. Riedl
e-mail: rene.riedl@jku.at

A. Auinger · R. Riedl
University of Applied Sciences Upper Austria, Steyr, Austria
e-mail: andreas.auinger@fh-steyr.at

F. D. Davis et al. (eds.), *Information Systems and Neuroscience*,
Lecture Notes in Information Systems and Organisation 29,
https://doi.org/10.1007/978-3-030-01087-4_31

it is viewed as "a loss of control" [2]. According to the American Psychological Association, in the US over two-thirds of the 2020 adults are supposed to be experiencing stress symptoms like fatigue. Already in 2005, every fifth person in the European Union was affected by work-related stress [3]. Expecting these numbers to rise, there is a great demand of research with the objective of developing stress-reducing methods. Moreover, a specific form of stress, referred to as technostress (i.e., stress caused by the use and ubiquity of information and communication technologies [4]), has become a major issue in organizations and in society in general.

In the last century, stress research has been typically divided into the areas of psychological, behavioral, or physiological studies [5]. Later, more specialized fields (i.e., psychoneuroimmunology) emerged [6]. Especially in scientific subfields like neuroscience, recently, there has been an increasing interest in human stress responses, including behavioral changes [1]. Stress is a "complex concept" often used in different ways with various meanings [7]. In general, a stress response should help people to manage challenging situations by keeping the organism in a state of homeostasis [1].

However, what does the term "stress" actually mean? Its physiological response was first shown empirically in 1936 by Hans Selye and at that time described as "a syndrome produced by diverse nocuous agents" [8]. In 1973, he stated: "Everybody knows what stress is and nobody knows what it is" [9]. Five years later, he defined stress as physiological non-specific reaction to exterior or interior demands [10]. The basic idea was that stress is a bodily response to environmental stimuli [5]. However, today we know that stress is not just a physiological reaction, but also related to psychological phenomena, such as a person's perception and cognitive reflection on a demanding situation, which, in turn, can result in physiological, neurophysiological, and behavioral responses [1, 2]. It generally describes a state that arouses the physiological or psychological homeostasis of our organism [11] and can even be activated by purely psychological states [6].

Importantly, stressor refers to the stimuli or challenging event, while stress response refers to the physical and emotional reaction or feedback caused by the stressor [12]. Whether and how humans respond to stress is determined by how we perceive a stimulus and how we react to it. Across paradigms, stress can be summarized as "a condition in which an individual is aroused by an aversive situation [...]. The magnitude of the stress and its physiological consequences are influenced greatly by the individual's perception of its ability to control the presence or intensity of the stimulation" [13].

Apart from subjective assessments like self-reports and standardized questionnaires, most commonly, the measurement of stress is conducted by recording cardiovascular or biochemical processes and/or states [5]. More precisely, these processes and/or states typically represent a heightened excitability, vigilance, or arousal, which can be monitored and evaluated using, for example, electrocardiography, electroencephalography, neurochemical levels, or behavioral activity [13]. Therefore, a variety of signals produced by the body or chemical substances in saliva, blood or urine samples are examined. We describe major signals and substances in the following.

The cardiovascular system responds to stress with increased heart rate (HR) and blood pressure (BP) [3, 5, 14]. Blood pressure can be measured as systolic blood pressure (SBP) or diastolic blood pressure (DBP). Past research has shown that SBP may be more strongly related to interpersonal stressors than DBP [15].

The endocrine system reacts to demanding situations by changing neurochemistry [13]. To restore homeostasis disturbed by environmental demands, a physiological reaction starts, involving autonomic nervous, neuroendocrine, metabolic, and immune system components. The hypothalamus-pituitary-adrenal (HPA) axis, including the hormone cortisol, plays a key role in this response system (for a simple description of the HPA axis processes, see [16]). Additionally, catecholamines (i.e., adrenaline and noradrenaline) are released into the circulation by the adrenal medulla [1, 11, 17]. It follows that cortisol and/or catecholamines are a reliable and valid marker of stress-related neurophysiological changes in our body. These substances are usually determined in blood or saliva samples [13, 15, 18]. Another sensitive salivary biomarker used for measuring stress-related changes in the sympathetic nervous system is salivary alpha-amylase (sAA) [14, 19, 20].

So far, the influence of stress on various physiological processes is well documented in the academic literature. In this paper, we give an interdisciplinary overview about the procedures and indicators typically used for measuring stress symptoms with a focus on the frequency of use. Consequently, we identify the most common and prevalent methods. However, in this paper we do not evaluate the reliability, validity, or accuracy of different measurement approaches. From an Information Systems (IS) perspective this paper has the goal to comprehensively, yet parsimoniously, inform technostress researchers about cardiovascular, neurophysiological, and biochemical stress indicators. Technostress has become a major topic in the IS literature in the last decade, and the topic has been investigated increasingly often based on neurobiological approaches [4, 16, 21–26]. The technostress topic is also highly important from a practical perspective because increasingly more people worldwide are negatively affected by the use and ubiquity of information and communication technologies (e.g., unreliability of systems such as computer breakdowns or long and variable systems response times, or permanent use of smartphones). Thus, studying (techno) stress measurement approaches is critical for the progress of the field, thereby also contributing to the development of effective countermeasures in practice.

In the next section, we describe the search process and underlying concept of the literature research, followed by a review of articles that used, criticized, or reviewed indicators for the measurement of stress. After briefly discussing the results, we close this paper by providing insight on future research directions.

2 Method

We searched for studies in which any approach for measuring stress was used including review articles and critiques without restrictions on publication date and application domain. The research was conducted using the online search service EBSCOhost, a collection of full-text databases, e-books, subject indexes, point-of-care medical references, and historical digital archives.

During the search process *stress* was considered as a master keyword, and was therefore combined as *title* element by default. The keyword list has been extended continuously by backward research (reviewing older literature cited in articles of previous search results) and forward research (identifying articles citing key articles of previous search results) [27]. The complete keyword list for the search field *title* is presented in footnote 1 (*AND* combinations listed in brackets).[1] The search by *title* elements and their combinations was limited by the following *subject terms* categories: physiology, neurology, neuroscience, psychology, behavioral science, biology, medicine, radiology, business, and information science.

We included articles in our review which met the following criteria: (i) source type: academic journal or conference paper; (ii) availability: full text; (iii) language: English; (iv) research subjects: humans; and (v) research purpose: review of, or study using, stress measurement approaches.

3 Results

In total, the keyword search yielded a result list of 676 articles. The first step of screening involved evaluating full titles and subjects to determine whether articles met the purpose of this literature review and fulfilled all defined criteria (e.g., articles dealing with animal research were excluded). A total of 164 full text articles were downloaded and considered for an abstract analysis.[2] Overall, databases covered publication dates from 1970 onwards (to 2017), all as indexed by March 2018, the month in which the queries were executed.

Next, the abstract of each article was examined, and all articles that matched the five inclusion criteria were checked to match the complete article against these criteria. Studies with vague abstracts to allow for immediate elimination also received a full text review. During this step, 36 articles were identified as ineligible and

---

[1]Physiological, neurological, neurophysiological, electrophysiological, neural, nerve-related, biological (indicator, measure, response, parameter, feedback, reaction), measurement, measuring, reading, record, logging, notes, testing, scan, report, monitoring, evaluation (device, tool, equipment, appliance, instrument, application, gadget, hardware, software, machine, utensil, implement, technology); pupil dilation; dilation, pupillary, pupil, constriction (response); miosis; mydriasis; eye tracking; eye tracker; eye movement; electrodermal activity; EDA; skin conductance (level, response); galvanic skin, electrodermal, sympathetic skin (response); GSR; EDR; SCR; psychogalvanic reflex; PGR; SCL; E-meter; biofeedback; heart, heart rate, HR, pulse rate (variability); HRV; heart activity; blood pressure; BP; electrocardiograph; ECG; EKG; electrocardiogram; brain; electroencephalography; EEG; magnetic resonance imaging; MRI; computed tomography scan; CT scan; CT; hormones; strain; load; workload; pressure; liability; level measurement; distress; nervous.

[2]The sources of the articles and databases are (number of downloads in brackets): PsycINFO (60), MEDLINE (47), Psychology and Behavioral Sciences Collection (30), SocINDEX (9), Business Source Premier (8), ERIC (7), PSYNDEX (2), GreenFILE (1).

not considered in further analyses. Finally, the full texts of 128 articles were analyzed to obtain detailed information about different stress indicators and measuring approaches. The focus of our analyses was on the methods section of the articles.

Table 1 shows the stress indicators used in the scientific literature (first column), the number of articles mentioning and/or using the indicator or method (second column), and the references (third column). Twelve papers (9.4%) were published prior to 1990, another twelve papers between 1990 and 1999, 43 (33.6%) papers were published between 2000 and 2009, and 61 papers (47.6%) in or after 2010.

Across all articles, a total of 21 different stress indicators emerged. Similar measurement approaches (e.g., SBP and DBP) are considered as one indicator. In the upper half of Table 1, which is more related to cardiovascular activity, the most frequently used indicators are HR/HRV (n = 50, 39.1%) and blood pressure (n = 40, 31.3%), followed by respiration rate (n = 8, 6.3%) and electrodermal activity (EDA) (n = 7, 5.5%). In the lower half of Table 1, which represents neurophysiological and biochemical indicators, the category of glucocorticoids including the hormone cortisol (n = 52, 40.6%) and the category of catecholamines including the hormone adrenalin (n = 13, 10.2%) outline the most commonly used stress indicators. Moreover, half of the articles (n = 64 or 50.0%) report about at least two different indicators, thus a combination of measurement approaches.

Apart from standard tools such as electrocardiography (ECG) and sphygmomanometers [29–31, 35, 43, 44, 55, 56, 75, 95, 117], also less frequently used tools were identified during full text analysis: Impedance cardiography (ICG) [55, 75], electrooculography (EOG) [117], electromyography (EMG) [30, 53, 62], high performance liquid chromatography (HPLC) [77] and microwave radar [56].

Furthermore, as a complement to cardiovascular, neurophysiological, and biochemical indicators, in almost half of all reviewed papers (n = 63, 49.2%) self-report approaches (specific surveys and questionnaires) are used to measure stress perceptions.[3]

[3] Standard questionnaire about participants subjective stress level (e.g., job tension, physical stress, psychological stress, strain) [44, 45, 59, 60, 63, 66, 74, 106, 118–126] or stress states and motivation (subjective work load, time urgency, state anxiety, involvement) [41, 44, 59, 89, 118, 120, 127], Trier Social Stress Test (TSST) [14, 36, 38, 39, 82, 92, 95, 103], Stress-Coping Scale (SCS), State-Trait Anxiety Inventory (STAI) or Iceberg Profile (IP) [38, 50, 64, 88, 117], Profile of Mood State (POMS) [64, 67, 88, 91, 128], Cleminshaw-Guidubaldi Parent Satisfaction Scale, Coping Resources Inventory of Stress (CRIS), Family Inventory of Life Events and Changes (FILE), Global Inventory of Stress (GIS), Parenting Stress Index (PSI), Parental Stress Scale or Perceived Stress Scale (PSS) [12, 40, 128–130], Health Opinion Survey, Population Health Perspective (PHP), General Health Questionnaire (GHQ) or Patient Health Questionnaire (PHQ) [12, 30, 131, 132], Hospital Anxiety and Depression Scale (HADS) [12, 30], Self-Assessment Manikin (SAM) [31, 48], Cox's Stress/Arousal Adjective Check List (SACL) [133, 134], NASA-Task Load Index (TLX) [135, 136], Social Stress Recall Task [130], Distress Tolerance Scale (DTS) [128], Montreal Imaging Stress Task (MIST) [91], Depression Anxiety Stress Scale (DASS) [3], State-Trait Anger Expression Inventory (STAXI) [88], Sports Anxiety Scale (SAS) [18], Pressure-Activation-Stress (PAS) scale [7], Parental Responsibility Scale (PRS) [12], Brief Symptom Inventory (BSI) [40], Perceived Stress Questionnaire (PSQ) [137], Affect-arousal Grid [42], Biographic Narrative Interpretative Method (BNIM) [138], Effort-Reward Imbalance Occupational Stress Scales [139], Anchoring-and-adjustment Questionnaire [75], Job Content Questionnaire (JCQ) [77], SWS-Survey [135],

Table 1 Stress measurement indicators and frequencies

Indicator	n^a	References
Heart/pulse rate or variability (HR, HRV)	50	[2, 3, 5, 12, 14, 28–72]
Systolic or diastolic blood pressure (SBP, DBP)	40	[2, 5, 12, 14, 28, 33–35, 42–45, 49, 52–54, 59, 61–65, 69–71, 73–87]
Respiration rate, respiratory frequency (RF)	8	[29, 30, 46, 47, 50, 53, 56, 72]
Electrodermal activity (EDA), skin conductance, galvanic skin response (GSR)	7	[30, 31, 46, 47, 70, 73, 88]
(Finger) temperature	4	[30, 46, 47, 53]
Event-related potential (ERP), electroencephalography (EEG)	4	[45, 53, 89, 90]
Respiratory sinus arrhythmia (RSA)	3	[3, 43, 48]
Pulse oximetry, oxygen saturation (SO_2) at finger, photoplethysmogram (PPG)	3	[3, 37, 46]
Magnetic resonance imaging (MRI)	3	[2, 60, 91]
End tidal CO_2 ($ETCO_2$)	1	[46]
Phosphorylase activation ratio	1	[71]
Corticosteroids/glucocorticoids, steroid hormones (e.g., cortisol)	52	[1, 2, 5–7, 11, 12, 14, 17–19, 28, 36, 38, 39, 48, 51, 52, 54, 58–60, 62, 64, 68, 74, 77, 82–84, 92–113]
Catecholamines (e.g. (nor) adrenalin, (nor) epinephrin)	13	[5, 11, 54, 62, 63, 77, 84–86, 103, 107–109]
Adrenocorticotropic hormone (ACTH)	8	[2, 11, 54, 94, 96, 106, 108, 114]
Blood lipids, cholesterol	6	[5, 43, 77, 83, 108, 115]
Salivary alpha amylase (sAA)	5	[14, 19, 97, 100, 116]
Sex hormones (e.g., estrogen, testosterone)	5	[2, 34, 96, 105, 110]
Blood sugar, insulin	5	[5, 43, 77, 108, 115]
Uric acid	1	[5]
Secretory immunoglobin A (slgA)	1	[68]
Peptic ulcer	1	[5]

[a] n = Number of articles mentioning, using, and/or reviewing the indicators or methods

4 Discussion

For more than eighty years, stress and its conceptual antecedent attract the attention of scientific research [8]. With the development of disciplines such as neuroscience and discoveries in modern brain research, academia achieved a broader understanding of the complex biological mechanisms related to stress. However, despite the importance of biology in human stress, there has also been an increasing interest in the changes of human behavior in regard to stress responses [1, 6, 11, 93]. To contribute to this research domain, we focused on stress measurement. We provided a short review of the literature and identified the most commonly used stress indicators. The list of stress measurement indicators and frequencies (see Table 1) constitutes a valuable basis for IS researchers, in particular those working in the field of technostress.

The majority of articles analyzed in this review (81.2%) have been published in this century, and almost half of the total number of papers was published in the current decade. This finding indicates that our list includes the most recent research in the stress domain. Interestingly, half of the articles report more than one indicator and therefore use a combination of measurement approaches, presumably to increase validity and/or explanatory power. Across all articles, the category of glucocorticoids including the hormone cortisol (52 of 128 papers, >40%), HR/HRV (50 papers, almost 40%), and blood pressure (40 papers, >30%) were used, criticized, or reviewed with the highest frequency. We surmise that this result is not only a function of the indicators' reliability, validity, and accuracy, but also a function of their ease of use and application costs (if compared to tools such as EEG or fMRI). Overall, we found that these indicators are the three most commonly used stress measurement approaches of the past fifty years. However, there is also a growing body of research for other indicators like the salivary biomarker sAA, likely because these methods are "readily accessible and easily obtained" [20]. As a next step, the focus could be on the relevance of stress indicators for different technological stressors. It will be rewarding to see what kind of measurement approaches IS researchers will use in future studies.

Acknowledgements This research was funded by the Upper Austrian Government as part of the project "Digitaler Stress in Unternehmen" (Basisfinanzierungsprojekt) at the University of Applied Sciences Upper Austria.

Survey of Health Care Professionals [140], Adjective Checklist on Emotions (EWL) [68], List of Cues for Determining Level of Stress [141], Pearlin and Schooler's List of Emotions [142], Taylor Manifest Anxiety Scale and Teaching Anxiety Scale [87], Role Strain Scale [143], and California Test of Personality (CTP) [132].

References

1. Sandi, C., Haller, J.: Stress and the social brain. Behavioural effects and neurobiological mechanisms. Nat. Rev. Neurosci. **16**(5), 290 (2015). https://doi.org/10.1038/nrn3918
2. Verma, R., Balhara, Y.P.S., Gupta, C.S.: Gender differences in stress response. Role of developmental and biological determinants. industrial. Psychiatry J. **20**(1), 4 (2011). https://doi.org/10.4103/0972-6748.98407
3. van der Zwan, J.E., de Vente, W., Huizink, A.C., Bögels, S.M., de Bruin, E.I.: Physical activity, mindfulness meditation, or heart rate variability biofeedback for stress reduction. A randomized controlled trial. Appl. Psychophysiol. Biofeedback **40**(4), 257–268 (2015). https://doi.org/10.1007/s10484-015-9293-x
4. Riedl, R.: On the biology of technostress. Literature review and research agenda. DATA BASE Adv. Inf. Syst. (ACM SIGMIS Database) **44**(1), 18–55 (2013)
5. Fried, Y., Rowland, K.M., Ferris, G.R.: The physiological measurement of work stress. Critique. Pers. Psychol. **37**, 583–615 (1984)
6. Sapolsky, R.M.: Stress and the brain. Individual variability and the inverted-U. Nat. Neurosci. **18**(10), 1344 (2015). https://doi.org/10.1038/nn.4109
7. Isaksson, J., Nilsson, K.W., Lindblad, F.: The pressure-activation-stress scale in relation to ADHD and cortisol. Eur. Child Adolesc. Psychiatry **24**(2), 153–161 (2015). https://doi.org/10.1007/s00787-014-0544-9
8. Selye, H.: A syndrome produced by diverse nocuous agents. Nature **138**(3479), 32 (1936). https://doi.org/10.1038/138032a0
9. Selye, H.: The evolution of the stress concept. The originator of the concept traces its development from the discovery in 1936 of the alarm reaction to modern therapeutic applications of syntoxic and catatoxic hormones. Am. Sci. **61**(6), 692–699 (1973)
10. Selye, H.: The Stress of Life. McGraw-Hill, New York (1976)
11. Krugers, H.J., Hoogenraad, C.C., Groc, L.: Stress hormones and AMPA receptor trafficking in synaptic plasticity and memory. Nat. Rev. Neurosci. **11**(10), 675 (2010). https://doi.org/10.1038/nrn2913
12. Foody, C., James, J.E., Leader, G.: Parenting stress, salivary biomarkers, and ambulatory blood pressure. A comparison between mothers and fathers of children with autism spectrum disorders. J. Autism Dev. Disord. **45**(4), 1084–1095 (2015). https://doi.org/10.1007/s10803-014-2263-y
13. Kim, J.J., Diamond, D.M.: The stressed hippocampus, synaptic plasticity and lost memories. Nat. Rev. Neurosci. **3**(6), 453 (2002). https://doi.org/10.1038/nrn849
14. Inagaki, T.K., Eisenberger, N.I.: Giving support to others reduces sympathetic nervous system-related responses to stress. Psychophysiology **53**(4), 427–435 (2016). https://doi.org/10.1111/psyp.12578
15. Lucas-Thompson, R.G., Granger, D.A.: Parent-child relationship quality moderates the link between marital conflict and adolescents' physiological responses to social evaluative threat. J. Fam. Psychol.: J. Div. Fam. Psychol. Am. Psychol. Assoc. (Division 43) (2014). https://doi.org/10.1037/a0037328
16. Riedl, R., Kindermann, H., Auinger, A., Javor, A.: Technostress from a neurobiological perspective. System breakdown increases the stress hormone cortisol in computer users. Bus Inf. Syst Eng **4**(2), 61–69 (2012). https://doi.org/10.1007/s12599-012-0207-7
17. Lupien, S.J., McEwen, B.S., Gunnar, M.R., Heim, C.: Effects of stress throughout the lifespan on the brain, behaviour and cognition. Nat. Rev. Neurosci. **10**(6), 434 (2009). https://doi.org/10.1038/nrn2639
18. Meland, A., Ishimatsu, K., Pensgaard, A.M., Wagstaff, A., Fonne, V., Garde, A.H., Harris, A.: Impact of mindfulness training on physiological measures of stress and objective measures of attention control in a military helicopter unit. Int. J. Aviat. Psychol. **25**(3–4), 191–208 (2015). https://doi.org/10.1080/10508414.2015.1162639

19. Kliewer, W.: Victimization and biological stress responses in urban adolescents. Emotion regulation as a moderator. J. Youth Adolesc. **45**(9), 1812–1823 (2016). https://doi.org/10.1007/s10964-015-0398-6
20. Nater, U.M., Rohleder, N.: Salivary alpha-amylase as a non-invasive biomarker for the sympathetic nervous system. Current state of research. Psychoneuroendocrinology **34**(4), 48–496 (2009). https://doi.org/10.1016/j.psyneuen.2009.01.014
21. Fischer, T., Riedl, R.: Technostress research. A nurturing ground for measurement pluralism? Commun. Assoc. Inf. Syst. **40**(17), 375–401 (2017)
22. Adam, M.T.P., Gimpel, H., Maedche, A., Riedl, R.: Design blueprint for stress-sensitive adaptive enterprise systems. Bus. Inf. Syst. Eng. **59**(4), 277–291 (2017). https://doi.org/10.1007/s12599-016-0451-3
23. Riedl, R., Kindermann, H., Auinger, A., Javor, A.: Computer breakdown as a stress factor during task completion under time pressure. Identifying gender differences based on skin conductance. Adv. Hum-Comput. Interact. (2013). https://doi.org/10.1155/2013/420169
24. Tams, S., Hill, K., Ortiz de Guinea, A., Thatcher, J., Grover, V.: NeuroIS-alternative or complement to existing methods? Illustrating the holistic effects of neuroscience and self-reported data in the context of technostress research. J. Assoc. Inf. Syst. (JAIS) **15**(Special Issue), 723–753 (2014)
25. Fischer, T., Halmerbauer, G., Meyr, E., Riedl, R.: Blood pressure measurement. A classic of stress measurement and its role in technostress research. In: Proceedings of the Gmunden Retreat on NeuroIS (2017)
26. Fischer, T., Riedl, R.: The status quo of neurophysiology in organizational technostress research. A review of studies published from 1978 to 2015. In: Proceedings of the Gmunden Retreat on NeuroIS (2015)
27. Vom Brocke, J., Simons, A., Niehaves, B., Niehaves, B., Reimer, K., Plattfaut, R., Cleven, A.: Reconstructing the giant. On the importance of rigour in documenting the literature search process. In: ECIS 2009 Proceedings **161** (2009)
28. Bishop-Fitzpatrick, L., Minshew, N.J., Mazefsky, C.A., Eack, S.M.: Perception of life as stressful, not biological response to stress, is associated with greater social disability in adults with autism spectrum disorder. J. Autism Dev. Disorders (2017). https://doi.org/10.1007/s10803-016-2910-6
29. Brindle, R.C., Ginty, A.T., Phillips, A.C., Fisher, J.P., McIntyre, D., Carroll, D.: Heart rate complexity. A novel approach to assessing cardiac stress reactivity. Psychophysiology (2016). https://doi.org/10.1111/psyp.12576
30. Greenberg, B.R., Grossman, E.F., Bolwell, G., Reynard, A.K., Pennell, N.A., Moravec, C.S., McKee, M.G.: Biofeedback assisted stress management in patients with lung cancer. A feasibility study. Appl. Psychophysiol. Biofeedback (2015). https://doi.org/10.1007/s10484-015-9277-x
31. Janka, A., Adler, C., Fischer, L., Perakakis, P., Guerra, P., Duschek, S.: Stress in crisis managers. Evidence from self-report and psychophysiological assessments. J. Behav. Med. (2015). https://doi.org/10.1007/s10865-015-9654-1
32. South, E.C., Kondo, M.C., Cheney, R.A., Branas, C.C.: Neighborhood blight, stress, and health. A walking trial of urban greening and ambulatory heart rate. Am. J. Public Health **105**(5), 909–913 (2015)
33. van der Vijgh, B., Beun, R.-J., van Rood, M., Werkhoven, P.: Meta-analysis of digital game and study characteristics eliciting physiological stress responses. Psychophysiology (2015). https://doi.org/10.1111/psyp.12431
34. Allen, A.J., McCubbin, J.A., Loveless, J.P., Helfer, S.G.: Effects of estrogen and opioid blockade on blood pressure reactivity to stress in postmenopausal women. J. Behav. Med. (2014). https://doi.org/10.1007/s10865-012-9468-3
35. Dragomir, A.I., Gentile, C., Nolan, R.P., D'antono, B.: Three-year stability of cardiovascular and autonomic nervous system responses to psychological stress. Psychophysiology (2014). https://doi.org/10.1111/psyp.12231

36. Klaperski, S., Dawans, B. von, Heinrichs, M., Fuchs, R.: Effects of a 12-week endurance training program on the physiological response to psychosocial stress in men. A randomized controlled trial. J. Behav. Med. (2014). https://doi.org/10.1007/s10865-014-9562-9
37. Kudo, N., Shinohara, H., Kodama, H.: Heart rate variability biofeedback intervention for reduction of psychological stress during the early postpartum period. Appl. Psychophysiol. Biofeedback (2014). https://doi.org/10.1007/s10484-014-9259-4
38. Polheber, J.P., Matchock, R.L.: The presence of a dog attenuates cortisol and heart rate in the trier social stress test compared to human friends. J. Behav. Med. (2014). https://doi.org/10.1007/s10865-013-9546-1
39. Saslow, L.R., McCoy, S., van der Löwe, I., Cosley, B., Vartan, A., Oveis, C., Keltner, D., Moskowitz, J.T., Epel, E.S.: Speaking under pressure. Low linguistic complexity is linked to high physiological and emotional stress reactivity. Psychophysiology (2014). https://doi.org/10.1111/psyp.12171
40. Whited, A., Larkin, K.T., Whited, M.: Effectiveness of emWave biofeedback in improving heart rate variability reactivity to and recovery from stress. Appl. Psychophysiol. Biofeedback (2014). https://doi.org/10.1007/s10484-014-9243-z
41. Prinsloo, G.E., Rauch, H.G.L., Lambert, M.I., Muench, F., Noakes, T.D., Derman, W.E.: The effect of short duration heart rate variability (HRV) biofeedback on cognitive performance during laboratory induced cognitive stress. Appl. Cognit. Psychol. (2011). https://doi.org/10.1002/acp.1750
42. Gordon, J.L., Ditto, B., D'antono, B.: Cognitive depressive symptoms associated with delayed heart rate recovery following interpersonal stress in healthy men and women. Psychophysiology (2012). https://doi.org/10.1111/j.1469-8986.2012.01397.x
43. Sauder, K.A., Johnston, E.R., Skulas-Ray, A.C., Campbell, T.S., West, S.G.: Effect of meal content on heart rate variability and cardiovascular reactivity to mental stress. Psychophysiology (2012). https://doi.org/10.1111/j.1469-8986.2011.01335.x
44. Sawyer, P.J., Major, B., Casad, B.J., Townsend, S.S.M., Berry Mendes, W.: Discrimination and the stress response. Psychological and physiological consequences of anticipating prejudice in interethnic interactions. Am. J. Public Health **102**(5), 1020–1026 (2012)
45. Weymar, M., Schwabe, L., Löw, A., Hamm, A.O.: Stress sensitizes the brain. Increased processing of unpleasant pictures after exposure to acute stress. J. Cogn. Neurosci. **24**(7), 1511–1518 (2012)
46. Hallman, D.M., Olsson, E.M.G., Schéele, B. von, Melin, L., Lyskov, E.: Effects of heart rate variability biofeedback in subjects with stress-related chronic neck pain. A pilot study. Appl. Psychophysiol. Biofeedback (2011). https://doi.org/10.1007/s10484-011-9147-0
47. Lin, H.-P., Lin, H.-Y., Lin, W.-L., Huang, A.C.-W.: Effects of stress, depression, and their interaction on heart rate, skin conductance, finger temperature, and respiratory rate. Sympathetic-parasympathetic hypothesis of stress and depression. J. Clin. Psychol. (2011). https://doi.org/10.1002/jclp.20833
48. Oldehinkel, A.J., Ormel, J., Bosch, N.M., Bouma, E.M.C., van Roon, A.M., Rosmalen, J.G.M., Riese, H.: Stressed out? Associations between perceived and physiological stress responses in adolescents. The TRAILS study. Psychophysiology (2011). https://doi.org/10.1111/j.1469-8986.2010.01118.x
49. Palomba, D., Ghisi, M., Scozzari, S., Sarlo, M., Bonso, E., Dorigatti, F., Palatini, P.: Biofeedback-assisted cardiovascular control in hypertensives exposed to emotional stress. A pilot study. Appl. Psychophysiol. Biofeedback (2011). https://doi.org/10.1007/s10484-011-9160-3
50. Prinsloo, G.E., Derman, W.E., Lambert, M.I., Laurie Rauch, H.G.: The effect of a single session of short duration biofeedback-induced deep breathing on measures of heart rate variability during laboratory-induced cognitive stress. A pilot study. Appl. Psychophysiol. Biofeedback (2013). https://doi.org/10.1007/s10484-013-9210-0
51. Kumsta, R., Sonuga-Barke, E.: The psychological and physiological stress response in ADHD. Eur. Child Adolesc. Psychiatry **19**(Suppl. 1), 80 (2010)

52. Lévesque, K., Moskowitz, D.S., Tardif, J.-C., Dupuis, G., D'antono, B.: Physiological stress responses in defensive individuals. Age and sex matter. Psychophysiology (2010). https://doi.org/10.1111/j.1469-8986.2009.00943.x
53. Ratanasiripong, P., Sverduk, K., Hayashino, D., Prince, J.: Setting up the next generation biofeedback program for stress and anxiety management for college students. A Simple and cost-effective approach. Coll. Student J. **44**(1) (2010)
54. Ulrich-Lai, Y.M., Herman, J.P.: Neural regulation of endocrine and autonomic stress responses. Nat. Rev. Neurosci. (2009). https://doi.org/10.1038/nrn2647
55. Wang, X., Ding, X., Su, S., Li, Z., Riese, H., Thayer, J.F., Treiber, F., Snieder, H.: Genetic influences on heart rate variability at rest and during stress. Psychophysiology (2009). https://doi.org/10.1111/j.1469-8986.2009.00793.x
56. Suzuki, S., Matsui, T., Imuta, H., Uenoyama, M., Yura, H., Ishihara, M., Kawakami, M.: A novel autonomic activation measurement method for stress monitoring. Non-contact measurement of heart rate variability using a compact microwave radar. Med. Biol. Eng. Comput. (2008). https://doi.org/10.1007/s11517-007-0298-3
57. Theurel, J., Offret, M., Gorgeon, C., Lepers, R.: Physiological stress monitoring of postmen during work. Work **31**, 229–236 (2008)
58. Floyd, K., Mikkelson, A.C., Tafoya, M.A., Farinelli, L., La Valley, A.G., Judd, J., Davis, K.L., Haynes, M.T., Wilson, J.: Human affection exchange:. XIV. Relational affection predicts resting heart rate and free cortisol secretion during acute stress. Behav. Med. **32**, 151–156 (2007)
59. Taylor, S.E., Welch, W.T., Kim, H.S., Sherman, D.K.: Cultural differences in the impact of social support on psychological and biological stress responses. Psychol. Sci. (2007). https://doi.org/10.1111/j.1467-9280.2007.01987.x
60. Wang, J., Rao, H., Wetmore, G.S., Furlan, P.M., Korczykowski, M., Dinges, D.F., Detre, J.A.: Perfusion functional MRI reveals cerebral blood flow pattern under psychological stress. In: Proceedings of the National Academy of Sciences of the United States of America (2005). https://doi.org/10.1073/pnas.0503082102
61. Chafin, S., Roy, M., Gerin, W., Christenfeld, N.: Music can facilitate blood pressure recovery from stress. Brit. J. Health Psychol. (2004). https://doi.org/10.1348/1359107041557020
62. Krantz, G., Forsman, M., Lundberg, U.: Consistency in physiological stress responses and electromyographic activity during induced stress exposure in women and men. Integr. Psych. Behav. (2004). https://doi.org/10.1007/bf02734276
63. Suarez, E.C., Saab, P.G., Llabre, M.M., Kuhn, C.M., Zimmerman, E.: Ethnicity, gender, and age effects on adrenoceptors and physiological responses to emotional stress. Psychophysiology (2004). https://doi.org/10.1111/j.1469-8986.00161.x
64. Rohrmann, S., Hennig, J., Netter, P.: Manipulation of physiological and emotional responses to stress in repressors and sensitizers. Psychol. Health **17**(5), 583–596 (2002)
65. Stein, F.: Occupational stress, relaxation therapies, exercise and biofeedback. Work **17**, 235–245 (2001)
66. Iwanaga, M., Yokoyama, H., Seiwa, H.: Effects of personal responsibility and latitude for type a and b individuals on psychological and physiological stress responses. Int. J. Behav. Med. (2000). https://doi.org/10.1207/s15327558ijbm0703_02
67. Scheufele, P.M.: Effects of progressive relaxation and classical music on measurements of attention, relaxation, and stress responses. J. Behav. Med. **23**(2), 207–228 (2000)
68. Huwe, S., Hennig, J., Netter, P.: Biological, emotional, behavioral, and coping reactions to examination stress in high and low state anxious subjects. Anxiety Stress Coping **11**, 47–65 (1998)
69. Muraoka, M.Y., Carlson, J.G., Chemtob, C.M.: Twenty-four-hour ambulatory blood pressure and heart rate monitoring in combat-related posttraumatic stress disorder. J. Trauma. Stress (1998). https://doi.org/10.1023/a:1024400628342
70. Bruning, N.S., Frew, D.R.: Effects of exercise, relaxation, and management skills training on physiological stress indicators. A field experiment. J. Appl. Psychol. **72**(4), 515–521 (1987)

71. Skinner, J.E., Beder, S.D., Entman, M.L.: Psychological stress activates phosphorylase in the heart of the conscious pig without increasing heart rate and blood pressure. Proc. Natl. Acad. Sci. U.S.A. **80**, 4513–4517 (1983)
72. Weiss, J.H.: Birth order and physiological stress response. Child Dev. **41**(2), 461–470 (1970)
73. Joshi, A., Kiran, R., Singla, H.K., Sah, A.N.: Stress management through regulation of blood pressure among college students. Work (2016). https://doi.org/10.3233/wor-162308
74. Hilmert, C.J., Ode, S., Zielke, D.J., Robinson, M.D.: Blood pressure reactivity predicts somatic reactivity to stress in daily life. J. Behav. Med. (2010). https://doi.org/10.1007/s10865-010-9256-x
75. Kassam, K.S., Koslov, K., Mendes, W.B.: Decisions under distress. Stress profiles influence anchoring and adjustment. Psychol. Sci. (2009). https://doi.org/10.1111/j.1467-9280.2009.02455.x
76. Peters, J.L., Kubzansky, L., McNeely, E., Schwartz, J., Spiro, A., Sparrow, D., Wright, R.O., Nie, H., Hu, H.: Stress as a potential modifier of the impact of lead levels on blood pressure. The normative aging study. Environ. Health Perspect. (2007). https://doi.org/10.1289/ehp.10002
77. Sun, J., Wang, S., Zhang, J.-Q., Li, W.: Assessing the cumulative effects of stress. The association between job stress and allostatic load in a large sample of Chinese employees. Work Stress (2007). https://doi.org/10.1080/02678370701742748
78. Rau, R.: The association between blood pressure and work stress. The importance of measuring isolated systolic hypertension. Work Stress (2006). https://doi.org/10.1080/02678370600679447
79. Gregg, M.E.D., Matyas, T.A., James, J.E.: Association between hemodynamic profile during laboratory stress and ambulatory pulse pressure. J. Behav. Med. (2005). https://doi.org/10.1007/s10865-005-9018-3
80. Hughes, B.M.: Study, examinations, and stress. Blood pressure assessments in college students. Educ. Rev. (2005). https://doi.org/10.1080/0013191042000274169
81. Vocks, S., Ockenfels, M., Jürgensen, R., Mussgay, L., Rüddel, H.: Blood pressure reactivity can be reduced by a cognitive behavioral stress management program. Int. J. Behav. Med. **11**(2), 63–70 (2004)
82. Brody, S., Preut, R., Schommer, K., Schürmeyer, T.H.: A randomized controlled trial of high dose ascorbic acid for reduction of blood pressure, cortisol, and subjective responses to psychological stress. Psychopharmacology (2002). https://doi.org/10.1007/s00213-001-0929-6
83. Elsass, P.: Stress and the workload of professional women in sweden. Acad. Manage. Perspect. (2000). https://doi.org/10.5465/ame.2000.26207946
84. Lovallo, W.R., al'Absi, M., Pincomb, G.A., Passey, R.B., Sung, B.H., Wilson, M.F.: Caffeine, extended stress, and blood pressure in borderline hypertensive men. Int. J. Behav. Med. (2000). https://doi.org/10.1207/s15327558ijbm0702_6
85. Evans, G.W., Bullinger, M., Hygge, S.: Chronic noise exposure and physiological response. A prospective study of children living under environmental stress. Psychol. Sci. **9**(1), 75–77 (1998)
86. James, G.D., Brown, D.E.: The biological stress response and lifestyle. Catecholamines and blood pressure. Annu. Rev. Anthropol. **26**, 313–335 (1997)
87. Lustman, P.J., Sowa, C.J.: Comparative efficacy of biofeedback and stress inoculation for stress reduction. J. Clin. Psychol. **39**(2), 191–197 (1983)
88. Ruiz-Robledillo, N., Moya-Albiol, L.: Lower electrodermal activity to acute stress in caregivers of people with autism spectrum disorder. An adaptive habituation to stress. J. Autism Dev. Disorders (2015). https://doi.org/10.1007/s10803-013-1996-3
89. Compton, R.J., Robinson, M.D., Ode, S., Quandt, L.C., Fineman, S.L., Carp, J.: Error-monitoring ability predicts daily stress regulation. Psychol. Sci. (2008). https://doi.org/10.1111/j.1467-9280.2008.02145.x
90. Gilbert, D.G., Estes, S.L., Welser, R.: Does noise stress modulate effects of smoking/nicotine? Mood, vigilance, and EEG responses. Psychopharmacology **129**, 382–389 (1997)

91. Ashare, R.L., Lerman, C., Cao, W., Falcone, M., Bernardo, L., Ruparel, K., Hopson, R., Gur, R., Pruessner, J.C., Loughead, J.: Nicotine withdrawal alters neural responses to psychosocial stress. Psychopharmacology (2016). https://doi.org/10.1007/s00213-016-4299-5
92. American Sociological Association: Relationships between self-rating stress scales and physiological stress measures. In: Conference Papers, 1–39 (2015)
93. McEwen, B.S., Bowles, N.P., Gray, J.D., Hill, M.N., Hunter, R.G., Karatsoreos, I.N., Nasca, C.: Mechanisms of stress in the brain. Nature Neurosci. (2015). https://doi.org/10.1038/nn.4086
94. Porcu, P., Morrow, A.L.: Divergent neuroactive steroid responses to stress and ethanol in rat and mouse strains. Relevance for human studies. Psychopharmacology (2014). https://doi.org/10.1007/s00213-014-3564-8
95. Quas, J.A., Rush, E.B., Yim, I.S., Nikolayev, M.: Effects of stress on memory in children and adolescents. Testing causal connections. Memory (2014). https://doi.org/10.1080/09658211.2013.809766
96. Sinclair, D., Purves-Tyson, T.D., Allen, K.M., Weickert, C.S.: Impacts of stress and sex hormones on dopamine neurotransmission in the adolescent brain. Psychopharmacology (2014). https://doi.org/10.1007/s00213-013-3415-z
97. van den Bos, E., Rooij, M. de, Miers, A.C., Bokhorst, C.L., Westenberg, P.M.: Adolescents' increasing stress response to social evaluation. Pubertal effects on cortisol and alpha-amylase during public speaking. Child Dev. (2014). https://doi.org/10.1111/cdev.12118
98. Oakley, R.H., Ren, R., Cruz-Topete, D., Bird, G.S., Myers, P.H., Boyle, M.C., Schneider, M.D., Willis, M.S., Cidlowski, J.A.: Essential role of stress hormone signaling in cardiomyocytes for the prevention of heart disease. In: Proceedings of the National Academy of Sciences of the United States of America (2013). https://doi.org/10.1073/pnas.1302546110
99. Taylor, C.J.: Physiological stress response to loss of social influence and threats to masculinity. In: Conf. Pap. Am. Sociol. Assoc. (2013). https://doi.org/10.1016/j.socscimed.2013.07.036
100. Rudolph, K.D., Troop-Gordon, W., Granger, D.A.: Individual differences in biological stress responses moderate the contribution of early peer victimization to subsequent depressive symptoms. Psychopharmacology (2011). https://doi.org/10.1007/s00213-010-1879-7
101. Taylor, C.J.: A gendered perspective on biological stress response to token status. Conf. Pap. Am. Sociol. Assoc. Ann. Meet., 1–42 (2011)
102. Rösler, U., Gebele, N., Hoffmann, K., Morling, K., Müller, A., Rau, R., Stephan, U.: Cortisol–ein geeigneter physiologischer Indikator für Belastungen am Arbeitsplatz? Zeitschrift für Arbeits- und Organisationspsychologie A&O (2010). https://doi.org/10.1026/0932-4089/a000011
103. Rimmele, U., Spillmann, M., Bärtschi, C., Wolf, O.T., Weber, C.S., Ehlert, U., Wirtz, P.H.: Melatonin improves memory acquisition under stress independent of stress hormone release. Psychopharmacology (2009). https://doi.org/10.1007/s00213-008-1344-z
104. Groc, L., Choquet, D., Chaouloff, F.: The stress hormone corticosterone conditions AMPAR surface trafficking and synaptic potentiation. Nat. Neurosci. (2008). https://doi.org/10.1038/nn.2150
105. Kajantie, E.: Physiological stress response, estrogen, and the male–female mortality gap. Curr. Dir. Psychol. Sci. (2008). https://doi.org/10.1111/j.1467-8721.2008.00604.x
106. McRae, A.L., Saladin, M.E., Brady, K.T., Upadhyaya, H., Back, S.E., Timmerman, M.A.: Stress reactivity. Biological and subjective responses to the cold pressor and trier social stressors. Hum. Psychopharmacol. (2006). https://doi.org/10.1002/hup.778
107. Alehagen, S., Wijma, B., Lundberg, U., Wijma, K.: Fear, pain and stress hormones during childbirth. J. Psychosom. Obstet. Gynecol. (2005). https://doi.org/10.1080/01443610400023072
108. de Kloet, E.R., Joëls, M., Holsboer, F.: Stress and the brain. From adaptation to disease. Nat. Rev. Neurosci. (2005). https://doi.org/10.1038/nrn1683
109. Gunnar, M.R., Cheatham, C.L.: Brain and behavior interface. Stress and the developing brain. Infant Ment. Health J. (2003). https://doi.org/10.1002/imhj.10052

110. Heinz, A., Hermann, D., Smolka, M.N., Rieks, M., Gräf, K.-J., Pöhlau, D., Kuhn, W., Bauer, M.: Effects of acute psychological stress on adhesion molecules, interleukins and sex hormones. Implications for coronary heart disease. Psychopharmacology (2003). https://doi.org/10.1007/s00213-002-1244-6
111. Morale, C., Brouwer, J., Testa, N., Tirolo, C., Barden, N., Dijkstra, C.D., Amor, S., Marchetti, B.: Stress, glucocorticoids and the susceptibility to develop autoimmune disorders of the central nervous system. Neurol. Sci. (2001). https://doi.org/10.1007/s100720170016
112. Licinio, J., Wong, M.-L.: The role of inflammatory mediators in the biology of major depression. Central nervous system cytokines modulate the biological substrate of depressive symptoms, regulate stress-responsive systems, and contribute to neurotoxicity and neuroprotection. Mol. Psychiatry **4**, 317–327 (1999)
113. Kelly, K.S., Hayslip Jr., B., Servaty, H.L., Ennis, M.P.: Physiological indicators of stress and intellectual performance among anxious older adults. Educ. Gerontol. **23**(5), 477–487 (1997)
114. Loving, T.J., Heffner, K.L., Kiecolt-Glaser, J., Glaser, R., Malarkey, W.B.: Stress hormone changes and marital conflict. spouses' relative power makes a difference. J. Marriage Fam. **66**, 595–612 (2004)
115. Bruenahl, C.A., Linden, M.: Common laboratory measures of global health may not be suited to assess, discriminate or predict chronic stress effects on biological systems. Nordic J. Psychiatry (2011). https://doi.org/10.3109/08039488.2010.542589
116. Monti, J.D., Abaied, J.L., Rudolph, K.D.: Contributions of socialization of coping to physiological responses to stress. Aust. J. Psychol. (2014). https://doi.org/10.1111/ajpy.12044
117. Hirokawa, K., Yagi, A., Miyata, Y.: Effects of stress coping strategies on psychological and physiological responses during speeches in Japanese and english. Soc. Behav. Pers. (2002). https://doi.org/10.2224/sbp.2002.30.2.203
118. Gärling, T., Gamble, A., Fors, F., Hjerm, M.: Emotional well-being related to time pressure, impediment to goal progress, and stress-related symptoms. J. Happiness Stud. (2016). https://doi.org/10.1007/s10902-015-9670-4
119. Pressman, R.M., Sugarman, D.B., Nemon, M.L., Desjarlais, J., Owens, J.A., Schettini-Evans, A.: Homework and family stress. With consideration of parents' self confidence, educational level, and cultural background. Am. J. Fam. Therapy (2015). https://doi.org/10.1080/01926187.2015.1061407
120. Dobrodolac, M., Marković, D., Čubranić-Dobrodolac, M., Denda, N.: Using work stress measurement to develop and implement a TQM programme. A case of counter clerks in Serbian post. Total Qual. Manag. Bus. Excellence (2014). https://doi.org/10.1080/14783363.2012.704280
121. Braunstein-Bercovitz, H.: Does stress enhance or impair selective attention? The effects of stress and perceptual load on negative priming. Anxiety, Stress, Coping (2003). https://doi.org/10.1080/10615800310000112560
122. Glaser, D.N., Tatum, B.C., Nebeker, D.M., Sorenson, R.C., Aiello, J.R.: Workload and social support. effects on performance and stress. Hum. Perform. **12**(2), 155–176 (1999)
123. Male, D., May, D.: Stress and health, workload and burnout in learning support coordinators in colleges of further education. Support Learn. (1998). https://doi.org/10.1111/1467-9604.00075
124. Korunka, C., Zauchner, S., Weiss, A.: New information technologies, job profiles, and external workload as predictors of subjectively experienced stress and dissatisfaction at work. Int. J. Hum. Comput. Interact. **9**(4), 407–424 (1997)
125. Kipnis, D., Schmidt, S.M.: Upward-influence styles. Relationship with performance evaluations, salary, and stress. Adm. Sci. Q. **33**, 528–542 (1988)
126. Kyriacou, C., Sutcliffe, J.: A note on teacher stress and locus of control. J. Occup. Psychol. **52**(3), 227–229 (1979). https://doi.org/10.1111/j.2044-8325.1979.tb00456.x
127. Friend, K.E.: Stress and performance. Effects of subjective work load and time urgency. Pers. Psychol. **35**, 623–633 (1982)
128. Gawrysiak, M.J., Leong, S.H., Grassetti, S.N., Wai, M., Shorey, R.C., Baime, M.J.: Dimensions of distress tolerance and the moderating effects on mindfulness-based stress reduction. Anxiety Stress Coping (2016). https://doi.org/10.1080/10615806.2015.1085513

129. Lessenberry, B.M., Rehfeldt, R.A.: Evaluating stress levels of parents of children with disabilities. Except. Child. **70**(2), 231–244 (2004)
130. Joshi, A., Kiran, R., Sah, A.N.: An experimental analysis to monitor and manage stress among engineering students using Galvanic Skin Response meter. Work **56**(3), 409–420 (2017). https://doi.org/10.3233/WOR-172507
131. Banks, J., Smyth, E.: 'Your whole life depends on it'. Academic stress and high-stakes testing in Ireland. J. Youth Stud. **18**(5), 598–616 (2015). https://doi.org/10.1080/13676261.2014.992317
132. Leigthon, D.C.: Measuring stress levels in school children as a program-monitoring device. Am. J. Public Health **62**(6), 799–806 (1972)
133. Cox, T., Mackay, C.: The measurement of self-reported stress and arousal. Br. J. Psychol. **76**, 183–186 (1985)
134. King, M.G., Burrows, G.D., Stanley, G.V.: Measurement of stress and arousal. Validation of the stress/arousal adjective checklist. Br. J. Psychol. **74**, 473–479 (1983)
135. González-Munoz, E.L., Guitérrez-Martínez, R.E.: Contribution of mental workload to job stress in industrial workers. Work **28**, 355–361 (2007)
136. MacDonald, W.: The impact of job demands and workload on stress and fatigue. Aust. Psychol. **38**(2), 102–107 (2003). https://doi.org/10.1080/00050060310001707107
137. Teufel, M., Stephan, K., Kowalski, A., Käsberger, S., Enck, P., Zipfel, S., Giel, K.E.: Impact of biofeedback on self-efficacy and stress reduction in obesity. A randomized controlled pilot study. Appl. Psychophysiol. Biofeedback **38**(3), 177–184 (2013). https://doi.org/10.1007/s10484-013-9223-8
138. Tucker, S.: An investigation of the stresses, pressures and challenges faced by primary school head teachers in a context of organisational change in schools. J. Soc. Work Pract. **24**(1), 63–74 (2010). https://doi.org/10.1080/02650530903532765
139. Tsutsumi, A., Iwata, N., Watanabe, N., de Jonge, J., Pikhart, H., Fernández-López, J.A., Xu, L., Peter, R., Knutsson, A., Niedhammer, I., Kawakami, N., Siegrist, J.: Application of item response theory to achieve cross-cultural comparability of occupational stress measurement. Int. J. Methods Psychiatr. Res. (2009). https://doi.org/10.1002/mpr.277
140. Barnes-Farrell, J.L., Rumery, S.M., Swody, C.A.: How do concepts of age relate to work and off-the-job stresses and strains? A field study of health care workers in five nations. Exp. Aging Res. **28**(1), 87–98 (2002). https://doi.org/10.1080/036107302753365577
141. Hinds, H., Burroughs, J.W.: How you know when you're stressed. Self-evaluations of stress. J. Gen. Psychol. **124**(1), 105–111 (1997)
142. Gigliotti, R.J., Huff, H.K.: Role-related conflicts, strains and stresses of older-adult college students. Sociol. Focus **28**(3), 329–342 (1995)
143. Van Meter, M.J.S., Agronow, S.J.: The stress of multiple roles. The case for role strain among married college women. Fam. Relat. **31**, 131–138 (1982)

The Neuroscience of Smartphone/Social Media Usage and the Growing Need to Include Methods from 'Psychoinformatics'

Christian Montag

Abstract The present work gives a brief overview of the current state of affairs in the investigation of the neuroscientific underpinnings of social media use. Such an overview is of importance because individuals spend significant amounts of time on these 'social' online channels. Despite several positive aspects of social media use, such as the ability to easily communicate with others across long distances, it is clear that detrimental effects on our brains and minds are possible. Given that much of the neuroscientific and psychological research conducted up to now relies solely on self-report measures to assess social media usage, it is argued that neuroscientists/psychologists need to include more digital traces resulting from human-machine/computer interaction, and/or information shared by individuals on social media, in their scientific analyses. In this realm, digital phenotyping can be achieved via methods of 'Psychoinformatics', a merger of the disciplines psychology and computer science/informatics.

Keywords Smartphone · Social media · Psychoinformatics · Digital phenotyping Nucleus accumbens · Anterior cingulate cortex

1 On Smartphone and Social Media Usage

Since the programming of the first HTML website by Tim Berners-Lee in the beginning of the 1990s, and particularly since the advent of Apple's IPhone in 2007, the digital revolution has dramatically changed the way we live in many areas of modern society. This groundbreaking movement towards a digital society is particularly

C. Montag (✉)
Department of Molecular Psychology, Institute of Psychology and Education, Ulm University, Ulm, Germany
e-mail: mail@christianmontag.de

C. Montag
MOE Key Laboratory for Neuroinformation, The Clinical Hospital of Chengdu Brain Science Institute, University of Electronic Science and Technology of China, Chengdu, China

F. D. Davis et al. (eds.), *Information Systems and Neuroscience*,
Lecture Notes in Information Systems and Organisation 29,
https://doi.org/10.1007/978-3-030-01087-4_32

evident in the ubiquity of the smartphone. This digital device strongly influences the way we communicate with others, entertain ourselves and travel around our environment. There are currently 2.5 billion smartphone users worldwide [1]. Much of our daily smartphone use is accounted for by the use of social media applications such as Facebook, WhatsApp and WeChat.

In line with the huge number of smartphone users worldwide, the number of users of these social media platforms is breathtaking. Facebook currently has almost two billion accounts [2], while the messenger application WhatsApp has an estimate of 1.5 billion users [3] and its Chinese cousin WeChat ('Wēixìn' literally, 'micro message') has about one billion users [4]. With its many features, the smartphone is not only omnipresent in terms of global user numbers, but also in terms of the length of daily usage. In work tracking more than 2400 participants, it became apparent that the typical smartphone user spends about 2.5 h each day on the phone [5]. Therefore, more than one work-day of a week is spent on the phone by many users. Self-report data indicates that most time spent on the phone is for leisure or private purposes, and not for business purposes [6].

Although the smartphone can help our productivity, our daily life can be negatively affected because of a constant influx of e-mails and messages via services such as WhatsApp, which can interrupt our work-processes [6, 7] and can lead to stress and lower well-being [8]. Moreover, excessive usage of the smartphone has been associated with ADHD like symptoms [9, 10]. This fits well with the observation that Internet addiction, with the specific form of (Internet) Gaming Disorder to be officially recognized in ICD-11 this year, has been repeatedly linked with tendencies towards ADHD [11, 12]. This all suggests that smartphone usage might exert beneficial, as well as detrimental, effects on our lives. It is therefore of importance to understand under which conditions interacting with the digital world enhances our lives, and to what extent smartphone usage poses problems [13, 14].

As mentioned above, the psychological literature has already shown that constant interruption and disruption of daily life activities might be the key to understanding when smartphone usage hinders our productivity and results in less happiness (namely, when the experience of flow at the work place is hindered by constant micro-breaks [15]). Taking a closer look at the usage of platforms such as Facebook or Instagram, it is clear that another source for negative emotionality can be found in the processes of social comparison and envy when consuming content from these social media platforms [16]. Social media is not always necessarily 'social', but may often be used to self-promote one's identity, personality and life events [17]. Typical images and messages posted often refer to the 'wonderful' and 'spectacular' life a person is living. Being confronted with these kinds of profiles could even result in depressive symptoms [18]. In light of this, it is not surprising that quitting Facebook for just one week can result in higher life satisfaction and well-being [19].

2 First Insights into the Neuroscientific Underpinnings of Social Media Usage

2.1 Neuroscientific Studies of Social Media Usage Are Still Scarce

Although many psychological mechanisms have been proposed for explaining why we are so attached to our smartphones (for an overview on conditioning principles please see Duke and Montag [7]) and why many use social media such as Facebook so extensively (see motives for Facebook usage [20, 21]), little is known about the neuroscientific underpinnings of social media usage (as is apparent in these reviews [22, 23]). These well-conducted reviews show that much of what we know about the neural underpinnings of social media usage has been derived from bordering scientific areas, such as the investigation of social dynamics developing from in- and exclusion of individuals; e.g. see the cyberball-paradigm [24]. In this regard, Meshi et al. [22] correctly summarized that "Neuroscience research with social media is still in its infancy", but that "there is great potential for future scientific discovery" (p. 9).

2.2 Understanding the Rewarding Aspects of Social Media Usage by Means of Functional Magnetic Resonance Imaging (FMRI)

A well understood phenomenon with respect to the usage of social media platforms such as Facebook or Instagram is the rewarding aspect of getting so-called 'Likes'. This also explains why a 'Like economy' developed on social media platforms, such as giving 'Like' for 'Like' [25], leading the concept of receiving a 'Like' for a creative or popular post of a photo or text ad absurdum. Several recent fMRI studies [26–29] have observed that the nucleus accumbens plays a pivotal role in understanding why humans are 'hunting' online for 'Likes'. In recent studies [28, 29] it was shown that processing pictures from one's own Instagram account with high versus low number of 'Likes' resulted in elevated activity in ventral striatal regions in the human brain (where the nucleus accumbens is located). In general, pictures with more 'Likes' (not necessarily one's own) elicited stronger activity in the accumbens region (see Fig. 1 on the next page).

Based on much research, it is well known that this brain region plays an important part in SEEKING activity, a term coined for an emotional/motivational system in Panksepp's Affective Neuroscience Theory [30, 31]. Activity in the SEEKING system describes a state of high energy accompanying motivated approach behavior in all mammals. Therefore, the expectation of getting 'Likes' might hijack the SEEKING system and urge Facebook users to return again and again to the social

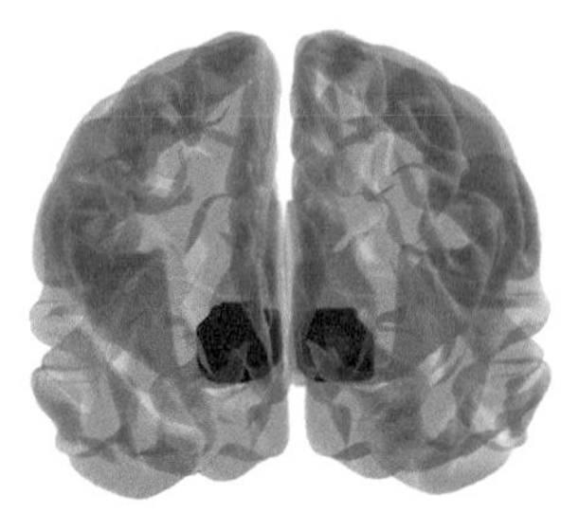
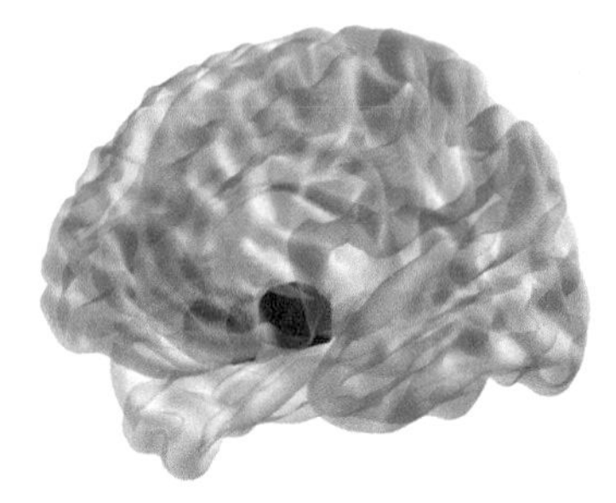

Fig. 1 The nucleus accumbens is involved in reward processing and represents a critical area to also understand the rewarding aspects of 'Likes' in social media research. Clearly, it only represents one of many brain areas involved to understand the neural basis of social media usage. For an overview on many other relevant brain areas such as the inferior frontal gyrus, the anterior temporal lobe or the temporoparietal junction (also known as the mentalizing network) see [22]. Thanks to Sebastian Markett for providing CM with the figure

media platform. The SEEKING system has been proposed to be anchored in the medial forebrain bundle [30]. These results illustrate how bottom-up processes (i.e. activity from evolutionary old regions of the human brain [31]) may explain why many social media users spend more and more time on Facebook, WhatsApp and similar applications. Clearly, striatal activity alone cannot explain this phenomenon, because typically prefrontal regions of the human cortex exert a tight grip on these older brain regions. Of note here, younger people with a less developed prefrontal cortex are more likely to be affected by subcortical activity (for age effects see [29] and also the following overview [23]). Activity in the PFC can result in powerful regulation of inner emotional urges, such that a person might think 'No, I don't have time to go on Facebook now. First, I have to finish my school homework'. In a study using electroencephalography (EEG), 'smartphone addicts' have been characterized by poor executive functions [32]. This means that they appear to have problems in exerting top-down control over the striatal regions, resulting in escalating social media usage, perhaps being characterized in extreme form as 'Internet Communication Disorder' (ICD) [33, 34].

2.3 *Social Media Usage/Internet Communication Disorder in Relation to Results from Studies Using Structural Brain Imaging (sMRI)*

Further insight into the neural underpinnings of social media usage stem from sMRI. In contrast to fMRI, structural imaging of the brain, amongst other features, allows researchers to get insights into individual differences in gray matter volume of the human brain. Such differences can be examined on a whole brain level, but also with respect to specific brain areas, such as the aforementioned nucleus accumbens. A recent study [35] observed that lower gray matter volume of the nucleus accumbens was associated with longer and more frequent usage of the Facebook application installed on smartphones. Of note, Facebook behavior was not assessed via self-report in this study, but directly tracked on the smartphones. This study therefore demonstrates the feasibility of combining real-world behavior with human neuroscience data. This will be discussed in more detail in the next section. Returning to the data linking lower gray matter volume of the nucleus accumbens to longer/higher frequent usage of Facebook on the smartphone, in the literature it has been observed that lower volume of the nucleus accumbens is associated with nicotine, alcohol and heroin addiction [36–38]. Of course, addictive behaviors in relation to these substances are not always directly comparable with ICD, but it could be suggested that some of the psychological mechanisms underlying excessive Facebook usage might partly resemble these other addictive behaviors. In line with this idea, a recently published study by Montag et al. [34] linked lower gray matter volume of the (subgenual) anterior cingulate cortex (ACC) to higher 'WeChat addiction' in a Chinese sample, a finding that fits with observations from both substance-dependency and Internet addiction (see discussion in [34]).

Although structural brain imaging gives no direct insights into the functionality of the human brain, the combination of results described in this section with those from the fMRI studies discussed further above reveal that subcortical areas react strongly in anticipation of social media usage (in particular in expectation of 'Likes'), and in 'social media addicts' dysfunction in top-down control mechanisms might lead to problems in the down-regulation of this increased activity from the ventral striatum [32]. Future studies will clearly need to paint a much more detailed picture on these many different activities and links with psychological processes such as mentalizing and self-referential cognition [23] while using Facebook and other social media applications. Much of what we currently know from a neuroscientific perspective on social media usage deals with the rewarding aspects of such platforms.

3 Adding Methods from 'Psychoinformatics' to the Neuroscientist's Toolbox

It has already been mentioned that a problem in much of the social media research conducted so far is the inclusion of self-report measures only. Without doubt, such measures will always be of importance because they give researchers valuable insights into subjectively perceived stress etc. due to social media usage, variables insufficiently operationalized by direct tracking of Internet or social media usage. Nevertheless, neglecting the possibilities of Psychoinformatics in the neuroscientific and psychological study of social media usage will ultimately lead to an incomplete understanding of why and how long people use these platforms [39]. Just consider the problems involved in estimating how long you spent on WhatsApp at a certain time last week? It has been reported in earlier work that people tend to have a distorted perception of time while using their phones, because they tend to get into 'the zone', and experience flow, on smartphones [40, 41]. Ironically, this kind of psychological state is often necessary for productive work.

Psychoinformatics describes the use of methods from computer science, such as machine learning or pattern classification, to study digital traces from human-computer interaction [42, 43] to do digital phenotyping [44]. Pioneering studies revealed the power of doing psychodiagnostics via analyzing the 'Like'-structure of a person's Facebook account (see meta-analysis [45]), and also via the study of Twitter accounts [46]. Amongst other things, it has been also shown that it is possible to predict socio-demographic variables, political attitudes, sexual orientation and personality traits from Facebook 'Likes' [47]. As a consequence, this kind of data, and in future all data from the Internet of Things (IoT), could be used to enhance one's own study design, not only in psychology, but also the neurosciences (see further examples with research on social network size [48, 49]). Nevertheless, many important obstacles have to be overcome in the near future. Among the most pressing might be addressing privacy issues, because digital phenotyping can clearly be misused for psychological targeting in sensitive areas [50, 51, 52].

4 Conclusion

In sum, the study of the neuroscientific/psychological underpinnings of social media use is still in its infancy and much additional work is needed to better understand the many psychological processes underlying the motivation to use platforms such as Facebook. With respect to the smartphone it becomes clear, that technology per se is neither good nor bad, but context matters. If used in the correct way, smartphones can makes us more productive, but there clearly exists a still to be defined point, where smartphones become detrimental for our mental health. Earlier it has been proposed that the link between productivity and smartphone usage can be described by an inverted U-function [15] with problems arising when a person is constantly interrupted by the smartphone and its social media channels. In addition and also in light of the recent Facebook-scandal [53], it will be necessary to re-think the design of the Facebook platform to make it indeed more social. Given the aforementioned possibilities to do psychodiagnostics via 'Likes', it will be also necessary to better protect the data of each user [54]. Finally, scientists have to come towards a consensus concerning the topic of smartphone addiction or smartphone use disorder. It is still not clear how these terms relate to Internet addiction/use disorder (4). Given that many smartphone users also use their phones for playing games and other applications, social media usage will not be sufficient to grasp and understand (excessive) smartphone usage.

Acknowledgements The position of CM is funded by a Heisenberg grant awarded to him by the German Research Foundation (MO 2363/3-2). CM thanks Andrew Cooper for language editing.

References

1. Statista.com https://www.statista.com/statistics/330695/number-of-smartphone-users-worldwide/
2. Statista.com https://www.statista.com/statistics/264810/number-of-monthly-active-facebook-users-worldwide/
3. Statistia.com https://www.statista.com/statistics/260819/number-of-monthly-active-whatsapp-users/
4. Statista.com https://www.statista.com/statistics/255778/number-of-active-wechat-messenger-accounts/
5. Montag, C., Błaszkiewicz, K., Sariyska, R., Lachmann, B., Andone, I., Trendafilov, B., Eibes, B., Markowetz, A.: Smartphone usage in the 21st century: who is active on WhatsApp? BMC Res. Notes **8**, 331 (2015)
6. Duke, É., Montag, C.: Smartphone addiction, daily interruptions and self-reported productivity. Addict. Behav. Rep. **6**, 90–95 (2017)
7. Duke, É., Montag, C.: Smartphone addiction and beyond: initial insights on an emerging research topic and its relationship to internet addiction. In: Montag, C., Reuter, M. (eds.) Internet Addiction, pp. 359–372. Springer, Berlin (2017)
8. Kushlev, K., Dunn, E.W.: Checking email less frequently reduces stress. Comput. Hum. Behav. **43**, 220–228 (2015)
9. Kushlev, K., Proulx, J., Dunn, E. W.: Silence your phones: smartphone notifications increase inattention and hyperactivity symptoms. In: Proceedings of the 2016 CHI Conference on Human Factors in Computing Systems, pp. 1011–1020 (2016) (ACM)
10. Hadar, A., Hadas, I., Lazarovits, A., Alyagon, U., Eliraz, D., Zangen, A.: Answering the missed call: Initial exploration of cognitive and electrophysiological changes associated with smartphone use and abuse. PLoS ONE **12**(7), e0180094 (2017)
11. Sariyska, R., Reuter, M., Lachmann, B., Montag, C.: Attention deficit/hyperactivity disorder is a better predictor for problematic Internet use than depression: evidence from Germany. J. Addict. Res. Ther. **6**, 1–6 (2015)
12. Yen, J.Y., Ko, C.H., Yen, C.F., Wu, H.Y., Yang, M.J.: The comorbid psychiatric symptoms of Internet addiction: attention deficit and hyperactivity disorder (ADHD), depression, social phobia, and hostility. J. Adolesc. Health **41**(1), 93–98 (2007)
13. Montag, C.: Homo Digitalis: Smartphones, soziale Netzwerke und das Gehirn. Springer, Berlin (2017)
14. Montag, C., Diefenbach, S.: Towards Homo Digitalis: Important Research Issues for Psychology and the Neurosciences at the Dawn of the Internet of Things and the Digital Society. Sustainability **10**, 415 (2018)
15. Montag, C., Walla, P.: Carpe diem instead of losing your social mind: beyond digital addiction and why we all suffer from digital overuse. Cogent. Psychol. **3**, 1157281 (2016)
16. Appel, H., Gerlach, A.L., Crusius, J.: The interplay between facebook use, social comparison, envy, and depression. Curr. Opin. Psychol. **9**, 44–49 (2016)
17. Bareket-Bojmel, L., Moran, S., Shahar, G.: Strategic self-presentation on Facebook: Personal motives and audience response to online behavior. Comput. Hum. Behav. **55**, 788–795 (2016)
18. Tandoc, E.C., Ferrucci, P., Duffy, M.: Facebook use, envy, and depression among college students: is facebooking depressing? Comput. Hum. Behav. **43**, 139–146 (2015)
19. Tromholt, M.: The facebook experiment: quitting facebook leads to higher levels of well-being. Cyberpsychol. Behav. Social. Networking **19**, 661–666 (2016)
20. Ryan, T., Chester, A., Reece, J., Xenos, S.: The uses and abuses of facebook: a review of facebook addiction. J. Behav. Addict. **3**, 133–148 (2014)

21. Sariyska, R., Lachmann, B., Cheng, C., Gnisci, A., Kaliszewska-Czeremska, K. Laconi, S., Zhong, S., Toraman, D., Montag, C.: The motivation for facebook use: Is it a matter of bonding or control over others? Evidence from a cross-cultural study. J. Individ. Differ. (in press)
22. Meshi, D., Tamir, D.I., Heekeren, H.R.: The emerging neuroscience of social media. Trends. Cogn. Sci. **19**, 771–782 (2015)
23. Crone, E.A., Konijn, E.A.: Media use and brain development during adolescence. Nat. Commun. **9**, 588 (2018)
24. Hartgerink, C.H., Van Beest, I., Wicherts, J.M., Williams, K.D.: The ordinal effects of ostracism: a meta-analysis of 120 Cyberball studies. PLoS ONE **10**, e0127002 (2015)
25. Gerlitz, C., Helmond, A.: The like economy: Social buttons and the data-intensive web. New Media. Soc. **15**, 1348–1365 (2013)
26. Meshi, D., Morawetz, C., Heekeren, H.R.: Nucleus accumbens response to gains in reputation for the self relative to gains for others predicts social media use. Frontiers. Hum. Neurosci. **7**, 439 (2013)
27. Turel, O., He, Q., Xue, G., Xiao, L., Bechara, A.: Examination of neural systems sub-serving facebook "addiction". Psychol. Rep. **115**, 675–695 (2014)
28. Sherman, L.E., Payton, A.A., Hernandez, L.M., Greenfield, P.M., Dapretto, M.: The power of the like in adolescence: effects of peer influence on neural and behavioral responses to social media. Psychol. Sci. **27**, 1027–1035 (2016)
29. Sherman, L.E., Greenfield, P.M., Hernandez, L.M., Dapretto, M.: Peer influence via instagram: effects on brain and behavior in adolescence and young adulthood. Child Dev. **89**, 37–47 (2018)
30. Ikemoto, S., Panksepp, J.: The role of nucleus accumbens dopamine in motivated behavior: a unifying interpretation with special reference to reward-seeking. Brain Res. Rev. **31**, 6–41 (1999)
31. Montag, C., Panksepp, J.: Primary emotional systems and personality: An evolutionary perspective. Frontiers. Psychol. **8**, 464 (2017)
32. Chen, J., Liang, Y., Mai, C., Zhong, X., Qu, C.: General deficit in inhibitory control of excessive smartphone users: evidence from an event-related potential study. Frontiers. Psychol. **7**, 511 (2016)
33. Wegmann, E., Brand, M.: Internet-communication disorder: It's a matter of social aspects, coping, and Internet-use expectancies. Frontiers. Psychol. **7**, 1747 (2016)
34. Montag, C., Zhao, Z., Sindermann, C., Xu, L., Fu, M., Li, J., Kendrick, K.M., Dai, J., Becker, B.: Internet communication disorder and the structure of the human brain: initial insights on WeChat addiction. Sci. Rep. **8**, 2155 (2018)
35. Montag, C., Markowetz, A., Blaszkiewicz, K., Andone, I., Lachmann, B., Sariyska, R., Trendafilov, B., Eibes, M., Kolb, J., Reuter, M., Weber, B., Markett, S.: Facebook usage on smartphones and gray matter volume of the nucleus accumbens. Behav. Brain Res. **329**, 221–228 (2017)
36. Das, D., Cherbuin, N., Anstey, K.J., Sachdev, P.S., Easteal, S.: Lifetime cigarette smoking is associated with striatal volume measures. Addict. Biol. **17**(4), 817–825 (2012)
37. Urošević, S., Collins, P., Muetzel, R., Schissel, A., Lim, K.O., Luciana, M.: Effects of reward sensitivity and regional brain volumes on substance use initiation in adolescence. Soc. Cogn. Affect. Neurosci. **10**, 106–113 (2014)
38. Seifert, C.L., Magon, S., Sprenger, T., Lang, U.E., Huber, C.G., Denier, N., Vogel, M., Schmidt, A., Radue, E.W., Borgwardt, S., Walter, M.: Reduced volume of the nucleus accumbens in heroin addiction. Eur. Arch. Psychiatry Clin. Neurosci. **265**, 637–645 (2015)
39. Montag, C., Duke, É., Markowetz, A.: Toward psychoinformatics: computer science meets psychology. Comput. Math. Methods. Med. (2016)
40. Lin, Y.H., Lin, Y.C., Lee, Y.H., Lin, P.H., Lin, S.H., Chang, L.R., Tseng, H.W., Yen, L.Y., Yang, C.C., Kuo, T.B.: Time distortion associated with smartphone addiction: identifying smartphone addiction via a mobile application (App). J. Psychiatr. Res. **65**, 139–145 (2015)

41. Montag, C., Błaszkiewicz, K., Lachmann, B., Sariyska, R., Andone, I., Trendafilov, B., Markowetz, A.: Recorded behavior as a valuable resource for diagnostics in mobile phone addiction: evidence from psychoinformatics. Behav. Sci. **5**, 434–442 (2015)
42. Yarkoni, T.: Psychoinformatics: New horizons at the interface of the psychological and computing sciences. Curr. Dir. Psychol. Sci. **21**, 391–397 (2012)
43. Markowetz, A., Błaszkiewicz, K., Montag, C., Switala, C., Schlaepfer, T.E.: Psychoinformatics: big data shaping modern psychometrics. Med. Hypotheses **82**, 405–411 (2014)
44. Onnela, J.P., Rauch, S.L.: Harnessing smartphone-based digital phenotyping to enhance behavioral and mental health. Neuropsychopharmacol.: Official Publ. Am. Coll. Neuropsychopharmacol. **41**, 1691–1696 (2016)
45. Azucar, D., Marengo, D., Settanni, M.: Predicting the Big 5 personality traits from digital footprints on social media: A meta-analysis. Pers. Individ. Differ. **124**, 150–159 (2018)
46. Quercia, D., Kosinski, M., Stillwell, D., Crowcroft, J.: Our twitter profiles, our selves: Predicting personality with twitter. In: 2011 IEEE Third International Conference on Privacy, Security, Risk and Trust (PASSAT) and 2011 IEEE Third Inernational Conference on Social Computing (SocialCom), pp. 180–185. IEEE (2011)
47. Kosinski, M., Stillwell, D., Graepel, T.: Private traits and attributes are predictable from digital records of human behavior. Proc. Natl. Acad. Sci. **110**, 5802–5805 (2013)
48. Von Der Heide, R., Vyas, G., Olson, I.R.: The social network-network: size is predicted by brain structure and function in the amygdala and paralimbic regions. Social cognitive and affective neuroscience **9**, 1962–1972 (2014)
49. Kanai, R., Bahrami, B., Roylance, R., Rees, G.: Online social network size is reflected in human brain structure. Proceedings of the Royal Society of London B **279**, 1327–1334 (2012)
50. Kshetri, N.: Big data's impact on privacy, security and consumer welfare. Telecommun. Policy. **38**, 1134–1145 (2014)
51. Matz, S.C., Kosinski, M., Nave, G., Stillwell, D.J.: Psychological targeting as an effective approach to digital mass persuasion. Proc. Natl. Acad. Sci. **114**, 12714–12719 (2017)
52. Chester, J., Montgomery, K.: The role of digital marketing in political campaigns. Internet. Policy. Rev. **6**, 1–20 (2017)
53. Bbc.com http://www.bbc.com/news/technology-43649018
54. Kosinski, M., Matz, S.C., Gosling, S.D., Popov, V., Stillwell, D.: Facebook as a research tool for the social sciences: opportunities, challenges, ethical considerations, and practical guidelines. Am. Psychol. **70**(6), 543–556 (2015)

Printed in the United States
By Bookmasters